# Carrier-Free Enzyme Immobilization

# FOUNDATIONS AND FRONTIERS IN ENZYMOLOGY

**Munishwar Nath Gupta, Series Editor**

Series Website: https://www.elsevier.com/en-gb/books-and-journals/book-series/foundations-and-frontiers-in-enzymology

**Foundations and Frontiers in Enzymology**

# Carrier-Free Enzyme Immobilization

## Fundamentals and Innovations in Biocatalysis


**Edited by**

**Mehdi Mohammadi**

Bioprocess Engineering Department, National Institute for Genetic Engineering and Biotechnology, (NIGEB) Tehran, Iran

**Maryam Yousefi**

Nanobiotechnology Research Center, Avicenna Research Institute, Academic Center for Education, Culture and Research (ACECR), Tehran, Iran



ELSEVIER

ACADEMIC PRESS

An imprint of Elsevier

elsevier.com/books-and-journals

Academic Press is an imprint of Elsevier

125 London Wall, London EC2Y 5AS, United Kingdom
50 Hampshire Street, 5th Floor, Cambridge, MA 02139, United States



ISBN: 978-0-443-44139-4

For Information on all Academic Press publications visit our website at https://www.elsevier.com/books-and-journals

*Publisher:* Megan Ball
*Acquisitions Editor:* Ali Afzal-Khan
*Editorial Project Manager:* Stuti Das
*Production Project Manager:* Anitha Sivaraj
*Cover Designer:* Bakyalakshmi S

Typeset by Aptara, New Delhi, India

# Contents

# Contributors

**Lina Ahlborn**
Institute for Biochemistry, Biotechnology and Bioinformatics, Technische Universität Braunschweig, Braunschweig, Germany

**Zofia Bukowska**
Institute of Chemical Technology and Engineering, Faculty of Chemical Technology, Poznan University of Technology, Poznan, Poland

**Jiandong Cui**
Laboratory of Industrial Fermentation Microbiology, Tianjin University of Science and Technology, Tianjin, China

**Yingjie Du**
Laboratory of Industrial Fermentation Microbiology, Tianjin University of Science and Technology, Tianjin, China

**Feihong Hou**
College of Biotechnology, Tianjin University of Science and Technology, Tianjin, China

**Rosli Md. Illias**
Department of Bioprocess & Polymer Engineering, Faculty of Chemical & Energy Engineering, Universiti Teknologi Malaysia, Skudai, Johor, Malaysia; The Innovation Centre in Agritechnology for Advanced Bioprocessing, Hub Pendidikan Tinggi Pagoh, Universiti Teknologi Malaysia, Pagoh, Johor, Malaysia

**Nardiah Rizwana Jaafar**
Department of Bioprocess & Polymer Engineering, Faculty of Chemical & Energy Engineering, Universiti Teknologi Malaysia, Skudai, Johor, Malaysia

**Nashriq Jailani**
Department of Bioprocess & Polymer Engineering, Faculty of Chemical & Energy Engineering, Universiti Teknologi Malaysia, Skudai, Johor, Malaysia

**Teofil Jesionowski**
Institute of Chemical Technology and Engineering, Faculty of Chemical Technology, Poznan University of Technology, Poznan, Poland

**Yanjun Jiang**
National-Local Joint Engineering Laboratory for Energy Conservation in Chemical Process Integration and Resources Utilization, Hebei University of Technology, Tianjin, China

**Yingjia Li**
Laboratory of Industrial Fermentation Microbiology, Tianjin University of Science and Technology, Tianjin, China

**Fufeng Liu**
College of Biotechnology, Tianjin University of Science and Technology, Tianjin, China

**Xingmiao Lu**
College of Biotechnology, Tianjin University of Science and Technology, Tianjin, China

**Cesar Mateo**
Department of Biocatalysis, Institute of Catalysis and Petrochemistry CSIC, Campus UAM, Madrid, Spain

**Mehdi Mohammadi**
Bioprocess Engineering Department, National Institute for Genetic Engineering and Biotechnology, Tehran, Iran

**Malwina Nowak**
Institute of Chemical Technology and Engineering, Faculty of Chemical Technology, Poznan University of Technology, Poznan, Poland

**Roshanida A. Rahman**
Department of Bioprocess & Polymer Engineering, Faculty of Chemical & Energy Engineering, Universiti Teknologi Malaysia, Skudai, Johor, Malaysia; The Innovation Centre in Agritechnology for Advanced Bioprocessing, Hub Pendidikan Tinggi Pagoh, Universiti Teknologi Malaysia, Pagoh, Johor, Malaysia

**Virendra Rathod**
Department of Chemical Engineering, Institute of Chemical Technology, Mumbai, Maharashtra, India

**Xiaofei Ren**
College of Biotechnology, Tianjin University of Science and Technology, Tianjin, China

**Agnieszka Rybarczyk**
Institute of Chemical Technology and Engineering, Faculty of Chemical Technology, Poznan University of Technology, Poznan, Poland

**Anett Schallmey**
Institute for Biochemistry, Biotechnology and Bioinformatics, Technische Universität Braunschweig, Braunschweig, Germany

**Santosh Sethi**
Department of Chemical Engineering, Institute of Chemical Technology, Mumbai, Maharashtra, India

**Mansour Shahedi**
Organic and Inorganic Chemistry, Shahid Beheshti University, Tehran, Iran

**Manman Shi**
Laboratory of Industrial Fermentation Microbiology, Tianjin University of Science and Technology, Tianjin, China

**Liying Wang**
College of Biotechnology, Tianjin University of Science and Technology, Tianjin, China

**Shilong Xing**
School of Chemical Engineering and Technology, Hebei University of Technology, Tianjin, China

**Maryam Yousefi**
Nanobiotechnology Research Center, Avicenna Research Institute, ACECR, Tehran, Iran

**Jakub Zdarta**
Institute of Chemical Technology and Engineering, Faculty of Chemical Technology, Poznan University of Technology, Poznan, Poland

**Yushuo Zhang**
College of Biotechnology, Tianjin University of Science and Technology, Tianjin, China

**Lixue Zhao**
Laboratory of Industrial Fermentation Microbiology, Tianjin University of Science and Technology, Tianjin, China; College of Biotechnology, Tianjin University of Science and Technology, Tianjin, China

SECTION

# Fundamentals of carrier-free and carrier-bound immobilization

# I

CHAPTER

# 1 Introduction to enzyme immobilization

**Malwina Nowak, Zofia Bukowska, Agnieszka Rybarczyk, Teofil Jesionowski and Jakub Zdarta**

*Institute of Chemical Technology and Engineering, Faculty of Chemical Technology, Poznan University of Technology, Poznan, Poland*

## 1.1 Enzyme immobilization

### 1.1.1 Biocatalysis - general information

Biocatalysis is defined as the use of natural or engineered biological catalysts to accelerate chemical reactions and has emerged as a central technology at the interface of chemistry, biology, and engineering sciences. Enzymes, which are defined as natural catalysts, are the primary agents of biocatalysis and usually offer exceptional chemo-, regio-, and stereoselectivity, enabling transformations that are often challenging to achieve through conventional synthetic methods (Mu et al., 2020). As a result, biocatalysis has gained significant attention in sectors such as pharmaceutical manufacturing, agrochemical production, food processing, and environmental remediation.

A key advantage of enzymes lies in their inherent specificity. Enzymes provide highly selective reaction pathways due to the three-dimensional arrangement of amino acids within their active sites. This structural precision allows them to differentiate between closely related substrates and control product stereochemistry with remarkable fidelity. Such capabilities are especially valuable in the synthesis of optically pure compounds, where traditional catalysis often requires extensive separations or the use of chiral auxiliaries.

Advances in molecular biology, protein engineering, and directed evolution have greatly expanded the landscape of possible biocatalytic reactions. Historically, enzymes were limited to the natural substrates they evolved to transform. Today, through iterative cycles of mutagenesis and selection, researchers can alter an enzyme's stability, activity, and substrate profile, effectively tailoring enzymes to industrial needs (Ndochinwa et al., 2024). Directed evolution, for example, has yielded enzymes capable of functioning in nonaqueous media, at elevated temperatures, or with completely novel substrates, thereby overcoming many of the constraints associated with early biocatalytic processes. In parallel, computational methods such as molecular modelling, machine learning–assisted sequence prediction, and quantum mechanical simulations have become essential tools in guiding enzyme design. Computational approaches allow researchers to predict beneficial mutations, analyze transition states,

**Carrier-Free Enzyme Immobilization: Fundamentals and Innovations in Biocatalysis.**
**DOI: https://doi.org/10.1016/B978-0-443-44139-4.00003-4**

and simulate enzyme–substrate interactions with increasing accuracy. The integration of computational strategies with laboratory evolution has accelerated the development of biocatalysts and reduced experimental cost and time.

From an environmental perspective, biocatalysis aligns closely with the principles of green chemistry. Enzymatic reactions typically occur under mild conditions, including ambient temperatures, neutral pH, and aqueous solvents, thus reducing the energy consumption and hazardous waste associated with traditional chemical syntheses (Sheldon, 2024; Sheldon & Brady, 2022). Additionally, the biodegradable nature of enzymes and their reliance on renewable resources further contribute to the sustainability of biocatalytic processes. In industrial settings, enzymatic cascades and multienzyme systems are increasingly used to streamline production, minimize by-products, and achieve high atom economy.

Despite these advantages, challenges remain. Enzyme stability under harsh industrial conditions, substrate inhibition, and high production costs can limit broader implementation. Nevertheless, biocatalysis continues to evolve as a transformative platform for sustainable and efficient chemical synthesis. Its combination of specificity, tunability, and environmental compatibility positions it as a vital component of future industrial biotechnology and green chemistry initiatives.

### 1.1.2 Enzyme immobilization - definition and significance

Enzymes, despite their remarkable catalytic efficiency and specificity, possess inherent limitations that restrict their direct application in many industrial and technological processes. As biological macromolecules, they are often sensitive to temperature fluctuations, extreme pH values, organic solvents, and mechanical stress, which can lead to rapid loss of activity or structural denaturation (Eğilmez & Haspolat, 2024). Additionally, free enzymes dispersed in solution cannot be easily recovered or reused, resulting in higher operational costs and increased waste generation. Their limited stability under harsh process conditions and their tendency to exhibit narrow substrate ranges further highlight the need for continuous improvement of enzymatic properties. To address these challenges, extensive research focuses on enhancing enzyme robustness, expanding their functional compatibility, and optimizing their performance for specific applications. One promising strategy is enzyme immobilization (Fig. 1.1), which involves confining enzymes in defined microenvironments that protect them from detrimental factors while maintaining their catalytic function.

The growing interest in enzyme immobilization reflects its profound conceptual and strategic significance within modern biotechnology. Immobilization is not merely a technical step applied to enzymatic systems, but rather an essential framework that shapes how enzymes are understood, engineered, and integrated into real-world applications (Tadesse & Liu, 2025; Wang et al., 2024). Its importance stems from the way it redefines the relationship between enzymes and their operational environment, transforming enzymes from freely moving molecular catalysts into components of structured, engineered systems. By situating enzymes within controlled

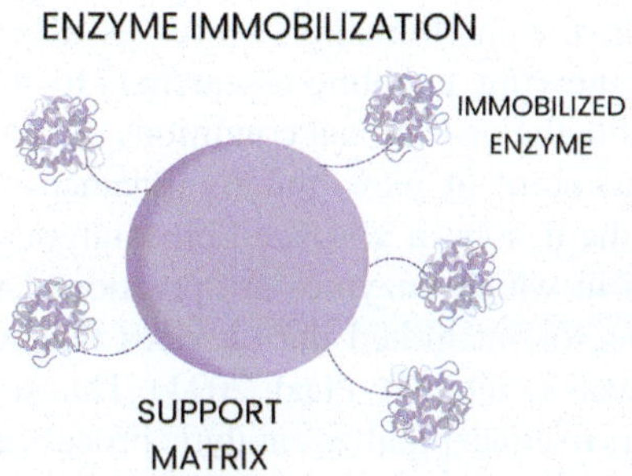


**FIGURE 1.1**

Concept of enzyme immobilization.

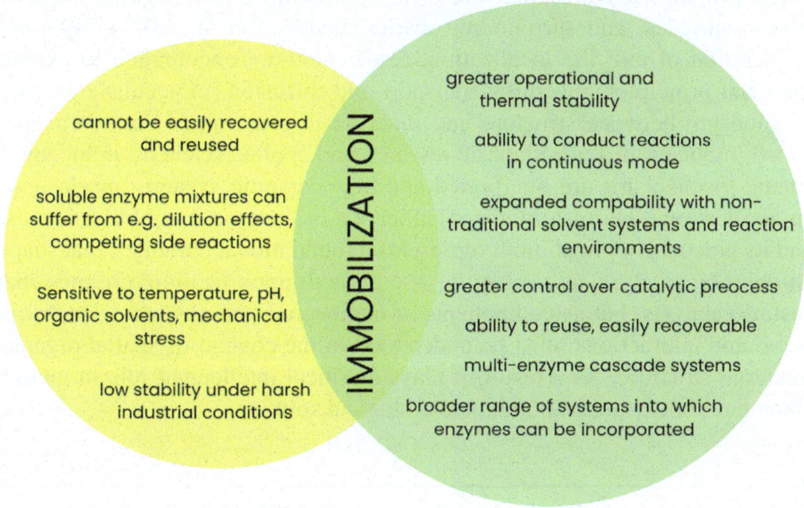


**FIGURE 1.2**

Advantages of immobilized enzymes.

spatial domains, immobilization introduces a new dimension to biocatalyst design, which emphasizes architecture, localization, and the interplay between biological function and material context (Maghraby et al., 2023) (Fig. 1.2). At a scientific level, immobilization represents a shift from viewing enzymes as isolated biochemical entities to considering them as functional elements within larger, multicomponent systems. This perspective is crucial for the advancement of bioprocess engineering, as it aligns enzymatic catalysis with principles of reactor design, materials science, and systems chemistry. The act of confining enzymes creates an interface between biology and engineered materials, prompting deeper exploration of how surface properties, microstructures, and physicochemical environments influence catalytic behavior

(Wang et al., 2023). Thus, immobilization serves as a bridge between molecular biology and applied engineering, enabling researchers to study enzymes in contexts that more closely resemble the operational conditions of industrial processes.

From a technological point of view, the significance of immobilization lies in its foundational role in the design of advanced biocatalytic platforms. It enables the construction of systems in which enzymes are positioned with precision, arranged in defined configurations, or integrated into devices that rely on spatial control at micro- and nanoscale levels (Žnidaršič-Plazl, 2021). This spatial organization allows researchers and engineers to conceptualize catalytic processes in new ways by moving from dispersed, unstructured reactions toward highly organized systems capable of sophisticated functionality. Immobilization, therefore, becomes a prerequisite for the development of modern biotechnologies, including modular reactors, integrated bioprocessing units, and multifunctional catalytic assemblies. Furthermore, immobilization provides a conceptual basis for rethinking how enzymes interact with substrates, surfaces, and surrounding media (Federsel et al., 2021). By forcing a reconsideration of enzyme localization, researchers are encouraged to explore the fundamental principles governing transport and diffusion, molecular ordering, and the relationship between structure and function in constrained environments. This has broad implications for both basic research and applied sciences, influencing how enzymatic mechanisms are interpreted and how catalytic systems are designed for future applications. Ultimately, the significance of enzyme immobilization extends beyond its practical application. It represents a foundational paradigm that shapes the evolution of biocatalysis as a scientific discipline. Immobilization reframes enzymes not just as catalysts, but as components of engineered systems whose performance, behavior, and interactions must be understood in the context of spatial organization and material structure. As a result, it plays a critical intellectual role in guiding the development of next-generation biotechnological solutions.

## 1.2 Advantages of enzyme immobilization

Immobilization is a key technique used in biotechnology, pharmaceuticals, the food industry, and environmental protection. Its growing popularity results from numerous advantages that significantly enhance the efficiency of biocatalytic processes and their economic feasibility (Fig. 1.2). In contrast to free-form catalysts, immobilized systems exhibit greater operational and thermal stability, meaning they are less susceptible to deactivation under harsh reaction conditions. Thanks to this, they can be used in processes carried out at elevated temperatures, extreme pH values, or in the presence of inhibitory substances.

One of the most important advantages of immobilization is the ability to conduct reactions in a continuous mode, for example, in flow bioreactors, where immobilized enzymes remain in the system while products are constantly removed. Such a configuration increases productivity and allows for more efficient use of resources. Immobilization also enables the development of processes that yield products of higher purity, as contamination related to enzyme proteins is significantly reduced (Attique et al., 2023).

Another important aspect of immobilization is the ability to reuse enzymes, which represents one of the most compelling advantages of converting soluble biocatalysts into heterogeneous, easily recoverable forms. Free enzymes, once mixed with substrates, typically become part of the reaction effluent and cannot be readily separated without complex, time-consuming, and costly purification procedures. Such downstream processing can significantly increase the overall expense of biocatalytic operations and generate large amounts of waste (Khan, 2021). Immobilized enzymes, in contrast, can be retained within reactors, removed by simple physical means such as filtration or centrifugation, or incorporated into fixed, structured systems that allow repeated or continuous use (Taheri-Kafrani et al., 2021). This reusability not only decreases catalyst consumption but also improves process economics by reducing labor, materials, and energy demands associated with catalyst replacement. In large-scale operations, where even minor efficiency gains may translate into substantial financial benefits, the economic advantages of enzyme immobilization are often decisive. Consequently, immobilization plays a pivotal role in enabling the transition from traditional chemical catalysis, which is frequently associated with harsh conditions and toxic reagents, to greener, more sustainable biological alternatives.

Immobilization also provides a powerful means of exerting greater control over catalytic processes, a level of manipulation that is difficult to achieve with free, soluble enzymes. By confining enzymes within defined microenvironments, it becomes possible to influence parameters such as substrate accessibility, local pH, surface charge, hydration state, and the structural characteristics of the surrounding matrix (Guisan et al., 2022). These microenvironmental effects can modulate reaction rates by enhancing or restricting substrate diffusion, altering enzyme–substrate interactions, and influencing the conformational flexibility of the protein. For example, immobilized enzymes situated within slightly hydrophobic microdomains may exhibit altered substrate specificity due to changes in binding affinities or restricted mobility of the catalytic residues. In certain cases, immobilization can even enhance catalytic activity. This phenomenon may arise from several factors, including improved enzyme orientation on solid surfaces, stabilization of the active conformation, or reduction in the formation of inactive aggregates (Caparco et al., 2022). While free enzymes in solution may undergo self-association or partial unfolding, immobilized enzymes are often held in conformations that favor catalytic turnover. The ability to fine-tune such conformational and environmental factors at the microscale is a distinguishing advantage of immobilized enzymes, enabling a level of catalytic optimization that extends beyond the constraints of the bulk reaction medium.

Beyond improved process control, immobilization directly supports the development of continuous reaction systems, which constitute the backbone of modern industrial bioprocessing. Continuous systems rely on catalysts that remain within the reactor while substrates and products are continuously fed in and removed. Free enzymes are unsuitable for these configurations because they would be washed out with the product stream, making it impossible to maintain a stable catalytic density (Chauhan et al., 2022). The loss of an active catalyst would compromise both productivity and product quality. Immobilized enzymes, however, remain fixed within the reactor system, including packed-bed columns, membrane reactors, fluidized beds, or other engineering configurations, and enable long-term, steady-state operation (Naramittanakul et al., 2021). The benefits of continuous bioprocessing include increased productivity, consistent product quality, reduced downtime associated with batch-to-batch transitions, and improved integration of upstream and downstream operations. This configuration is especially relevant for pharmaceutical manufacturing, where consistent reaction conditions are essential for meeting regulatory requirements and ensuring batch reproducibility (Rajaraman, 2024). Continuous systems employing immobilized enzymes also enable more efficient heat and mass transfer, providing a higher degree of process intensification. As the global industry increasingly shifts toward continuous manufacturing paradigms, immobilized enzymes will continue to play a key enabling role, bridging the gap between biological catalysis and large-scale industrial throughput.

Immobilization is also essential for constructing multienzyme cascade systems, where multiple enzymatic steps occur sequentially within a single reactor or within coordinated modules (Cai et al., 2025). In natural metabolic pathways, enzymes operate cooperatively, transferring unstable intermediates with remarkable speed and precision. Recreating such efficient, multistep processes in engineered systems poses significant challenges when using free enzymes. Soluble enzyme mixtures can suffer from dilution effects, intermediate instability, competing side reactions, and suboptimal spatial arrangements (Babaei et al., 2023). Immobilization overcomes these limitations by enabling the co-localization of enzymes within defined structures. When enzymes are immobilized in close proximity, intermediates can be transferred directly from one active site to the next, minimizing diffusion losses and preventing degradation. This close arrangement, often referred to as substrate channeling, enhances reaction efficiency and allows more complex biotransformation to be performed with greater control (Selvaraj et al., 2022). Such multienzyme cascades have become increasingly important in the synthesis of pharmaceuticals, high-value chemicals, flavors, fragrances, and chiral intermediates. These cascades may significantly reduce the number of purification steps required, thereby lowering costs and minimizing waste. Moreover, immobilized cascades can be engineered to mimic entire enzymatic pathways, offering an elegant and sustainable alternative to multistep chemical synthesis (Zhang et al., 2021).

Environmental sustainability represents another domain in which immobilization plays an essential role. Because immobilized enzymes can be reused repeatedly, they reduce the frequency with which new catalyst batches must be produced or disposed of. Moreover, many immobilized systems operate effectively under mild, environmentally benign conditions, such as ambient temperature, neutral pH, and aqueous media, thus reducing energy consumption and eliminating the need for hazardous solvents (Najim et al., 2024). These considerations make immobilized enzymes particularly attractive for industries committed to adopting greener, more sustainable manufacturing practices.

The extended operational lifetimes of immobilized enzymes also help minimize waste generation, while the biodegradability of many enzyme-support systems ensures compatibility with environmentally responsible disposal methods.

From a materials and engineering perspective, immobilization broadens the range of systems into which enzymes can be incorporated, thus enabling the development of smart and multifunctional biocatalytic devices. Immobilized enzymes can be integrated into membranes, porous beads, nanofibers, sol–gel matrices, hydrogels, microreactors, biosensors, and bioelectronic devices (Rather et al., 2022). These engineered platforms allow precise control over reaction kinetics, catalytic accessibility, and environmental responsiveness. For example, immobilized enzymes are central components of biosensors used for medical diagnostics. Glucose biosensors, which are the most widely used biosensors worldwide, rely on immobilized glucose oxidase (GOx) to achieve consistent, reproducible, and long-term performance (Sreenivasulu & Rao, 2024). The immobilized enzyme provides rapid response times, stable signal generation, and reliable functioning over repeated measurement cycles, all of which are critical for clinical accuracy.

Immobilization also expands the compatibility of enzymes with nontraditional solvent systems and reaction environments. Free enzymes typically exhibit limited stability in organic solvents, ionic liquids, deep eutectic solvents, or high-salt solutions, all of which are commonly used in industrial processes (Khan, 2025). Immobilized enzymes, however, often display enhanced tolerance to such conditions due to the stabilization provided by the immobilization matrix. In some cases, immobilized enzymes remain active in nearly anhydrous environments, enabling biocatalysis in solvent systems where free enzymes would rapidly denature (Asaduzzaman & Salmon, 2022). These developments broaden the operational scope of enzyme catalysis and create opportunities for applications in synthetic organic chemistry, polymer modification, fine chemical production, and other fields where traditional enzymatic processes were previously unsuitable.

Despite the numerous advantages of immobilization, it is important to recognize that immobilization should not be viewed merely as a method for fixing enzymes in place. Rather, it is a strategic design philosophy that allows for the rational engineering of biocatalytic systems with enhanced robustness, efficiency, and controllability. By tailoring the physicochemical environment of immobilized enzymes, researchers can modulate key performance metrics such as turnover frequency, thermal and chemical stability, substrate affinity, inhibition resistance, and overall catalytic lifetime. Immobilization thus provides an essential tool for aligning enzymatic properties with industrial requirements, enabling biocatalysts to perform reliably under conditions that might otherwise compromise their functionality (Hassan et al., 2025). As biotechnology continues to evolve, immobilization will remain central in bridging disciplines such as molecular biology, protein engineering, materials science, chemical engineering, and environmental technology. The emergence of new support materials, including metal–organic frameworks, polymer–inorganic hybrids, nanostructured surfaces, and stimuli-responsive polymers, offers exciting possibilities for next-generation immobilized systems with unprecedented functionality. Similarly, advances in computational modeling, enzyme engineering, and system biology provide new insights that can guide the design of optimized immobilized biocatalysts.

## 1.3 Carrier bound versus carrier free - techniques of enzyme immobilization

The wide range of enzyme immobilization techniques reflects the dynamic progress of biotechnology and the growing demands of industry, medicine, and research. Modern approaches to anchoring biocatalysts encompass a broad spectrum of strategies that differ in the durability of enzyme attachment, the nature of the supporting materials, and their overall impact on enzyme activity and stability (Mirsalami et al., 2024). This diversity makes it possible to tailor immobilization methods to specific process requirements, including reaction conditions, substrate type, or the desired longevity of the catalytic system (Yuan et al., 2023). The variety of available methodologies fosters innovation as well, supporting the development of complex biocatalytic systems, integration with advanced materials, and the creation of novel technological platforms. Regardless of the chosen strategy, all immobilization techniques share the common objective of optimizing enzyme performance under practical conditions while improving the efficiency of biotechnological processes. As a result, immobilization continues to be a key tool for translating enzymatic reactions from laboratory settings into industrial applications, while maintaining the precision and specificity characteristic of biocatalysis. Enzyme immobilization methods are typically grouped by the immobilization mechanism—i.e., the type of interactions involved. However, the division into carrier-free and carrier-based techniques is also frequently encountered (Fig. 1.3).

### 1.3.1 Carrier-based immobilization techniques

Carrier-based immobilization represents one of the most established and versatile strategies for stabilizing enzymes and adapting them for repeated or continuous use in industrial, analytical, and biomedical processes. In this approach, enzymes are attached to solid supports (often referred to as carriers) through physical or chemical interactions that limit their mobility while preserving their catalytic function. The carrier thus becomes an integral structural component of the biocatalytic system, offering mechanical stability, protection from denaturation, and improved handling properties (Mohidem et al., 2023). The diversity of carriers, combined with the range of possible binding mechanisms, has enabled significant progress in designing tailored immobilized enzyme systems capable of withstanding demanding operational conditions.

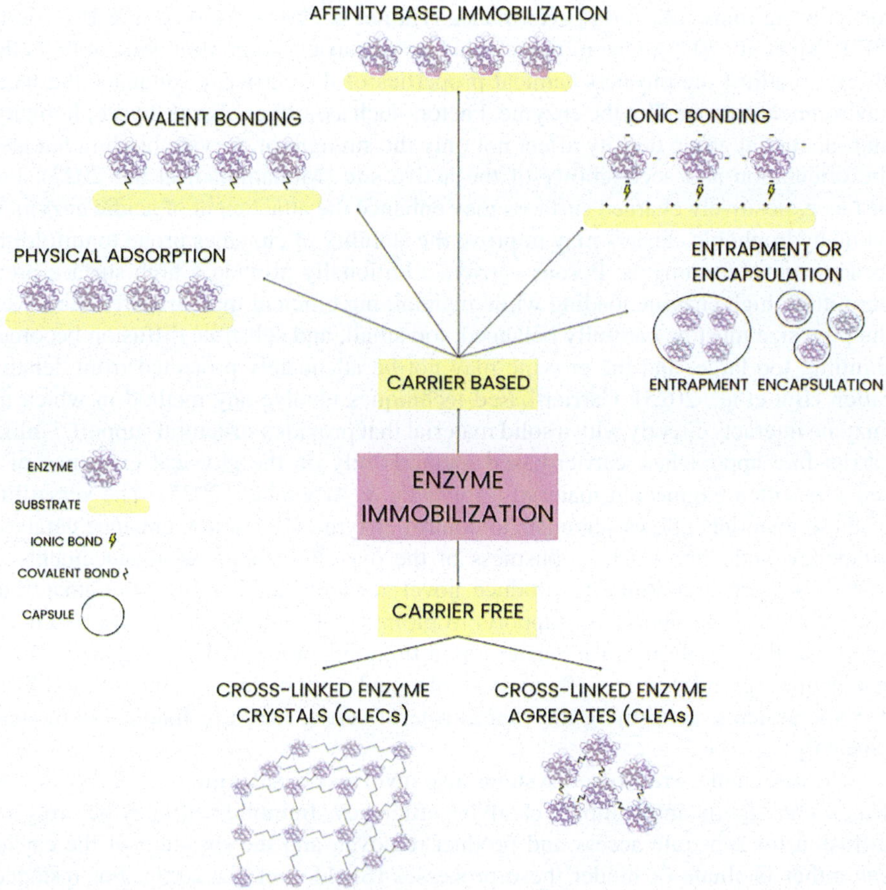


**FIGURE 1.3**

Different techniques of enzyme immobilization.

The selection of an appropriate carrier is a decisive step in designing a successful immobilized enzyme system. Carriers vary widely in chemical composition, structure, porosity, hydrophobicity, and surface reactivity. Traditional materials include natural polymers such as cellulose, agarose, chitosan, and alginate, which are valued for their biocompatibility and mild surface chemistries. Synthetic polymers, including polyacrylamide, polystyrene, and polymethacrylates, offer greater tunability, allowing precise control over pore size, particle shape, and chemical functionality. Inorganic materials, such as silica, glass beads, and various metal oxides, are chosen for applications requiring high mechanical strength or operation under harsh chemical conditions. More recently, advanced materials such as nanostructured carriers, magnetic particles, and hybrid organic–inorganic matrices have expanded the possibilities for tailoring immobilization performance (Hao et al., 2024; Matveeva & Bronstein, 2021; Xu et al., 2025). One of the advantages of carrier-bound immobilization is the ability to adjust the physicochemical properties of the carrier to influence the local environment surrounding the enzyme. Factors such as surface charge, hydrophobicity, and functional group density affect not only the strength of enzyme binding but also the orientation and accessibility of the active site (Mohammadi et al., 2023). For instance, positively charged surfaces may enhance the attachment of acidic enzymes, while hydrophobic carriers may improve the stability of enzymes prone to unfolding in aqueous environments. Porous carriers additionally provide a high surface area, permitting high enzyme loading while maintaining efficient mass transfer. However, the pore size must be carefully balanced: too small, and substrate diffusion becomes limiting; too large, and the enzyme may not be adequately protected from denaturation (Bai et al., 2025). Carrier-based techniques involve any method in which an enzyme interacts directly with a solid material that provides structural support. Unlike carrier-free approaches, carrier-based methods rely on the physical existence of a separate, often engineered material (Holyavka & Artyukhov, 2025). The versatility of these materials allows scientists to control enzyme orientation, enhance catalytic efficiency, and improve the robustness of the overall system. As advancements in materials science continue to produce novel carriers, ranging from nanomaterials and hybrid organic-inorganic supports to bioinspired scaffolds, it should be underlined that the possibilities for carrier-based immobilization will only expand. These innovations promise more efficient, stable, and application-specific immobilized enzyme systems, confirming the central role of carrier-based techniques in modern biocatalysis.

The design of carrier-based systems also involves considerations of mass transfer, which can significantly impact catalytic efficiency. Immobilized enzymes rely on diffusion for substrate access and product removal, and the structure of the carrier can either facilitate or hinder these processes (Sheldon et al., 2021). For instance, highly porous materials improve substrate accessibility, but excessively complex pore structures may slow down diffusion. To address such limitations, researchers often tailor carrier morphology using macroporous beads, hierarchical pore systems, or surface-immobilized thin layers to optimize reaction rates while maintaining enzyme stability (Ran et al., 2025). In contrast, nanoparticle-based carriers, including magnetic nanoparticles, offer exceptionally high surface-area-to-volume ratios that can significantly reduce diffusion limitations. Moreover, their small size allows them to remain suspended during reaction while enabling rapid and efficient magnetic recovery.

Within the broad category of carrier-based immobilization, several distinct techniques can be identified based on the type of interaction responsible for binding the enzyme to the carrier. Although each method has its own advantages, limitations, and areas of application, all share the common characteristic of anchoring the enzyme to a solid phase. The primary techniques included in this group are physical adsorption, covalent binding, affinity-based immobilization, ionic binding, and entrapment or encapsulation within porous carriers (Alvarado-Ramírez et al., 2021). Together, these methods cover a wide range of interactions, from weak and reversible to strong and permanent, enabling the design of immobilization strategies tailored to specific requirements.

One of the most straightforward carrier-based methods is physical adsorption, which relies on weak, noncovalent interactions between an enzyme and the surface of a carrier (Silva Almeida et al., 2024). These forces may include van der Waals interactions, hydrogen bonding, hydrophobic interactions, or electrostatic attraction. Adsorption is attractive due to its simplicity, low energy requirements, and minimal impact on enzyme structure. For immobilization, the enzyme is typically mixed with the carrier in an aqueous environment, allowing spontaneous attachment driven by surface forces. Because this method does not involve chemical modification, enzyme activity is usually well preserved (de Andrade Silva et al., 2022). However, the weakness of the interactions also means that desorption can occur under conditions of changing pH, ionic strength, or the presence of competing molecules. Despite this drawback, adsorption remains a widely used carrier-based technique, especially in cases where mild immobilization conditions are necessary. A related, but more selective category is affinity-based immobilization, which exploits specific biological interactions between the enzyme and a ligand attached to the carrier. These ligands may include substrates, inhibitors, cofactors, antibodies, metal ions, or engineered recognition tags such as His-tags (Sardar & Mirza, 2023). The highly specific nature of the interaction allows for precise orientation and strong yet often reversible binding. Affinity-based methods are particularly valuable for applications in biosensors and analytical devices, in which the fidelity and consistency of enzyme activity are critical (Zarini et al., 2025). Although this approach may require more complex carrier preparation and higher material costs, its ability to provide controlled immobilization with minimal loss of catalytic function makes it an important part of the carrier-based technique family. Closely related to adsorption and affinity interactions is ionic binding, another prominent technique included within the carrier-based immobilization group. Many carriers are engineered to contain charged functional groups, such as quaternary ammonium groups, carboxylates, or phosphates, that can interact with complementary charges on the enzyme's surface (Badgujar et al., 2022). These ionic interactions are stronger than simple adsorption but still reversible under certain conditions. Ionic binding offers a balance between stability and mild immobilization, making it useful in processes where enzyme recycling is needed but extreme chemical resistance is not required. Because surface charge distribution varies among enzymes, this method can also help orient the enzyme favorably on the carrier surface, enhancing accessibility of the active site.

A second major technique within carrier-based immobilization is covalent binding, in which enzymes form strong, irreversible bonds with reactive functional groups on the carrier surface. Carriers are often chemically activated using reagents that introduce groups such as aldehydes, epoxides, or carbodiimide-derived intermediates (Prabhakar et al., 2025). These groups can react with amino, carboxyl, thiol, or hydroxyl moieties on the enzyme, forming stable covalent linkages. Because covalent attachment provides exceptional stability, immobilized enzymes created through this method exhibit resistance to harsh operational conditions, including changes in pH, temperature, and solvent composition (Almulaiky et al., 2024). This technique is especially advantageous for industrial systems that require long-term use and minimal enzyme leaching. However, inappropriate coupling chemistry or excessive modification may damage the enzyme's active site, so optimization is usually required. Nevertheless, covalent immobilization is a cornerstone of carrier-based techniques due to its robustness and reliability.

A special subset of carrier-based immobilization includes entrapment and encapsulation within porous carriers. Although these methods differ mechanistically from binding to surface functional groups, they still qualify as carrier-based because the enzyme is physically confined within a solid matrix. In entrapment, the enzyme is incorporated into the internal structure of a porous carrier, such as a gel, polymer network, or inorganic matrix (Imam et al., 2021). By contrast, encapsulation often involves enclosing the enzyme within microcapsules or hollow particles (Weng et al., 2023). In both cases, the enzyme does not form direct chemical bonds with the carrier; instead, it is retained through size exclusion or physical confinement. These techniques offer enhanced protection from environmental stresses, but mass-transfer limitations must be considered, especially when dealing with large substrates or high-viscosity reaction media.

Taken together, these methods all rely on the presence of a solid phase that supports the enzyme, controls its microenvironment, and enables its reuse. The diversity of these approaches allows researchers and engineers to tailor immobilization strategies to specific reaction requirements and operational conditions. Whether stability, cost, enzyme orientation, or mass-transfer efficiency is the primary concern, carrier-based methods offer a versatile toolbox for optimizing biocatalytic processes.

Carrier-bound immobilization also facilitates the engineering of multifunctional biocatalytic systems (Ölçücü et al., 2021). By controlling the spatial arrangement of enzymes on a carrier, it is possible to create cascades in which multiple enzymes act sequentially to convert a substrate into a final product. Co-immobilization improves reaction efficiency by minimizing diffusion losses between steps and has proven especially valuable in synthesizing complex chemicals or degrading recalcitrant pollutants. In some advanced designs, carriers are engineered to provide microenvironments tailored to the differing requirements of each enzyme, enabling synergistic processes that would be challenging to achieve in solution (Zhang et al., 2025).

In industrial applications, carrier-bound immobilized enzymes are indispensable in processes such as food production, pharmaceutical synthesis, biofuel processing, and wastewater treatment. For example, immobilized lipases are widely used in esterification and transesterification reactions, where their enhanced stability and reusability significantly reduce production costs. For instance, immobilized glucose isomerase remains a cornerstone of high-fructose corn syrup production, operating continuously for extended periods (Jin et al., 2021). In clinical settings, carriers are used to immobilize enzymes for biosensors, diagnostic devices, or therapeutic applications, where stability and controlled activity are essential (Prakash et al., 2024).

Carrier-bound systems also play a role in bioremediation, allowing enzymes to withstand pollutants or extreme environmental conditions while degrading contaminants efficiently (Osuoha & Nwaichi, 2021).

The development of novel carriers continues to drive innovation in immobilization technology. Advances in materials science, from metal–organic frameworks to graphene derivatives and bioinspired polymers, offer unprecedented control over surface chemistry, geometry, and mechanical properties. Such materials can be designed to respond to stimuli (e.g., pH, temperature, magnetic fields), enabling dynamic regulation of enzyme activity. Furthermore, progress in nanotechnology has led to carriers that facilitate intracellular delivery, targeted catalysis, or integration into microfluidic systems (Wang et al., 2021a). These evolving capabilities are expanding the role of immobilized enzymes beyond traditional industrial uses into areas such as synthetic biology, green chemistry, and advanced biomedical engineering. Overall, carrier-bound immobilization techniques offer a powerful set of tools for enhancing enzyme performance, extending their operational lifetime, and enabling implementation in diverse technological environments. By carefully selecting carriers, optimizing binding strategies, and engineering appropriate microenvironments, researchers can design highly efficient, robust, and application-specific biocatalytic systems. As new materials and hybrid approaches emerge, carrier-bound immobilization will continue to play an important role in modern biocatalysis and the development of sustainable, enzyme-driven technologies.

### 1.3.2 Carrier free immobilization techniques

Carrier-free immobilization represents an important class of methods in which enzymes are stabilized and immobilized without the use of an external solid support. Unlike carrier-based approaches, these techniques eliminate the need for a separate matrix or scaffold, relying instead on the self-assembly or cross-linking of enzymes into stable, insoluble structures (Chauhan et al., 2022). Carrier-free systems have gained considerable attention in recent years due to their high catalytic density, reduced diffusional limitations, and lower production costs, making them particularly attractive for industrial biocatalysis and processes requiring high volumetric productivity (Ölçücü et al., 2023).

The crucial principle of carrier-free immobilization is that the enzyme itself becomes the structural element of the immobilized system. This contrasts with traditional immobilization methods, where the carrier may account for 90%–99% of the total mass of the biocatalyst, significantly diluting catalytic activity (Bisht et al., 2023). By removing the carrier, these techniques yield preparations with exceptional enzyme loading, often resulting in improved reaction rates and more compact reactor designs. Furthermore, eliminating the carrier reduces the environmental footprint and cost of immobilized biocatalyst production, particularly important for large-scale industrial applications. Among the various carrier-free techniques, the most widely used are cross-linked enzyme aggregates (CLEAs) and cross-linked enzyme crystals

(CLECs) (Chen et al., 2023). CLEAs are formed by precipitating the enzyme using agents such as ammonium sulfate, organic solvents, or polymers and subsequently cross-linking the aggregated enzyme molecules with bifunctional agents like glutaraldehyde (Costa et al., 2023). This process produces stable, insoluble particles that retain high catalytic activity. CLEAs offer several advantages: ease of preparation, compatibility with crude enzyme extracts, enhanced stability, and tunability through co-aggregation with additives or other enzymes. Multienzyme CLEAs, for example, enable cascade reactions within a single immobilized system, improving overall efficiency. CLECs, by comparison, start with the crystallization of the enzyme, followed by cross-linking the ordered crystalline structure. This method yields highly pure, mechanically robust catalysts with uniform properties (Kubiak et al., 2022). Although crystallization can be challenging for some enzymes, CLECs exhibit exceptional stability and long operational lifetimes. Their well-defined structure often provides enhanced resistance to temperature, organic solvents, and proteolytic degradation, making them valuable for specialty chemical synthesis and pharmaceutical applications.

Another emerging approach in the carrier-free category is enzyme self-assembly, in which enzymes spontaneously organize into macromolecular structures through engineered interfaces or naturally occurring interaction domains (Lou et al., 2023). This method can produce ordered assemblies with controlled geometry, orientation, and catalytic function. Although still under active development, self-assembled enzyme structures show promise for nanoscale catalysis, biosensing, and synthetic biology.

It might be briefly summarized that carrier-free immobilization also reduces diffusional limitations typically associated with porous carriers. Substrates can access catalytic sites more easily, increasing turnover rates and allowing efficient processing of high-molecular-weight compounds. However, because there is no protective matrix, mechanical stability can be lower compared to supported systems, and careful optimization of cross-linking and particle morphology is required. Overall, carrier-free immobilization techniques provide powerful alternatives to carrier-based methods, offering high catalytic density, improved sustainability, and the potential for enhanced enzyme performance. As research advances and new cross-linking strategies and self-assembly methods emerge, carrier-free systems are poised to play an increasingly significant role in industrial biocatalysis and advanced biotechnological applications. Nevertheless, a detailed description of the carrier-free approach will be presented in the next chapter of this book.

### 1.3.3 Carrier free versus carrier based immobilization techniques

Among the major immobilization approaches, *carrier-based* and *carrier-free* techniques represent two contrasting but complementary philosophies. Carrier-based methods involve attaching enzymes to an external solid support, whereas carrier-free methods rely solely on the enzyme molecules themselves, typically structured through aggregation, crystallization, or self-assembly. Understanding the strengths of each approach is essential for designing efficient biocatalytic systems tailored to specific operational needs.

Carrier-based immobilization is the most established class of techniques and offers several practical and engineering advantages. As stated above, the use of a solid support allows immobilized enzymes to be easily handled, separated from reaction mixtures, and integrated into continuous reactors such as packed-bed or fluidized-bed configurations. This makes carrier-based systems suitable for large-scale industrial processes where catalyst recovery and reusability are critical. The diversity of available carriers also provides flexibility to design microenvironments that enhance enzyme stability. Surface chemistry can be tailored to promote favorable enzyme orientation, increase resistance to denaturation, or modulate hydrophobicity and charge (Ranimol & Sunkar, 2022). Furthermore, covalent immobilization onto carriers offers exceptional operational durability, preventing enzyme leaching even under harsh process conditions. For applications requiring long-term stability, mechanical robustness, and ease of process integration, carrier-based immobilization remains the preferred choice.

In contrast, carrier-free immobilization focuses on maximizing catalytic density by eliminating the mass contribution of an external support. Methods such as CLEAs and CLECs produce compact, insoluble biocatalysts composed almost entirely of enzyme molecules. As a result, carrier-free preparations exhibit extremely high volumetric activity, making them especially advantageous in systems where space efficiency or high reaction rates are priorities (Patti et al., 2025). The absence of an inert carrier also reduces production costs, both by lowering material expenses and by enabling the use of crude enzyme extracts in some cases. Moreover, carrier-free immobilization often yields improved mass transfer properties, as substrates do not need to diffuse through the pores of a carrier. This is particularly beneficial for reactions involving high-molecular-weight substrates or viscous media. Additionally, the structural rigidity introduced through cross-linking frequently enhances thermal, pH, and solvent stability, extending the lifespan of the biocatalyst.

Despite these differences, carrier-based and carrier-free approaches should not be viewed as competing technologies but rather as complementary strategies with distinct advantages (Fig. 1.4). Carrier-based systems showed great mechanical stability and compatibility with continuous operations, which makes them ideal for extensive industrial use. They also allow for versatile designs, including co-immobilization, surface patterning, and integration with advanced materials. Carrier-free systems, meanwhile, offer high catalytic efficiency and sustainability by minimizing waste and maximizing active enzyme content. They are particularly useful when high productivity per unit volume is needed, when carriers would introduce diffusion limitations, or when reducing costs is essential. Choosing between these techniques depends heavily on the intended application. For large-scale continuous bioprocessing, carrier-based immobilization typically offers superior performance due to its robustness and ease of handling. For batch reactions requiring high enzyme loading or minimal diffusional resistance, carrier-free systems may provide significant benefits. In emerging fields such as synthetic biology, microreactor technology, and green chemistry, both

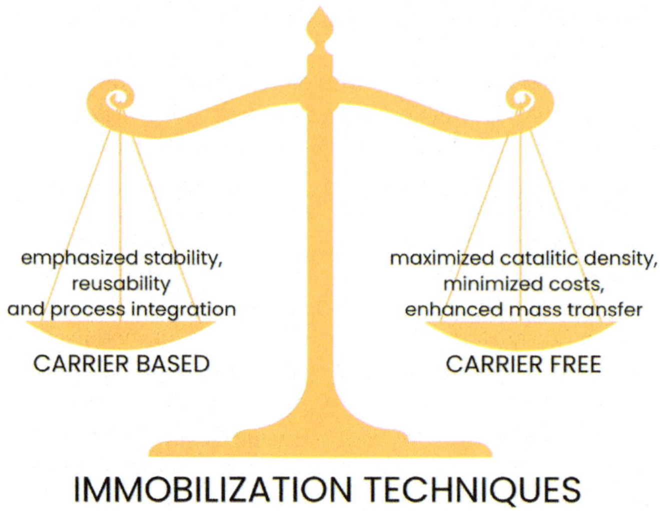


**FIGURE 1.4**

Comparison of carrier based and carrier free techniques.

strategies continue to evolve, often being combined to exploit the strengths of each - including hybrid systems where enzymes are first structured through carrier-free methods and then integrated with supporting materials. Both carrier-free and carrier-based immobilization offer unique and valuable advantages. Carrier-based techniques emphasize stability, reusability, and process integration, while carrier-free approaches maximize catalytic density, minimize costs, and enhance mass transfer. A thoughtful comparison of these strengths enables researchers and engineers to design optimized biocatalytic systems tailored to modern industrial and technological needs.

## 1.4 Applications of immobilized enzymes

Immobilized enzymes, practically from all catalytic groups, are finding more and more practical applications in many branches of industry and everyday life, as shown in Fig. 1.5.

Oxidoreductases (EC1) constitute one of the most frequently immobilized classes of enzymes. The data summarized in Table 1.1 indicate that immobilization of oxidoreductases generally results in moderate to high retention of enzymatic activity, accompanied by a significant improvement in operational and thermal stability. In the case of GOx immobilized on polyester nonwoven fabric using polyethyleneimine, 78% of the activity of the free enzyme was retained. Importantly, immobilization enabled repeated reuse of the enzyme for at least 15 cycles and preserved its activity at elevated temperatures (Zhang et al., 2021b). A similar trend was observed for

**Table 1.1** Overview of immobilized enzymes from different EC classes, including enzyme type, support material, activity recovery, stability and applications.

| EC | Immobilized enzyme | Support material | Activity recovery | Reusability | Application | References |
|---|---|---|---|---|---|---|
| EC1 | Glucose oxidase | Polyester nonwoven fabric, polyethyleneimine | 78% | 78% | Biocatalysis research, and heterogeneous bio-Fenton system | Zhang et al. (2021b) |
| | Horseradish peroxidase | $Fe_3O_4$–$NH_2$/hNF | 45% | 94% | Development of the textile industry and the protection of the environment, dye wastewater decolorization | Bakar et al. (2024) |
| | Laccase | Magnetic amino-functionalized metal–organic framework ($Fe_3O_4$-$NH_2$@MIL-101(Cr)) | 88% | 87% | Dye removal form wastewater | Amari et al. (2021) |
| | Oxidoreductase | Epoxy functionalized $Fe_3O_4$ nanoparticles | 80% | 90% | Promising as a nano-biocatalyst carrying out in many industries and different fields | Ali et al. (2022) |
| | B-Glucosidase | Bio-graphene | >98.5% | 72% | The development of more sustainable catalysts for a variety of biological applications. | Alatzoglou et al. (2023) |
| | Chloroperoxidase | Pore structure of MOFs | >95.2% | 70.7% | Potential in wastewater treatment | Yang et al. (2023) |

*(continued on next page)*

**Table 1.1** Overview of immobilized enzymes from different EC classes, including enzyme type, support material, activity recovery, stability and applications—cont'd

| EC | Immobilized enzyme | Support material | Activity recovery | Reusability | Application | References |
|---|---|---|---|---|---|---|
| EC2 | Glycosyltransferase | Chitosan beads | 72.2% | 72% (after 4 cycles) 53% (after 8 cycles) | Potentially value for the production of Reb M | Wang et al. (2021b) |
| | Transaminases | Polypropylene (PP) membrane | 30%–50% | 85% | Future use in flow mode hybrid processes | Meersseman Arango et al. (2024) |
| | Fructose transferase | Eucalyptus biochar functionalized with glutaraldehyde | 96.51% | 48% | FOS production | da Silva et al. (2024) |
| | P. fluorescens $\omega$-transaminase | Cross-linked enzyme aggregates | 29% | 74% | Biocatalytic applications | Velasco-Lozano et al. (2020) |
| | Glycogen synthase kinase-3$\beta$ | 316L stainless steel plates | Enzyme inhibition activity retained after immobiliza-tion | 85% | Cardiovascular biomaterials (stents), endothelialization enhancement | Zhang et al. (2021c) |

(continued on next page)

| | | | | | | |
|---|---|---|---|---|---|---|
| EC3 | α-glucosidase | Biocompatible super porous p(HEMA) cryogel | 90% | >50% | Biotechnological use in real industrial applications | Demirci et al. (2020) |
| | α-chymotrypsin | Zwitterionic microgel synthesized from poly (carboxybetaine methyl methacrylate) | 72% | 72% | Applications in organic chemistry | Erfani et al. (2021) |
| | *Aspergillus niger* cellulase | $Fe_3O_4$/graphene oxide/chitosan nanocomposite | 86% | 89% | Hydrolysis of pretreated cellulose in wheat and rice bran | Ashwini John et al. (2023) |
| | *Trichoderma reesei cellulase* | Chitosan modified $Fe_3O_4$/graphene oxide nanocomposite activated with 2% GDA | 78% | 66% | Saccharification of microcrystalline cellulose | Asar et al. (2020) |
| | Xylanase | $SiO_2$ nanoparticles | 84% | 84% | Production of xylooligosaccharides for various applications in the food, paper, animal sectors and for the bioconversion of lignocellulosic waste | Pagolu et al. (2021) |

*(continued on next page)*

**Table 1.1** Overview of immobilized enzymes from different EC classes, including enzyme type, support material, activity recovery, stability and applications—cont'd

| EC | Immobilized enzyme | Support material | Activity recovery | Reusability | Application | References |
|---|---|---|---|---|---|---|
| EC4 | Carbonic anhydrase | CS/PVA Hydrogel-MOF Composites | 50% | 86% | $CO_2$ catalytic conversion | Cao et al. (2025) |
| | Phenylalanine ammonia-lyase | Electrospun nanofibrous membrane | 36% | 84% | *Not reported* | Hsieh et al. (2023) |
| | Fructose-6-phosphate aldolase | Cobalt-chelated agarose (Co-IDA) | 93.8% | 80% | Synthesis of a precursor of D-fagomine | Masdeu et al. (2021) |
| | Fumarase | Ceramic beads | 86% | 86% | Used in the industrial production of L-malate | Yamane et al. (2024) |
| EC5 | *D*-allulose 3-epimerase | Zeolitic imidazolate framework | 80% | 80% | Industrial-scale manufacturing of D-allulose | Tan et al. (2023) |
| | *D*-allulose 3-epimerase | Epoxy resin | 75% | 80% | Production of D-allulose | Tan et al. (2023) |
| | Glucose isomerase | *E. coli* cells expressing glucose isomerase | 55% | 85% | Biocatalyst for HFCS-55 production | Jin et al. (2017) |

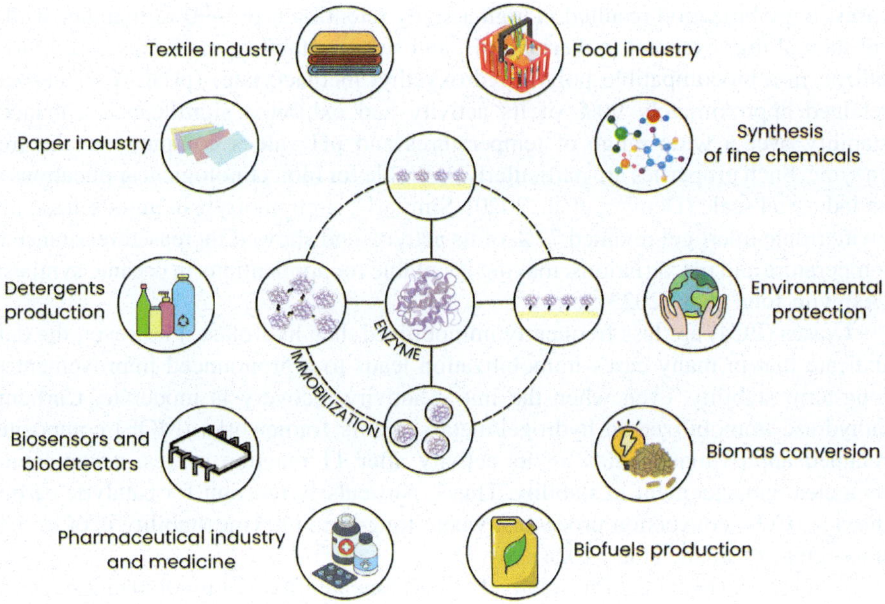


**FIGURE 1.5**

Application of immobilized enzymes.

horseradish peroxidase (HRP) immobilized on magnetic $Fe_3O_4$–$NH_2$/hNF nanocomposites. Although the retained activity decreased to approximately 45% after six reaction cycles, the immobilized enzyme exhibited higher thermal stability compared to its free counterpart. Such properties are particularly advantageous in dye wastewater decolorization processes in the textile industry (Bakar et al., 2024).

Transferases (EC2) are considerably more demanding with respect to immobilization than oxidoreductases or hydrolases. This is mainly due to their complex catalytic mechanisms, the requirement for cofactors, and their higher sensitivity to diffusional limitations. The data presented in Table 1.1 clearly reflect these characteristics, showing a wider variation in retained activity after immobilization. Glycosyltransferase immobilized on chitosan beads exhibited a gradual decrease in activity with successive reuse cycles, retaining 72.5% activity after four cycles and 53.1% after eight cycles. Despite this decline, the enzyme maintained high structural stability (Wang et al., 2021b). In contrast, $\omega$-transaminase immobilized in the form of CLEAs showed a relatively low activity recovery (29%), but significantly enhanced thermostability at temperatures of 60°C–70°C. This example clearly illustrates the typical trade-off observed in the immobilization of transferases, where reduced initial activity is compensated by markedly improved stability (Velasco-Lozano et al., 2020).

Hydrolases (EC3) represent the most extensively studied and most frequently immobilized class of enzymes, which is clearly reflected in the presented data. In most cases, immobilization resulted in high activity retention (70%–90%) together with a substantial improvement in thermal, pH, and storage stability. α-Glucosidase immobilized in a biocompatible poly(2-hydroxyethyl methacrylate) (p(HEMA)) cryogel retained approximately 90% of its activity and exhibited significantly enhanced stability over a wide range of temperatures and pH values compared to the free enzyme. Such properties are particularly desirable for biotechnological applications at an industrial scale (Demirci et al., 2020). Similarly, α-chymotrypsin immobilized in a zwitterionic microgel retained 72% of its activity and showed increased resistance to temperature and pH variations, making it suitable for applications in organic synthesis (Ashwini John et al., 2023).

Lyases (EC4) are less frequently immobilized than hydrolases; however, the data indicate that in many cases immobilization leads to a pronounced improvement in long-term stability, even when the initial activity recovery is moderate. Carbonic anhydrase immobilized in hydrogel/metal–organic frameworks (MOF) composites retained approximately 50% of its activity after 11 reaction cycles, accompanied by a clear enhancement in stability. This is particularly relevant for catalytic carbon dioxide ($CO_2$) conversion processes, where long-term enzyme stability is of critical importance (Yamane et al., 2024).

Isomerases (EC5) exhibit relatively good amenability to immobilization, as evidenced by the high levels of retained activity reported in the table. D-allulose 3-epimerase immobilized in both zeolitic imidazolate framework (ZIF) structures and epoxy resins retained between 75% and 80% of its activity, while simultaneously showing high storage and operational stability (Jin et al., 2017). On the other hand, ligases (EC6) belong to the most challenging enzyme classes with respect to immobilization, which is reflected in the limited number of publications reporting their successful immobilization. Ligases catalyze bond-forming reactions, which significantly complicates their immobilization and application in heterogeneous systems. Further, translocases (EC7) represent the least explored enzyme class in the context of immobilization, and the number of available literature reports remains very limited. These enzymes catalyze the transport of ions or molecules across biological membranes, making their activity strictly dependent on membrane integrity and electrochemical gradients.

## 1.5 Conclusions

Enzyme immobilization represents a key strategy in modern biotechnology, enabling the effective use of enzymes under industrial conditions where stability and consistent performance are critical. Analysis of different approaches has shown that both carrier-based and carrier-free techniques offer unique advantages and can be tailored to specific applications. Immobilization enhances thermal and chemical stability, allows for enzyme reuse, and enables continuous operation, significantly improving the efficiency of biotechnological processes while reducing operational costs. Moreover, controlling the enzyme microenvironment in immobilized systems allows modulation of catalytic parameters, such as substrate accessibility and activity, which is difficult to achieve with free enzymes in solution. Comparison of carrier-based and carrier-free methods indicates that the choice of strategy depends on process requirements: carrier-based techniques are advantageous in continuous systems and applications demanding high mechanical stability, while carrier-free approaches maximize catalytic density and minimize diffusional limitations. Hybrid strategies that combine both methods can further exploit the strengths of each approach. Applications of immobilized enzymes span a wide range of industries, including food production, pharmaceuticals, environmental remediation, biosensors, and synthetic biology. Different enzyme classes exhibit varying susceptibility to immobilization: hydrolases and oxidoreductases are the most stable and commonly used, whereas ligases and translocases require further research and optimization. In conclusion, enzyme immobilization is not merely a stabilization technique but a foundational concept in modern biocatalyst design, enabling the development of efficient, sustainable, and versatile biotechnological systems. Its continued advancement, together with progress in enzyme engineering and materials science, will be critical for introducing innovative and environmentally friendly industrial processes.

## Acknowledgments

This work was supported by the National Science Centre, Poland under the research Grant number 2024/54/E/ST8/00089.

## References

Alatzoglou, C., Patila, M., Giannakopoulou, A., Spyrou, K., Yan, F., Li, W., Chalmpes, N., Polydera, A. C., Rudolf, P., Gournis, D., & Stamatis, H. (2023). Development of a multi-enzymatic biocatalytic system through immobilization on high quality few-layer bio-graphene. *Nanomaterials, 13*(1), 127. doi:10.3390/nano13010127.

Ali, A. O., Ali, K. A., Shahein, Y. E., Shokeer, A., Sharada, H. M., & Abdalla, M. S. (2022). Epoxy functionalized iron oxide magnetic nanoparticles for catalase enzyme covalent immobilization. *Chemical Papers, 76*(7), 4431–4441. www.springer.com/11696. doi:10.1007/s11696-022-02180-6.

Almulaiky, Y. Q., Alkabli, J., & El-Shishtawy, R. M. (2024). Improving enzyme immobilization: A new carrier-based magnetic polymer for enhanced covalent binding of laccase enzyme. *International Journal of Biological Macromolecules, 282*, 137362. https://www.sciencedirect.com/science/journal/01418130. doi:10.1016/j.ijbiomac.2024.137362.

Alvarado-Ramírez, L., Rostro-Alanis, M., Rodríguez-Rodríguez, J., Castillo-Zacarías, C., Sosa-Hernández, J. E., Barceló, D., Iqbal, H. M. N., & Parra-Saldívar, R. (2021). Exploring current tendencies in techniques and materials for immobilization of laccases—A review. *International Journal of Biological Macromolecules, 181*, 683–696. www.elsevier.com/locate/ijbiomac. doi:10.1016/j.ijbiomac.2021.03.175.

Amari, A., Alzahrani, F. M., Alsaiari, N. S., Katubi, K. M., Rebah, F. B., & Tahoon, M. A. (2021). Magnetic metal organic framework immobilized laccase for wastewater decolorization. *Processes, 9*(5), 774. https://www.mdpi.com/2227-9717/9/5/774/pdf. doi:10.3390/pr9050774.

Asaduzzaman, F., & Salmon, S. (2022). Enzyme immobilization: Polymer-solvent-enzyme compatibility. *Molecular Systems Design and Engineering, 7*(11), 1385–1414. www.rsc.org/journals-books-databases/about-journals/molecular-systems-design-engineering/. doi:10.1039/d2me00140c.

Asar, M. F., Ahmad, N., & Husain, Q. (2020). Chitosan modified $Fe_3O_4$/graphene oxide nanocomposite as a support for high yield and stable immobilization of cellulase: Its application in the saccharification of microcrystalline cellulose. *Preparative Biochemistry and Biotechnology, 50*(5), 460–467. www.tandf.co.uk/journals/titles/10826068.asp. doi:10.1080/10826068.2019.1706562.

Ashwini John, J., Samuel, M. S., & Selvarajan, E. (2023). Immobilized cellulase on $Fe_3O_4$/GO/CS nanocomposite as a magnetically recyclable catalyst for biofuel application. *Fuel, 333*, 126364. doi:10.1016/j.fuel.2022.126364.

Attique, S. A., Qurat ul Ain, Hussain, N., Bilal, M., & Iqbal, H. M. N. (2023). Enzyme immobilization approaches. *Biocatalyst immobilization* (pp. 37–54). Academic Press. doi:10.1016/B978-0-323-91317-1.00007-4.

Babaei, H., Ghobadi Nejad, Z., Yaghmaei, S., & Farhadi, F. (2023). Co-immobilization of multi-enzyme cascade system into the metal–organic frameworks for the removal of Bisphenol A. *Chemical Engineering Journal, 461*, 142050. doi:10.1016/j.cej.2023.142050.

Badgujar, K. C., Badgujar, V. C., & Bhanage, B. M. (2022). Recent update on use of ionic liquids for enzyme immobilization, activation, and catalysis: A partnership for sustainability. *Current Opinion in Green and Sustainable Chemistry, 36*, 100621. http://www.journals.elsevier.com/current-opinion-in-green-and-sustainable-chemistry. doi:10.1016/j.cogsc.2022.100621.

Bai, H., Yu, D., & Du, X. (2025). Review of porous microspheres for enzyme immobilization: Strategies, applications, and prospects. *International Journal of Biological Macromolecules, 295*, 139627. doi:10.1016/j.ijbiomac.2025.139627.

Bakar, B., Akbulut, M., Ulusal, F., Ulu, A., Özdemir, N., & Ateş, B. (2024). Horseradish peroxidase immobilized onto mesoporous magnetic hybrid nanoflowers for enzymatic decolorization of textile dyes: A highly robust bioreactor and boosted enzyme stability. *ACS Omega, 9*(23), 24558–24573. pubs.acs.org/journal/acsodf. doi:10.1021/acsomega.4c00703.

Bisht, M., Thayallath, S. K., Bharadwaj, P., Franklin, G., & Mondal, D. (2023). Biomass-derived functional materials as carriers for enzymes: Towards sustainable and robust biocatalysts. *Green Chemistry, 25*(12), 4591–4624. http://pubs.rsc.org/en/journals/journal/gc. doi:10.1039/d2gc04792f.

Cai, X., Huang, Y., & Zhu, C. (2025). Immobilized multi-enzyme/nanozyme biomimetic cascade catalysis for biosensing applications. *Advanced Healthcare Materials, 14*(8), e2401834. http://onlinelibrary.wiley.com/journal/10.1002/(ISSN)2192-2659. doi:10.1002/adhm.202401834.

Cao, Y., Yang, P., Zhao, R., & Wang, F. (2025). Recent progress in enzyme immobilization to metal–organic frameworks to enhance the $CO_2$ conversion efficiency. *Molecules, 30*(2), 251. doi:10.3390/molecules30020251.

Caparco, A. A., Dautel, D. R., & Champion, J. A. (2022). Protein mediated enzyme immobilization. *Small, 18*(19), e2106425. http://onlinelibrary.wiley.com/journal/10.1002/(ISSN)1613-6829. doi:10.1002/smll.202106425.

Chauhan, V., Kaushal, D., Dhiman, V. K., Kanwar, S. S., Singh, D., Dhiman, V. K., & Pandey, H. (2022). An insight in developing carrier-free immobilized enzymes. *Frontiers in Bioengineering and Biotechnology, 10*, 794411. http://journal.frontiersin.org/journal/bioengineering-and-biotechnology#archive. doi:10.3389/fbioe.2022.794411.

Chen, N., Chang, B., Shi, N., Yan, W., Lu, F., & Liu, F. (2023). Cross-linked enzyme aggregates immobilization: Preparation, characterization, and applications. *Critical Reviews in Biotechnology, 43*(3), 369–383. http://www.tandfonline.com/loi/ibty20. doi:10.1080/07388551.2022.2038073.

Costa, I. O., Morais, J. R. F., de Medeiros Dantas, J. M., Gonçalves, L. R. B., dos Santos, E. S., & Rios, N. S. (2023). Enzyme immobilization technology as a tool to innovate in the production of biofuels: A special review of the cross-linked enzyme aggregates (CLEAs) strategy. *Enzyme and Microbial Technology, 170*, 110300. www.elsevier.com/locate/enzmictec. doi:10.1016/j.enzmictec.2023.110300.

da Silva, T. F., de Oliveira, K. M., Pinto, R. R., Saraiva, L. S., Marchesan, R., Perna, R. F., Morales, S. A. V., & Xavier, M. D. C. A. (2024). Immobilization of a microbial fructosyltransferase on biochar and biochar functionalized with glutaraldehyde aiming at fructooligossacharides production. *DESAFIOS – Revista Interdisciplinar da Universidade Federal do Tocantins, 11*(3). doi:10.20873/2024_v3_17.

de Andrade Silva, T., Keijok, W. J., Guimarães, M. C. C., Cassini, S. T. A., & de Oliveira, J. P. (2022). Impact of immobilization strategies on the activity and recyclability of lipases in nanomagnetic supports. *Scientific Reports, 12*(1), 6448. www.nature.com/srep/index.html. doi:10.1038/s41598-022-10721-y.

Demirci, S., Sahiner, M., Yilmaz, S., Karadag, E., & Sahiner, N. (2020). Enhanced enzymatic activity and stability by in situ entrapment of $\alpha$-Glucosidase within super porous p(HEMA) cryogels during synthesis. *Biotechnology Reports, 28*, e00534. doi:10.1016/j.btre.2020.e00534.

Eğilmez, H. İ., & Haspolat, E. (2024). Temperature-Dependent Parameters in Enzyme Kinetics: Impacts on Enzyme Denaturation. *Fundam. J. Math. Appl., 7*, 226–235. https://doi.org/10.33401/fujma.1517334. doi:10.33401/fujma.1517334.

Erfani, A., Zarrintaj, P., Seaberg, J., Ramsey, J. D., & Aichele, C. P. (2021). Zwitterionic poly(carboxybetaine) microgels for enzyme (chymotrypsin) covalent immobilization with extended stability and activity. *Journal of Applied Polymer Science, 138*(22), e50545. http://onlinelibrary.wiley.com/journal/10.1002/(ISSN)1097-4628. doi:10.1002/app.50545.

Federsel, H.-J., Moody, T. S., & Taylor, S. J. C. (2021). Recent Trends in Enzyme Immobilization—Concepts for Expanding the Biocatalysis Toolbox. *Molecules, 26*, 2822. https://doi.org/10.3390/molecules26092822. doi:10.3390/molecules26092822.

Guisan, J. M., Fernandez-Lorente, G., Rocha-Martin, J., & Moreno-Gamero, D. (2022). Enzyme immobilization strategies for the design of robust and efficient biocatalysts. *Current Opinion in Green and Sustainable Chemistry, 35*, 100593. doi:10.1016/j.cogsc.2022.100593.

Hao, X., Liu, P., & Chu, X. (2024). Recent advances in the strategies of simultaneous enzyme immobilization accompanied by nanocarrier synthesis. *Applied Sciences, 14*(9), 3702. doi:10.3390/app14093702.

Hassan, M. E., Zhu, X., de Souza, E. F., Elnashar, M. M., & Lu, F. (2025). Enzyme immobilization advances: A key to unlocking renewable bioenergy potential. *Green Chemistry, 27*(37), 11289–11311. http://pubs.rsc.org/en/journals/journal/gc. doi:10.1039/d5gc03388h.

Holyavka, M. G., & Artyukhov, V. G. (2025). Prospects and problems in enzyme immobilization methodology: Comprehensive review. *Biophysical Reviews, 17*(5), 1471–1509. https://www.springer.com/journal/12551. doi:10.1007/s12551-025-01360-4.

Hsieh, C. Y., Hong, P. Y., & Hsieh, L. S. (2023). Immobilization of BoPAL3 phenylalanine ammonia-lyase on electrospun nanofibrous membranes of polyvinyl alcohol/nylon 6/chitosan crosslinked with dextran polyaldehyde. *Polymers, 15*(18), 3699. http://www.mdpi.com/journal/polymers. doi:10.3390/polym15183699.

Imam, H. T., Marr, P. C., & Marr, A. C. (2021). Enzyme entrapment, biocatalyst immobilization without covalent attachment. *Green Chemistry, 23*(14), 4980–5005. http://pubs.rsc.org/en/journals/journal/gc. doi:10.1039/d1gc01852c.

Jin, L. Q., Chen, X. X., Jin, Y. T., Shentu, J. K., Liu, Z. Q., & Zheng, Y. G. (2021). Immobilization of recombinant *Escherichia coli* cells expressing glucose isomerase using modified diatomite as a carrier for effective production of high fructose corn syrup in packed bed reactor. *Bioprocess and Biosystems Engineering, 44*(8), 1781–1792. https://rd.springer.com/journal/449. doi:10.1007/s00449-021-02560-4.

Jin, L. Q., Xu, Q., Liu, Z. Q., Jia, D. X., Liao, C. J., Chen, D. S., & Zheng, Y. G. (2017). Immobilization of recombinant glucose isomerase for efficient production of high fructose corn syrup. *Applied Biochemistry and Biotechnology, 183*(1), 293–306. http://www.springer.com/humana+press/journal/12010. doi:10.1007/s12010-017-2445-0.

Khan, M. F. (2025). Enhancing stability of enzymes for industrial applications: Molecular insights and emerging approaches. *World Journal of Microbiology and Biotechnology, 41*(10), 310. https://www.springer.com/journal/11274. doi:10.1007/s11274-025-04568-4.

Khan, M. R. (2021). Immobilized enzymes: A comprehensive review. *Bulletin of the National Research Centre, 45*, 207.

Kubiak, M., Kampen, I., & Schilde, C. (2022). Structure-based modeling of the mechanical behavior of cross-linked enzyme crystals. *Crystals, 12*(4), 441. doi:10.3390/cryst12040441.

Lou, Y., Zhang, B., Ye, X., & Wang, Z.-G. (2023). Self-assembly of the de novo designed peptides to produce supramolecular catalysts with built-in enzyme-like active sites: A review of structure–activity relationship. *Materials Today Nano, 21*, 100302. doi:10.1016/j.mtnano.2023.100302.

Maghraby, Y. R., El-Shabasy, R. M., Ibrahim, A. H., & Azzazy, H. M. E.-S. (2023). Enzyme Immobilization Technologies and Industrial Applications. *ACS Omega, 8*, 5184–5196. https://doi.org/10.1021/acsomega.2c07560. doi:10.1021/acsomega.2c07560.

Masdeu, G., Vázquez, L. M., López-Santín, J., Caminal, G., Kralj, S., Makovec, D., Álvaro, G., Guillén, M., & Marr, A. C. (2021). Synthesis of a precursor of D-fagomine by immobilized fructose-6-phosphate aldolase. *PLoS One, 16*(4), e0250513. doi:10.1371/journal.pone.0250513.

Matveeva, V. G., & Bronstein, L. M. (2021). Magnetic nanoparticle-containing supports as carriers of immobilized enzymes: Key factors influencing the biocatalyst performance. *Nanomaterials, 11*(9), 2257. https://www.mdpi.com/2079-4991/11/9/2257/pdf. doi:10.3390/nano11092257.

Meersseman Arango, H., Nguyen, X. D. L., Luis, P., Leyssens, T., Roura Padrosa, D., Paradisi, F., & Debecker, D. P. (2024). Membrane-immobilized transaminases for the synthesis of enantiopure amines. *RSC Sustainability, 2*(10), 3139–3152. https://www.rsc.org/journals-books-databases/about-journals/rsc-sustainability/#scope. doi:10.1039/d4su00293h.

Mirsalami, S. M., Mirsalami, M., & Ghodousian, A. (2024). Techniques for immobilizing enzymes to create durable and effective biocatalysts. *Results in Chemistry, 7*, 101486. doi:10.1016/j.rechem.2024.101486.

Mohammadi, M., Shahedi, M., Ahrari, F., Mostafavi, M., Habibi, Z., & Yousefi, M. (2023). Isocyanide-based multi-component reactions for carrier-free and carrier-bound covalent immobilization of enzymes. *Nature Protocols, 18*(5), 1641–1657. https://www.springer.com/journal/41596. doi:10.1038/s41596-023-00812-z.

Mohidem, N. A., Mohamad, M., Rashid, M. U., Norizan, M. N., Hamzah, F., & Mat, H. B. (2023). Recent advances in enzyme immobilisation strategies: An overview of techniques and composite carriers. *Journal of Composites Science, 7*(12), 488. www.mdpi.com/journal/jcs. doi:10.3390/jcs7120488.

Mu, R., Wang, Z., Wamsley, M. C., Duke, C. N., Lii, P. H., Epley, S. E., … Roberts, P. J. (2020). Application of enzymes in regioselective and stereoselective organic reactions. *Catalysts, 10*(8), 832. https://doi.org/10.3390/catal10080832. doi:10.3390/catal10080832.

Najim, A. A., Radeef, A. Y., al-Doori, I., & Jabbar, Z. H. (2024). Immobilization: The promising technique to protect and increase the efficiency of microorganisms to remove contaminants. *Journal of Chemical Technology and Biotechnology, 99*(8), 1707–1733. http://onlinelibrary.wiley.com/journal/10.1002/(ISSN)1097-4660. doi:10.1002/jctb.7638.

Naramittanakul, A., Buttranon, S., Petchsuk, A., Chaiyen, P., & Weeranoppanant, N. (2021). Development of a continuous-flow system with immobilized biocatalysts towards sustainable bioprocessing. *Reaction Chemistry and Engineering, 6*(10), 1771–1790. www.rsc.org/journals-books-databases/about-journals/reaction-chemistry-engineering/. doi:10.1039/d1re00189b.

Ndochinwa, O. G., Wang, Q.-Y., Amadi, O. C., Nwagu, T. N., Nnamchi, C. I., Okeke, E. S., & Moneke, A. N. (2024). *Current status and emerging frontiers in enzyme engineering: An industrial perspective, Heliyon, 10.* https://doi.org/10.1016/j.heliyon.2024.e32673. doi:10.1016/j.heliyon.2024.e32673.

Ölçücü, G., Klaus, O., Jaeger, K. E., Drepper, T., & Krauss, U. (2021). Emerging solutions for *in vivo* biocatalyst immobilization: Tailor-made catalysts for industrial biocatalysis. *ACS Sustainable Chemistry and Engineering, 9*(27), 8919–8945. http://pubs.acs.org/journal/ascecg. doi:10.1021/acssuschemeng.1c02045.

Ölçücü, G., Krauss, U., Jaeger, K. E., & Pietruszka, J. (2023). Carrier-free enzyme immobilizates for flow chemistry. *Chemie-Ingenieur-Technik, 95*(4), 531–542. http://onlinelibrary.wiley.com/journal/10.1002/(ISSN)1522-2640. doi:10.1002/cite.202200167.

Osuoha, J. O., & Nwaichi, E. O. (2021). Enzymatic technologies as green and sustainable techniques for remediation of oil-contaminated environment: State of the art. *International Journal of Environmental Science and Technology, 18*(5), 1299–1322. doi:10.1007/s13762-020-02876-w.

Pagolu, R., Singh, R., Shanmugam, R., Kondaveeti, S., Patel, S. K. S., Kalia, V. C., & Lee, J. K. (2021). Site-directed lysine modification of xylanase for oriented immobilization onto silicon dioxide nanoparticles. *Bioresource Technology, 331*, 125063. www.elsevier.com/locate/biortech. doi:10.1016/j.biortech.2021.125063.

Patti, S., Magrini Alunno, I., Pedroni, S., Riva, S., Ferrandi, E. E., & Monti, D. (2025). Advances and challenges in the development of immobilized enzymes for batch and flow biocatalyzed processes. *ChemSusChem, 18*(8), e202402007. http://onlinelibrary.wiley.com/journal/10.1002/(ISSN)1864-564X. doi:10.1002/cssc.202402007.

Prabhakar, T., Giaretta, J., Zulli, R., Rath, R. J., Farajikhah, S., Talebian, S., & Dehghani, F. (2025). Covalent immobilization: A review from an enzyme perspective. *Chemical Engineering Journal, 503*, 158054. doi:10.1016/j.cej.2024.158054.

Prakash, O., Verma, D., & Singh, P. C. (2024). Exploring enzyme-immobilized MOFs and their application potential: Biosensing, biocatalysis, targeted drug delivery and cancer therapy. *Journal of Materials Chemistry B, 12*(40), 10198–10214. http://pubs.rsc.org/en/journals/journal/tb. doi:10.1039/d4tb01556h.

Rajaraman, R. (2024). Beyond conventional models: Integer and fractional order analysis of nonlinear Michaelis-Menten kinetics in immobilised enzyme reactors. *Engineering Computations, 41*(8/9), 1987–2025. doi:10.1108/ec-03-2024-0238.

Ran, L., Lu, Y., Chen, L., He, M., & Deng, Z. (2025). Design, synthesis, and application of immobilized enzymes on artificial porous materials. *Advanced Science, 12*(20), e2500345. http://onlinelibrary.wiley.com/journal/10.1002/(ISSN)2198-3844. doi:10.1002/advs.202500345.

Ranimol, G., & Sunkar, S. (2022). A comparative study of free laccase and laccase immobilized in copper alginate. *Indian Journal of Biochemistry and Biophysics, 59*(2), 214–223. doi:10.56042/ijbb.v59i2.27154.

Rather, A. H., Khan, R. S., Wani, T. U., Beigh, M. A., & Sheikh, F. A. (2022). Overview on immobilization of enzymes on synthetic polymeric nanofibers fabricated by electrospinning. *Biotechnology and Bioengineering, 119*(1), 9–33. http://onlinelibrary.wiley.com/journal/10.1002/(ISSN)1097-0290. doi:10.1002/bit.27963.

Sardar, M., & Mirza, K. (2023). Affinity immobilization and affinity layers. In M. L Ferreira, & M. L. Foresti (Eds.), *Biocatalyst immobilization* (pp. 269–290). Academic Press. doi:10.1016/B978-0-323-91317-1.00003-7.

Selvaraj, C., Rudhra, O., Alothaim, A. S., Alkhanani, M., & Singh, S. K. (2022). Structure and chemistry of enzymatic active sites that play a role in the switch and conformation mechanism. *Advances in Protein Chemistry and Structural Biology, 130*, 59–83. https://www.elsevier.com/books/advances-in-protein-chemistry-and-structural-biology/eisenberg/978-0-12-374827-0. doi:10.1016/bs.apcsb.2022.02.002.

Sheldon, R. A. (2024). Green chemistry and biocatalysis: Engineering a sustainable future. *Catalysis Today, 431*, 114571. https://doi.org/10.1016/j.cattod.2024.114571. doi:10.1016/j.cattod.2024.114571.

Sheldon, R. A., Basso, A., & Brady, D. (2021). New frontiers in enzyme immobilisation: Robust biocatalysts for a circular bio-based economy. *Chemical Society Reviews, 50*(10), 5850–5862. http://pubs.rsc.org/en/journals/journal/cs. doi:10.1039/d1cs00015b.

Sheldon, R. A., & Brady, D. (2022). Green Chemistry, Biocatalysis, and the Chemical Industry of the Future. *ChemSusChem, 15*, e202102628. https://doi.org/10.1002/cssc.202102628. doi:10.1002/cssc.202102628.

Silva Almeida, C., Simão Neto, F., da Silva Sousa, P., da Silva Aires, F. I., de Matos Filho, J. R., Gama Cavalcante, A. L., Sousa Junior, P. G. D., Melo, R. L. F., & José, J. C. (2024). Enhancing lipase immobilization via physical adsorption: Advancements in stability, reusability, and industrial applications for sustainable biotechnological processes. *ACS Omega, 9*(47), 46698–46732. pubs.acs.org/journal/acsodf. doi:10.1021/acsomega.4c07088.

Sreenivasulu, G., & Rao, R. R. (2024). Advanced techniques for glucose oxidase immobilization: Evolution, computational integration, and biomedical applications. *Advances in Engineering and Intelligence Systems, 3*(3), 82–93. https://aeis.bilijipub.com/article_206708_c1ee919a62c3d9434fe94da25f165bf3.pdf. doi:10.22034/aeis.2024.473823.1216.

Tadesse, M., & Liu, Y. (2025). Recent Advances in Enzyme Immobilization: The Role of Artificial Intelligence, Novel Nanomaterials, and Dynamic Carrier Systems. *Catalysts, 15*, 571. https://doi.org/10.3390/catal15060571. doi:10.3390/catal15060571.

Taheri-Kafrani, A., Kharazmi, S., Nasrollahzadeh, M., Soozanipour, A., Ejeian, F., Etedali, P., Mansouri-Tehrani, H. A., Razmjou, A., Yek, S. M. G., & Varma, R. S. (2021). Recent developments in enzyme immobilization technology for high-throughput processing in food industries. *Critical Reviews in Food Science and Nutrition, 61*(19), 3160–3196. https://www.tandfonline.com/loi/bfsn20. doi:10.1080/10408398.2020.1793726.

Tan, J. H., Chen, A., Bi, J., Lim, Y. H., Wong, F. T., & Ow, D. S. W. (2023). The engineering, expression, and immobilization of epimerases for D-allulose production. *International Journal of Molecular Sciences, 24*(16), 12703. http://www.mdpi.com/journal/ijms. doi:10.3390/ijms241612703.

Velasco-Lozano, S., Jackson, E., Ripoll, M., López-Gallego, F., & Betancor, L. (2020). Stabilization of ω-transaminase from *Pseudomonas fluorescens* by immobilization techniques. *International Journal of Biological Macromolecules, 164*, 4318–4328. www.elsevier.com/locate/ijbiomac. doi:10.1016/j.ijbiomac.2020.09.003.

Wang, H., Kou, X., Gao, R., Huang, S., Chen, G., & Ouyang, G. (2024). Enzyme-Immobilized Porous Crystals for Environmental Applications. *Environ. Sci. Technol., 58*, 11869–11886. https://doi.org/10.1021/acs.est.4c01273. doi:10.1021/acs.est.4c01273.

Wang, K., Hui, K. N., San Hui, K., Peng, S., & Xu, Y. (2021a). Recent progress in metal-organic framework/graphene-derived materials for energy storage and conversion: Design, preparation, and application. *Chemical Science, 12*(16), 5737–5766. http://pubs.rsc.org/en/journals/journal/sc. doi:10.1039/d1sc00095k.

Wang, K.-Y., Zhang, J., Hsu, Y.-C., Lin, H., Han, Z., Pang, J., … Zhou, H.-C. (2023). Bioinspired Framework Catalysts: From Enzyme Immobilization to Biomimetic Catalysis. *Chem. Rev., 123*, 5347–5420. https://doi.org/10.1021/acs.chemrev.2c00879. doi:10.1021/acs.chemrev.2c00879.

Wang, Z., Liu, W., Liu, W., Ma, Y., Li, Y., Wang, B., Wei, X., Liu, Z., & Song, H. (2021b). Co-immobilized recombinant glycosyltransferases efficiently convert rebaudioside A to M in cascade. *RSC Advances, 11*(26), 15785–15794. http://pubs.rsc.org/en/journals/journal/ra. doi:10.1039/d0ra10574k.

Weng, Y., Ranaweera, S., Zou, D., Cameron, A. P., Chen, X., Song, H., & Zhao, C. X. (2023). Improved enzyme thermal stability, loading and bioavailability using alginate encapsulation. *Food Hydrocolloids, 137*, 108385. https://www.journals.elsevier.com/food-hydrocolloids. doi:10.1016/j.foodhyd.2022.108385.

Xu, X., Wan, J., Sun, J., Wu, L., & Lei, J. (2025). Tailored metal-organic framework-based enzyme hybrids: Immobilization strategies, improved performance, and biological applications. *Small Science, 5*(10), e2500222. https://onlinelibrary.wiley.com/journal/26884046. doi:10.1002/smsc.202500222.

Yamane, M., Iwazumi, K., & Osanai, T. (2024). Immobilization of fumarase from thermophilic eukaryotic red alga *Cyanidioschyzon merolae*on ceramic carrier. *The Journal of General and Applied Microbiology, 70*(2). doi:10.2323/jgam.2024.02.003.

Yang, P., Yang, W., Zhang, H., & Zhao, R. (2023). Metal–organic framework for the immobilization of oxidoreductase enzymes: Scopes and perspectives. *Materials, 16*(19), 6572. doi:10.3390/ma16196572.

Yuan, Y., Shen, J., & Salmon, S. (2023). Developing enzyme immobilization with fibrous membranes: Longevity and characterization considerations. *Membranes, 13*(5), 532. doi:10.3390/membranes13050532.

Zarini, L., Carzaniga, T., Pirotta, M., Damin, F., Brambilla, D., Chiari, M., Bassanini, I., Gagni, P., Mussida, A., Casiraghi, L., Buscaglia, M., & Sola, L. (2025). Affinity-based copolymer coating for oriented protein immobilization in biosensor development. *Biosensors, 15*(10), 670. doi:10.3390/bios15100670.

Zhang, J., Lovell, J. F., Shi, J., & Zhang, Y. (2025). Nanomaterials for co-immobilization of multiple enzymes. *BMEMat, 3*(1), e12080. https://onlinelibrary.wiley.com/journal/27517446. doi:10.1002/bmm2.12080.

Zhang, M., Shi, X., Sun, H., Xu, D., Gao, Y., Wu, X., Zhang, J., & Zhang, J. (2021c). Immobilization of glycogen synthase kinase-3$\beta$ inhibitor on 316L stainless steel via polydopamine to accelerate endothelialization. *Frontiers in Bioengineering and Biotechnology, 9*, 806151. http://journal.frontiersin.org/journal/bioengineering-and-biotechnology#archive. doi:10.3389/fbioe.2021.806151.

Zhang, X., Li, G., Chen, G., Wu, D., Wu, Y., & James, T. D. (2021a). Enzyme mimics for engineered biomimetic cascade nanoreactors: Mechanism, applications, and prospects. *Advanced Functional Materials, 31*(50), e2106139. http://onlinelibrary.wiley.com/journal/10.1002/(ISSN)1616-3028. doi:10.1002/adfm.202106139.

Zhang, Y., Li, X., Wang, J., & Wang, Q. (2021b). Immobilizing redox enzyme on amino functional group-integrated tailor-made polyester textile: High loading, stability, and application in a bio-Fenton system. *ACS Sustainable Chemistry & Engineering, 9*(37), 12463–12473. Retrieved December 17, 2025 https://pubs.acs.org/doi/10.1021/acssuschemeng.1c03775 .

Žnidaršič-Plazl, P. (2021). Biocatalytic process intensification *via* efficient biocatalyst immobilization, miniaturization, and process integration, Current Opinion in Green and Sustainable. *Chemistry, 32*, 100546. https://doi.org/10.1016/j.cogsc.2021.100546. doi:10.1016/j.cogsc.2021.100546.

CHAPTER

# 2 Principles of carrier-free immobilization

**Mehdi Mohammadi**[a] **and Maryam Yousefi**[b]

[a]*Bioprocess Engineering Department, National Institute for Genetic Engineering and Biotechnology, Tehran, Iran,* [b]*Nanobiotechnology Research Center, Avicenna Research Institute, ACECR, Tehran, Iran*

## 2.1 Introduction

Carrier-free immobilization of enzymes is most simply defined as the intermolecular cross-linking of enzyme molecules to each other, thereby achieving immobilization without the incorporation of an external support or matrix. To fully grasp the concept, it is essential to thoroughly understand physiochemical principles underlying the transition from enzymes in their soluble homogeneous forms to heterogeneous catalytically active biocatalysts. Along this pathway, three key techniques govern carrier-free enzyme immobilization: protein precipitation, protein crystallization, and cross-linking technologies. Together, these processes facilitate creating various immobilized enzyme forms, including cross-linked enzymes (CLEs), cross-linked enzyme aggregates (CLEAs), cross-linked enzyme crystals (CLECs), and related assemblies. The specific details of each technique greatly influence the size, shape, and functional properties of the resulting immobilized biocatalyst.

## 2.2 Protein aggregation/precipitation

Protein precipitation is an essential biochemical technique used to isolate proteins from a solution by decreasing their solubility. This causes proteins to aggregate and separate from the solution as a solid phase. The process of protein precipitation is governed by various physicochemical principles that influence protein solubility and intermolecular interactions. To fully understand these mechanisms, it is important first to understand the factors that affect protein solubility.

## 2.3 Protein solubility

Protein solubility is a fundamental concept that describes the extent to which a protein can dissolve in a solvent, typically water, under specified conditions. Protein solubility is intimately connected to aggregation behavior that can be either reversible or

**Carrier-Free Enzyme Immobilization: Fundamentals and Innovations in Biocatalysis.**
**DOI: https://doi.org/10.1016/B978-0-443-44139-4.00004-6**

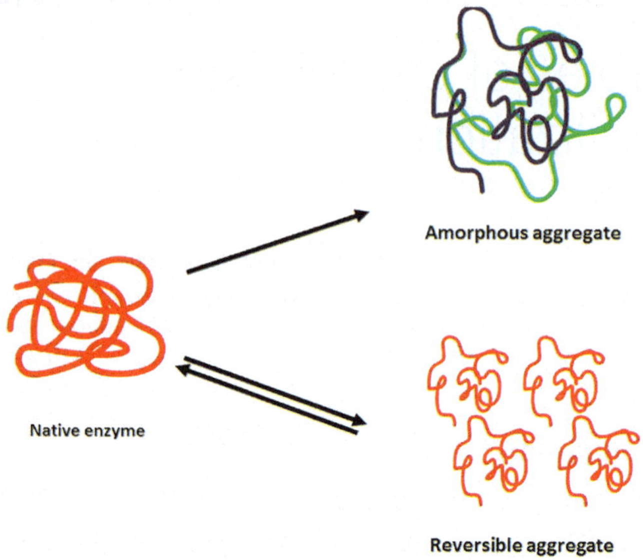


**FIGURE 2.1**

Enzyme aggregation forms.

Various enzyme aggregation types, including reversible aggregates with ordered structures and irreversible amorphous aggregates formed from disordered protein assemblies.

irreversible (Fig. 2.1). Under controlled environmental conditions, reversible aggregation (precipitation) can occur without significant structural changes, allowing proteins to be resolubilized (Roberts, 2014; Schein, 1990).

This type of aggregation is of critical importance in the enzyme immobilization industry and is widely used to produce CLEAs. In contrast, irreversible aggregation often results from protein unfolding, exposing hydrophobic patches that drive strong intermolecular interactions to form large, insoluble aggregates. Irreversible aggregation can be detrimental to protein function and stability and is considered a vital concern in enzyme applications and pharmaceutical formulations (Wong et al., 2018). It is governed by a complex interaction between the protein's intrinsic characteristics and the surrounding environmental factors. Intrinsic properties include the amino acid sequence and composition, the three-dimensional structure, and the distribution of hydrophobic and hydrophilic residues on the protein surface (Trevino et al., 2008). The amino acid sequence and composition have a profound effect on protein solubility because they determine the physicochemical properties of the protein surface that governs interactions with water molecules. Proteins with a higher proportion of charged and polar protic residues on their surface tend to have increased solubility by forming strong hydrogen bonds or electrostatic interactions, while those enriched in hydrophobic amino acids often display reduced solubility due to a greater tendency to aggregate (Gao et al., 2024). Experimental substitution analyses at solvent-accessible

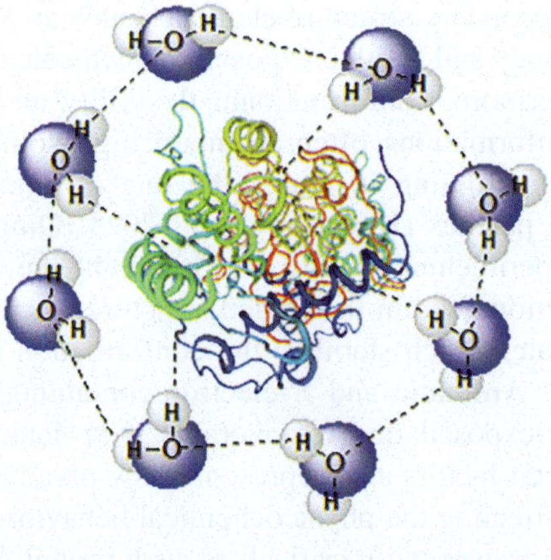


**FIGURE 2.2**

Protein hydration shell: Water H-bonds to backbone C = O/N–H, side-chain $-COO^-$, $-CONH_2$, –OH, $-NH_3^+$.

Water molecules forming a hydration shell around the protein surface influence protein structure, dynamics, and solubility.

sites have demonstrated that aspartic acid, glutamic acid, and serine contribute most favorably to protein solubility, especially at higher net charges. These residues improve water binding (hydration) by forming strong hydrogen bonds and effectively prevent aggregation through electrostatic and polar interactions (Fig. 2.2). Other hydrophilic residues like asparagine, glutamine, and threonine contribute less favorably, likely due to their amide groups interacting differently with solvent molecules. Lysine and arginine, as positively charged residues, have complex effects that depend on overall protein charge; their solubility contributions can be favorable or unfavorable depending on pH and net charge conditions.

Protein three-dimensional conformation also plays a key role; globular proteins with compact and well-folded structures generally exhibit higher solubility compared to unfolded or partially folded proteins that expose internal hydrophobic residues, which favor aggregation (Niwa et al., 2009). Additionally, the conformation of proteins profoundly influences their solubility in aqueous environments due to the way the three-dimensional structure dictates the distribution and accessibility of polar, charged, and hydrophobic amino acid residues on the protein surface. In a properly folded, globular protein, hydrophobic residues tend to be buried in the interior, forming a compact core, while hydrophilic and charged residues are predominantly exposed on the surface (Pimtawong et al., 2025; Vihinen, 2020). This arrangement promotes favorable interactions with water molecules primarily through hydrogen bonding and electrostatic forces, facilitating protein solubility. Conversely, when proteins are unfolded or partially misfolded, internal hydrophobic residues become exposed to the solvent, increasing the propensity for intermolecular hydrophobic interactions that lead to aggregation and reduced solubility (Vihinen, 2020). Certain structural motifs and secondary structure elements, such as $\beta$-strands and $\alpha$-helices, influence surface topology and residue exposure, which can also modulate solubility (Schein, 1990). Furthermore, conformational flexibility and dynamics play roles; rigid, well-packed conformations often maintain high solubility, while increased flexibility or dynamic unfolding increases the rate of aggregation by transiently exposing hydrophobic patches (Havlásek et al., 2025; Qing et al., 2022). Recent computational and experimental studies also highlight the importance of specific residue-residue interactions within the folded structure, such as salt bridges involving lysine and glutamate, which stabilize the conformation and promote solubility (Trevino et al., 2007). Aromatic and $\pi$-electron-containing residues can promote aggregation if surface-exposed due to favorable $\pi$–$\pi$ interactions, which reduce solubility. Environmental factors also represent a key parameter influencing protein solubility, effectively affecting the physicochemical behavior of proteins in solution. The pH of the solution is especially critical, as each protein has an isoelectric point (pI) at which its net charge is zero, minimizing electrostatic repulsion and enhancing the tendency to aggregate and precipitate (Qing et al., 2022). Similarly, ionic strength affects solubility in a biphasic manner: low to moderate salt concentrations can shield charged groups and improve solubility, while high salt levels promote "salting out," where water availability for protein hydration decreases, leading to precipitation (Duong-Ly & Gabelli, 2014). Temperature is a well-established factor influencing the solubility of chemical compounds, typically promoting increased solubility with rising temperatures. In proteins, moderate temperature elevations enhance solubility by increasing molecular mobility and disrupting intermolecular interactions, whereas excessive thermal stress induces conformational destabilization, resulting in denaturation and irreversible aggregation (Qing et al., 2022).

## 2.4 Principles of methods for reversible protein aggregation

Reversible protein aggregation is a dynamic and regulated process in which proteins temporarily assemble into larger size complexes in response to environmental changes without compromising their functional properties. Several methods have been introduced to induce precipitation in different biological applications of proteins, including the preparation of CLEAs. These precipitation techniques generally alter enzyme solubility by manipulating factors such as salt concentration (salting out), pH levels (isoelectric precipitation), or the addition of organic solvents, enabling controlled enzyme aggregation that facilitates their subsequent immobilization through cross-linking.

### 2.4.1 Salting out

Salting out is a widely used protein precipitation technique that capitalizes on the decreased solubility of enzymes in solutions with high ionic strength (Duong-Ly & Gabelli, 2014). When salts are added to an enzyme solution, they dissociate into

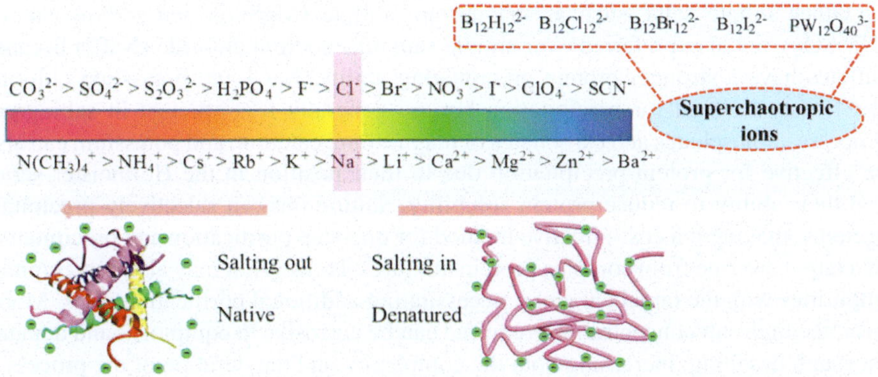


**FIGURE 2.3**

The Hofmeister series:

Illustrating ion-specific effects on protein stability and solubility.

*From: Kang, B., Tang, H., Zhao, Z. & Song, S. (2020). Hofmeister series: Insights of ion specificity from amphiphilic assembly and interface property.* ACS Omega, 5*(12), 6229–6239. https://doi.org/10.1021/acsomega.0c00237.*

ions that interact strongly with water molecules, effectively competing with the enzymes for hydration. This competition decreases the number of free water molecules available to solvate enzyme molecules. Since enzymes have both hydrophilic (water-attracting) and hydrophobic (water-repelling) regions, the reduction in solvent availability weakens the interaction between water and the enzyme's hydrophilic amino acid residues. Consequently, enzyme molecules tend to aggregate through hydrophobic interactions as these regions cluster together to minimize contact with the ion-rich environment. This leads to enzyme aggregation and subsequent precipitation, after which the enzyme can be collected by centrifugation or filtration, or used directly for the preparation of CLEAs. The aggregation process is further facilitated as salt ions neutralize the charge repulsion between enzyme molecules. Since each enzyme has a different threshold for salt concentration, selective fractionation and partial purification of enzyme mixtures can also be achieved by carefully controlling the type and amount of salt used. Salting out is reversible and generally preserves enzyme activity because it avoids harsh chemical denaturation. Ammonium sulfate ($(NH_4)_2SO_4$) is the most commonly used salt for protein precipitation due to its high solubility, low toxicity, and affordability (Wingfield, 1998). It is highly efficient because both its anion and cation rank high in the Hofmeister series (Fig. 2.3). The Hofmeister series ranks ions according to their distinct effects on the solubility, stability, and conformation of proteins and macromolecules in aqueous solutions (Lo Nostro & Ninham, 2012; Tadeo et al., 2009). It classifies ions from kosmotropes, which enhance structural stability, to chaotropes, which induce structural destabilization, based on their impact on water structure, protein hydration, and molecular interactions that regulate protein folding, aggregation, and precipitation.

Other sulfate salts, such as magnesium sulfate ($MgSO_4$) and sodium sulfate ($Na_2SO_4$), are also preferred over simple salts like sodium chloride (NaCl) because sulfates have a stronger protein precipitating ability. NaCl and potassium chloride (KCl) may be used for precipitation but are generally less efficient in salting-out processes. Phosphates and carbonates of magnesium, calcium, and potassium can also be effective for protein precipitation due to their position in the Hofmeister series and their ability to reduce protein solubility. Salting out can selectively precipitate proteins, making it a cost-effective method for one-step purification and precipitation of a target enzyme from complex protein mixtures. However, it may also coprecipitate impurities with the target enzymes, necessitating additional purification steps. Moreover, the high salt concentrations required can be corrosive to equipment and demand thorough desalting, increasing both the complexity and the duration of the process.

### 2.4.2 Solvent-assisted protein aggregation

The use of organic solvents or polar hydrophilic agents like polyethylene glycol (PEG) in protein reversible precipitation is another widely employed technique for protein precipitation (Atha & Ingham, 1981; Ingham, 1984). Organic solvents such as ethanol, acetone, methanol, and acetonitrile are commonly used because they alter the solubility of proteins by disrupting the hydration shell surrounding protein molecules (Burgess, 2009). The disruption of the protein hydration shell involves altering the structured layer of water molecules that normally surrounds and stabilizes the protein in aqueous solution. Protein molecules in water are enveloped by this layer, which interacts with the protein's surface through hydrogen bonds, electrostatic interactions, and van der Waals forces (Fig. 2.4). The hydration shell stabilizes the native protein structure and maintains its solubility by preventing hydrophobic contacts between protein molecules. Organic solvents, particularly those capable of forming hydrogen bonds, interfere with the hydrogen-bonding network between water molecules and the polar or charged groups on the protein surface. They can replace some of the tightly bound water molecules in the hydration shell or reduce their number by altering the local environment, leading to partial dehydration of the protein surface. As water is removed or its interactions are altered, hydrophobic patches on the protein surface become exposed, facilitating protein–protein hydrophobic interactions and subsequent aggregation (Nickerson et al., 2022).

The process typically requires careful optimization of solvent type, concentration, temperature, and pH to maximize precipitation efficiency while minimizing protein denaturation and loss of activity (Akyüz et al., 2024). For instance, acetone is known for rapid protein precipitation because of its high miscibility with water and its ability to form hydrogen bonds, allowing it to effectively compete with and replace water molecules in the protein's hydration shell. Acetone also has a low freezing point, which allows precipitation to be conducted at low temperatures that help maintain protein stability and prevent denaturation. Its volatility facilitates rapid evaporation, making the removal of acetone from protein precipitates easier and faster, thus simplifying downstream processing.

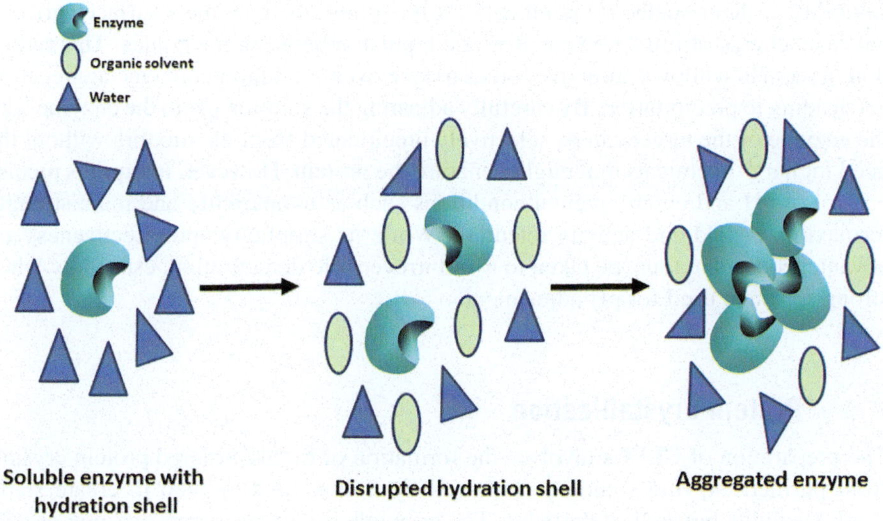


**FIGURE 2.4**

Enzyme aggregation by organic solvents.

Organic solvents disrupt the hydration shell surrounding protein molecules.

In solvent-based precipitation, the percentage of solvent is a critical parameter, as excessive concentrations can lead to conformational changes in protein structure by replacing the water molecules inside the protein structure. This few numbers of water molecules act as molecular lubricant within the protein tertiary structure and are essential for maintaining conformational flexibility required for functional activity (Klibanov, 2001). Organic solvents, particularly those with log *P* values below 2, can strongly disrupt this crucial water–enzyme interaction, resulting in enzyme inactivation or irreversible denaturation. Precipitation efficiency can also vary depending on solvent type, concentration, temperature, and protein properties, sometimes causing incomplete precipitation and reduced recovery. Impurities may coprecipitate, complicating downstream purification, while the flammability, toxicity, and disposal requirements of many solvents pose safety and environmental challenges, often requiring additional steps to completely remove residual solvent.

### 2.4.3 Isoelectric point-induced protein aggregation

Although isoelectric precipitation is a widely used technique for enzyme isolation and purification, it is rarely used for CLEAs preparation due to the typical preference for salt or solvent precipitants that maintain better enzyme activity and provide more controlled aggregate formation (Novák & Havlíček, 2016). Isoelectric precipitation takes advantage of the unique property of proteins and enzymes to have minimum solubility at their isoelectric point (pI). At this point, the enzyme's surface carries a balanced charge, minimizing electrostatic repulsion between molecules. This reduction in repulsion allows attractive forces to take over, facilitating enzyme aggregation and leading to precipitation. By carefully adjusting the solution pH to the enzyme's pI, the enzyme of the target can be selectively precipitated from the mixture without the need for harsh chemicals that might denature the protein. However, it requires precise control of pH and environmental conditions such as temperature and ionic strength to maximize yield and activity retention. While its simplicity and effectiveness are advantageous, care must be taken to avoid irreversible denaturation, especially when strong acids are used for pH adjustment.

### 2.4.4 Protein crystallization

The preparation of CLECs involves the formation of highly ordered protein crystals from purified enzyme solutions, followed by a cross-linking step to create stabilized, insoluble biocatalyst particles. The principle is similar to conventional protein crystallization: a supersaturated protein solution promotes nucleation and crystal growth, typically using methods such as batch crystallization, vapor diffusion, and counter-diffusion (Chayen, 1998). Key factors controlling crystallization include protein concentration, precipitant type and concentration, pH, and temperature. At the molecular level, protein crystallization is guided by the balance between enthalpic gains from specific intermolecular interactions (e.g., hydrogen bonds, van der Waals forces) that stabilize the crystal contacts, and entropic costs due to the loss of freedom when proteins become ordered (Schmit & Dill, 2010). To favor crystallization thermodynamically, the decrease in free energy (Gibbs free energy) must be negative, which happens under supersaturated conditions. The formation of protein crystals chemically goes through the creation of a supersaturated protein solution, followed by a process that involves two main steps: nucleation and growth. Initially, purified protein molecules are dissolved in an aqueous solution. Conditions such as pH, temperature, ionic strength, and the presence of precipitating agents are adjusted to reduce protein solubility, leading to a supersaturated state. Afterward in the nucleation step, the protein molecules start to come together and form small aggregates or clusters to initiate crystallization. Nucleation requires protein molecules to orient in the correct direction to form crystal-like contacts and overcome the energy barrier associated with ordering the molecules. When nuclei are stable, additional protein molecules sequentially add to the nuclei, growing the crystal lattice in an ordered, repeating pattern. Growth often occurs by layer-by-layer addition, where molecules attach to specific crystal faces, creating a highly ordered three-dimensional array. Once crystals are formed under optimized conditions, they can be treated with cross-linking agents to covalently bond protein molecules within the crystal lattice, thereby converting fragile protein crystals into robust CLECs with enhanced thermal and operational stability (Zelinski & Waldmann, 1997).

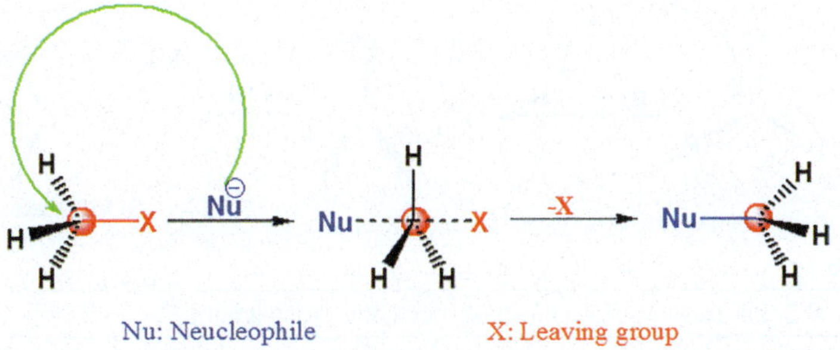


**FIGURE 2.5**

Nucleophilic substitution reaction.

Formation of stable covalent cross-links between enzyme molecules proceeds via nucleophilic substitution.

## 2.5 Principles of cross-linking in carrier-free immobilization

Proteins structure has several nucleophilic groups on their surface including amine moieties which are particularly important and reactive in a nucleophilic reaction. A nucleophilic reaction occurs when an atom possessing an available accessible lone pair of electrons attacks an electron-deficient atom (the electrophile), often resulting in the formation of a new covalent bond (Fig. 2.5). In chemistry, a covalent bond forms when two atoms share electron pairs, creating a strong and stable attachment. In the case of proteins, the covalent bonds can be formed between the nucleophilic groups on the protein surface and suitable reactive partners in the reaction medium. The most reactive amine groups of protein surfaces mainly come from the side chains of the amino acid lysine, as well as the N-terminal end of the protein chain. Since amines have a nonbonding lone electron pair, they act as strong nucleophiles that can readily donate electrons to react and form strong covalent bonds. The reactive amine groups are often exposed on the outer surface of proteins, making them easily accessible for chemical reactions.

The abundance of amine groups on the protein surface and their nucleophilicity make them the most common reactive sites when forming covalent linkages on proteins (Spicer & Davis, 2014). Amine reactivity underlies many methods for enzyme immobilization or cross-linking, as they react with several electrophiles without significantly disturbing the protein's natural structure or function. Other nucleophiles, such as carboxylic acid groups from glutamic acid and aspartic acid and thiol groups from cysteine residues, also exist on protein surfaces, but amine groups are typically more abundant and therefore play the leading role in surface chemistry.

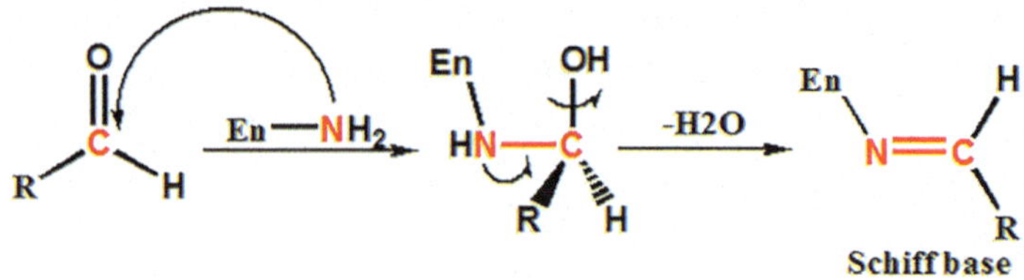


**FIGURE 2.6**

Formation of Schiff base in aldehyde-based enzyme cross-linking.

Schiff base formation between enzyme amino groups and aldehyde cross-linkers enables covalent enzyme coupling.

### 2.5.1 Nucleophilic attachment to aldehyde linkers

Compounds with aldehyde functional groups are widely used as cross-linking agents in cross-linking strategies for the preparation of carrier-free enzyme biocatalysts such as CLEAs and CLECs. These compounds react with protein surfaces to generate covalently joined enzyme molecules and produce robust, insoluble biocatalyst networks. Aldehyde-activated cross-linking is favored for CLEAs and CLECs because it typically occurs under mild, enzyme-compatible conditions and allows direct transformation of active enzymes from reversible aggregates or crystals into solid, heterogeneous, stable biocatalysts, often preserving a large fraction of enzymatic activity. The chemistry of aldehyde-amine nucleophilic attachment is the fundamental principle of enzyme cross-linking. Aldehyde functional groups of the active cross-linking agents are highly electrophilic, meaning that their carbonyl carbon atom is electron-deficient and thus chemically reactive toward nucleophiles like amines present on protein surfaces (Cheung & Nimni, 1982). When an aldehyde compound is offered to an enzyme aggregate or crystal, primary amines act as nucleophiles, and the lone pair of nitrogen attacks the carbonyl carbon of an aldehyde to initiate a nucleophilic addition reaction and generate a transient, tetrahedral intermediate (Fig. 2.6). Subsequent elimination of water from this intermediate leads to the formation of an imine, also known as a Schiff base, which has a C = N double bond linking the enzyme to the aldehyde. Schiff bases are generally more stable in nonaqueous environments and can hydrolyze back to the starting aldehyde and amine in water, especially under acidic conditions. For enhanced stability, especially in bioconjugation and immobilization, the imine linkage is further reduced to a secondary amine by reaction with mild reducing agents, making the cross-link irreversible and permanent. Cross-linking of enzyme aggregates/crystals by bifunctional aldehyde linkers generally targets lysine residues as they are widely distributed on enzyme surfaces. However, the efficiency and specificity of aldehyde-amine coupling depend on factors such as pH, protein structure, aldehyde type and concentration, and steric accessibility of the lysine residues (Cheung & Nimni, 1982).

Among aldehyde-based cross-linkers, glutaraldehyde (GA) is by far the most commonly applied in preparing CLEAs or CLECs. GA is a linear five-carbon dialdehyde ($OHC–(CH_2)_3–CHO$) and owes its effectiveness to having two reactive aldehyde groups per molecule, allowing it to simultaneously cross-link multiple protein molecules. GA can exist as a monomer or as part of oligomeric/polymeric forms in solution, which has important consequences for its reactivity and the structure of the resulting protein network (Barbosa et al., 2014). At dilute levels, GA predominantly remains in its simple monomeric forms, such as the free aldehyde, its hydrate, or a hemiacetal species. When the concentration increases, it undergoes self-association, giving rise to oligomeric hemiacetals (Migneault et al., 2004). However, the primary pathway in GA's reaction with proteins begins with the nucleophilic attack by a lysine amine on the carbonyl carbon of GA, leading to Schiff base formation. Multiple lysine side chains from different molecules can be joined in this fashion, yielding a complex three-dimensional network. While the initial imine bonds are reversible, ongoing reaction steps and increasing complexity of the aggregate shift the equilibrium toward highly stable, often irreversible linkages. Under physiological conditions (neutral pH, ambient temperature), GA cross-linking is typically rapid and yields high immobilization efficiency. GA can react with not only primary amines but also, to a lesser degree, other nucleophiles like thiol groups of cysteine and phenolic hydroxyls of tyrosine, further enhancing the cross-linking density of the network (Modenez et al., 2018). The two aldehyde groups in GA can preassemble into polymeric forms, meaning that GA-based cross-linking is capable of forming both simple one-to-one and complex many-to-many covalent structures, depending on the reaction conditions. GA cross-linking also provides self-limiting aggregation, meaning that particles do not grow indefinitely but instead reach a critical size determined by enzyme concentration, cross-linker amount, and mixing dynamics. This feature is particularly beneficial for forming CLEAs, as it enables the controlled assembly of insoluble, recoverable enzyme clusters with high activity. The mechanical strength, water resistance, and long-term stability of GA-mediated CLEAs and CLECs greatly exceed those of noncovalently aggregated enzymes (Sheldon, 2011). Nonetheless, GA's high reactivity can also present challenges. If lysine residues are located near or within the enzyme's active site, partial or full inactivation may occur. Thus, careful optimization of reaction conditions, concentration of cross-linker, choice of buffer, reaction temperature, and the cross-linking time is essential for maximizing both yield and retained catalytic function (Migneault et al., 2004). Alternative protocols may involve preblocking sensitive residues, using lower concentrations, or including protective additives.

Dextran aldehyde is a renewable, nontoxic macromolecular polyaldehyde widely used as an alternative to GA in protein cross-linking, especially in the preparation of CLEAs and CLECs (Mateo et al., 2004). Derived from the oxidation of dextran, a natural polysaccharide composed of glucose units linked mainly by $\alpha$-1,6-glycosidic bonds, dextran aldehyde contains multiple aldehyde groups per polymer chain, allowing extensive cross-linking with amino groups on the protein surface (Fig. 2.7).

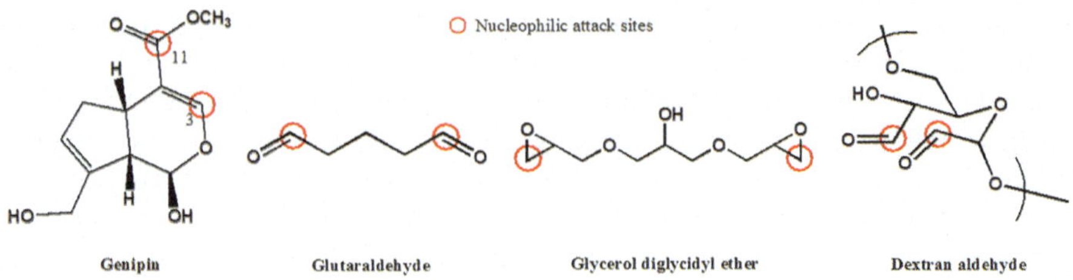


**FIGURE 2.7**

Molecular structure of enzyme cross-linking agents:

Illustrating reactive moieties responsible for forming covalent bonds with enzymes.

This multisite reactivity results in a high density of cross-linking that can stabilize the enzyme's structure without aggressively reacting with the enzyme. Compared with short-chain, smaller aldehydes like GA, the larger size and polymeric nature of dextran aldehyde reduce the potential inactivation of the final biocatalyst due to its steric hindrance and limited diffusion into enzyme cores. The polymeric dextran chains can form multiple cross-link points simultaneously, leading predominantly to intramolecular cross-linking within the same enzyme molecule or intermolecular cross-linking between neighboring enzyme molecules, depending on polymer concentration and chain length. This dual capability of dextran aldehyde, together with its larger molecular size and multialdehyde functionality, allows adjustment of protein mobility and flexibility, which can lead to enhanced thermal and operational stability of CLEs compared to those cross-linked with smaller aldehydes. The use of dextran aldehyde has been successfully demonstrated to increase the stability of various lipases and other enzymes, with documented increases in thermal stability and resistance to denaturing agents, making it a favorable choice for sensitive enzymes or applications requiring gentle immobilization. Additionally, dextran aldehyde's polymeric nature can create enzyme coatings, enhancing surface coverage and protection of immobilized enzymes (Liu & Smith, 2023; Mateo et al., 2004). However, its use often requires optimization of the concentration and oxidation degree of dextran to achieve the desired stability and activity balance.

## 2.5.2 Nucleophilic attachment to epoxy linkers

Bifunctional epoxy cross-linkers such as glycerol diglycidyl ether (GDE) can also be used in enzyme cross-linking/immobilization due to their ability to form stable covalent bonds with enzymes under mild conditions (Fig. 2.7) (Weiser et al., 2014). Epoxides contain highly reactive oxirane rings that act as electrophiles and can readily open upon reaction with nucleophilic groups present on protein surfaces, such as amines, thiols, and hydroxyl groups of residues of lysine, cysteine, tyrosine, aspartate, or glutamate. This reaction leads to the stable covalent attachment of enzyme molecules (Mateo et al., 2013; Mohammadi et al., 2016). The cross-linking process with epoxy groups typically involves nucleophilic attack on the strained three-membered oxirane ring by nucleophiles such as the primary amine groups of lysine residues on the protein surface. This ring-opening reaction results in covalent bond formation via a hydroxylated linker, effectively cross-linking enzyme molecules together. Epoxy compounds are comparatively stable in aqueous environments, allowing longer reaction times and easier handling than other more reactive functional groups. However, their slower reactivity can be a limitation, necessitating optimized conditions such as longer incubation periods to achieve efficient coupling, which in turn may reduce enzyme activity. Although the capability of epoxy groups to react with a broader range of nucleophilic residues than GA is particularly advantageous for proteins with only a limited number of surface-exposed lysine residues, this high variability may result in a nonuniform particle size distribution of the cross-linked biocatalyst.

### 2.5.3 Genipin-mediated enzyme cross-linking

Genipin is a naturally derived cross-linking agent, recognized as a less toxic alternative to conventional chemical cross-linkers like GA. Its application in protein cross-linking has expanded due to its biocompatibility, low cytotoxicity, and ability to form stable covalent bonds with primary amine groups in proteins under mild conditions (Tacias-Pascacio et al., 2019; Wahba, 2024). However, the multistep extraction and purification, and its limited commercial availability, contribute to the higher market price of genipin compared with synthetic alternatives such as GA. Consequently, this elevated cost represents a significant limitation for large-scale or industrial applications. Nevertheless, genipin can be more appropriately employed in specialized systems where its superior biocompatibility and reduced cytotoxicity provide distinct advantages over conventional cross-linkers. The cross-linking mechanism involves a multistep chemical reaction primarily targeting primary amines, mainly the amino groups of lysine residues. Initially, the ester group at the C11 of genipin undergoes a nucleophilic substitution reaction with the amine groups of lysine residues located on the surface of an enzyme molecule, leading to the formation of genipin-protein complex (Fig. 2.8).

The occurrence of this reaction has been confirmed by the detection of methanol released into the reaction mixture (Butler et al., 2003). Then the lone pair of electrons on an amine group from the second enzyme molecule initiates a nucleophilic attack on the unsaturated ether bond in the six-membered cyclic alkene at the C3 position of genipin. This reaction results in the cleavage of the ring structure and the formation of an aldehyde intermediate. Subsequently, this aldehyde intermediate reacts with the same amine to yield a nitrogen-containing cyclic intermediate (Butler et al., 2003).

## 2.6 Principles of enzyme cross-linking in soluble form

After the introduction of CLECs in the early 1960s, advances in protein insolubilization technology made it possible to directly cross-link soluble enzymes using

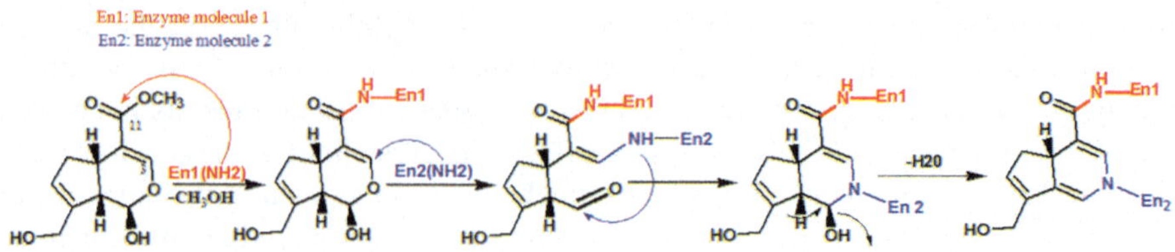


**FIGURE 2.8**

Mechanism of genipin-mediated cross-linking.

The reaction involves nucleophilic substitution of the ester group by lysine amino groups forming a genipin-enzyme complex, followed by nucleophilic attack of a second enzyme molecule on the cyclic ether ring leading to nitrogen-containing cyclic intermediate formation.

bifunctional agents such as GA (Quiocho & Richards, 1964). Habeeb pioneered this approach by generating insoluble trypsin through direct GA treatment (Habeeb, 1967). Despite using a high enzyme concentration (50 mg/mL), the cross-linking efficiency was low, reaching only up to 10%. Nevertheless, the resulting insoluble trypsin retained 80% of its enzymatic activity, and the reaction was found to be pH-dependent, with the greatest activity retention at lower pH levels. To increase the degree of cross-linking, Habeeb incorporated solid ammonium sulfate, pioneering what we now recognize as CLEA preparation. In Jansen and Olson reported that GA could be used to insolubilize a crystalline suspension of papain, a sulfhydryl-containing enzyme, resulting in an insoluble enzyme that retained 48% of its ester hydrolyzing activity (Jansen & Olson, 1969). Adding high concentrations of protein feeders, such as bovine serum albumin (BSA), enabled another approach that resulted in the one-step formation of CLEs of $\beta$-galactosidase with up to 64% efficiency. However, the large amount of BSA required diminishes the overall advantage of this method (Khare & Gupta, 1990).

Although enhanced thermostability was one of the earliest reported benefits of CLEs, these systems frequently exhibited significant limitations, including reduced activity retention, poor reproducibility, and low mechanical stability. Additional challenges included the handling of gelatinous CLEs preparations, low cross-linking yield, and the lack of control over the geometric properties of the CLEs (Cao et al., 2003). The limitations of CLEs have prompted a shift in focus toward more advanced and efficient enzyme cross-linking methods, such as CLEAs and CLECs. Following these developments, there was a significant gap in publications on producing CLEs until 2020, till the innovative approach of using isocyanide-based multicomponent reactions (IMCRs) to produce CLEs has reignited interest and spurred renewed attention in this field (Ashjari et al., 2020; Mohammadi et al., 2023; Shahedi et al., 2025). IMCRs are a prominent class of MRCs in organic chemistry where an isocyanide reagent reacts simultaneously with two or more other components to form complex molecules in a single step (Dömling, 1998). These reactions are highly valued for their efficiency and ability to rapidly generate diverse compound libraries. The classic examples include the Passerini three-component reaction and the Ugi four-component

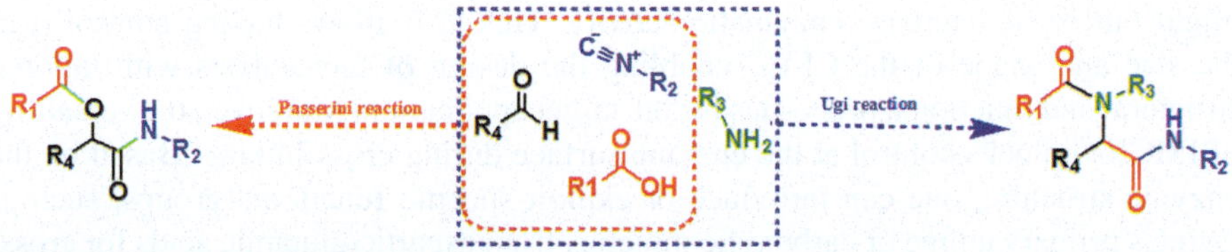


**FIGURE 2.9**

Isocyanide-based multicomponent reactions.

The reaction enables efficient one-step covalent cross-linking of soluble enzymes.

reaction, both of which yield products valuable in medicinal chemistry and drug discovery (Fig. 2.9) (Shahedi et al., 2025).

The unique reactivity of isocyanides, due to their unusual valence structure allowing both electrophilic and nucleophilic behavior, enables the formation of various bonds and scaffolds. IMCRs have broad compatibility with various functional groups and facilitate sustainable and efficient reactions, often under mild conditions. Today, they are widely applied in fields like polymer science, material science, and biological chemistry due to their versatility and robustness. IMCRs have also shown great promise in the rapid and efficient covalent immobilization of enzymes on solid carriers (Mohammadi et al., 2015; Shahedi et al., 2025). Building on this, the application of IMCRs in enzyme cross-linking has been expanded for the preparation of CLEs, offering an advantage by effectively bypassing the initial and often challenging step of enzyme aggregation or crystallization (Mohammadi et al., 2023). The formation of enzyme crystals or aggregates requires carefully controlled chemical or physical conditions, such as precipitation agents, temperature shifts, or prolonged incubation times, which are often time-consuming, labor-intensive, and may lead to inconsistent particle size and partial enzyme inactivation. Besides, the preparation of CLECs typically requires highly pure enzymes, which adds complexity and cost to the process. By employing IMCRs, the entire process can be streamlined into a one-step cross-linking reaction that directly captures and binds enzyme molecules without the need for prior aggregation or crystallization. This approach not only reduces the preparation time but also minimizes the use of potentially damaging chemicals or physical treatments, simplifying the workflow and improving reproducibility. The remarkable flexibility of IMCRs in accepting different functional groups empowers researchers to choose among a wide range of commercially available linkers featuring diverse functional groups, toxicity profiles, hydrophobicity, and solubility characteristics Fig. 2.9). IMCRs, such as the well-known Ugi or Passerini reactions, are compatible with amines, carboxylic acids, aldehydes, and epoxides, allowing for the straightforward incorporation of several compounds into the enzyme cross-linking design (Ahrari et al., 2023, 2022, 2024). This flexibility means that one can select linkers best suited for the specific application, whether aiming for biocompatibility, minimizing toxicity, or tailoring hydrophobicity to influence enzyme behavior. By carefully selecting the proper size of the linkers, it is possible to avoid concerns related to small molecules penetrating into the enzyme's internal structure, which might otherwise decrease enzymatic activity. This also allows having control over the size and shape of the CLEs, enabling the design of biocatalysts with tailored structural and functional properties for the target application. Moreover, the versatility of IMCRs unlocks control at the enzyme surface during cross-linking. Based on the enzyme structure, one can introduce or exploit specific functional groups, such as lysine's primary amine or carboxylic moieties from aspartic/glutamic acids for cross-linking reactions. This targeted approach can preserve enzyme activity, allow for fine-tuning of immobilization, and facilitate the creation of catalytically active enzyme preparations.

### 2.6.1 Mechanism of cross-linking in isocyanide-based multicomponent reactions

When employing IMCRs for cross-linking, with GA as the cross-linking agent, the process can proceed via the Ugi reaction. The Ugi reaction is a classic four-component condensation reaction involving an amine, a carbonyl compound (typically an aldehyde), a carboxylic acid, and an isocyanide. In this mechanism, GA serves as the carbonyl compound, reacting with an amine group and a carboxylic acid group in the presence of an isocyanide, which occurs under mild aqueous conditions, preserving enzyme activity. The mechanism begins with the condensation of the amine and GA to form an imine or Schiff base intermediate. Then, the isocyanide adds to the imine carbon, followed by the nucleophilic attack of the carboxylate group, resulting in the formation of an amide linkage (Fig. 2.9). This approach differs fundamentally from conventional GA-amine reactions, by creating a stable, multifunctional covalent network. In the context of enzyme cross-linking, this mechanism ensures that both the amine and carboxylic groups on proteins are involved, creating strong covalent cross-links between enzyme molecules. Chemistry permits the introduction of several bifunctional aldehyde compounds as cross-linkers to modulate the structural and functional properties of the final biocatalyst. When employing a bifunctional diacid linker in place of GA for cross-linking, the mechanism is proposed to proceed via a three-component reaction pathway. The cross-linking initiates with the reaction of the carboxyl groups from the linker with two isocyanide molecules, generating highly reactive intermediates, which are subsequently subjected to nucleophilic attack by either free amine groups or carboxylic acid groups exposed on the enzyme surface. The ensuing reactions lead to the formation of a stable cross-linked network, effectively and covalently bridging enzyme molecules.

## 2.7 Conclusion

Carrier-free immobilization relies on a delicate balance between protein solubility, reversible aggregation, and covalent cross-linking, and this chapter shows that successful design of a carrier-free immobilized system is only possible when these elements are treated as a unified physicochemical problem. The discussion of precipitation methods and crystallization principles underlines that aggregation must remain controlled, preserving native structure and hydration while preorganizing enzymes into architectures that are amenable to efficient cross-linking. On this foundation, the comparison of aldehyde-, epoxy-, polysaccharide-, and genipin-based linkers, together with newer isocyanide multicomponent strategies, demonstrates that linker chemistry is not merely a technical detail but a powerful handle to tune particle morphology, mass transfer, stability, and activity retention in carrier-free biocatalysts. The principles summarized here argue for a rational, structure-guided approach in which enzyme surface chemistry, aggregation route, and cross-linker design are co-optimized, enabling the next generation of robust, sustainable, and application-specific carrier-free immobilized enzymes for industrial and pharmaceutical processes.

## AI disclosure

During the preparation of this work the authors used perplexity in order to polish the language. After using this tool, the authors reviewed and edited the content as needed and takefull responsibility for the content of the publication.

## References


Modenez, I. A., Sastre, D., Moraes, F. C., & Netto, C. M. (2018). Influence of glutaraldehyde cross-linking modes on the recyclability of immobilized lipase B from *Candida antarctica* for transesterification of soybean oil. *Molecules, 23*(9), 2230. doi:10.3390/molecules23092230.

Ahrari, F., Yousefi, M., Habibi, Z., & Mohammadi, M. (2022). Application of undecanedicarboxylic acid to prepare cross-linked enzymes (CLEs) of *Rhizomucor miehei* lipase (RML): Selective enrichment of polyunsaturated fatty acids. *Molecular Catalysis, 520*, 112172. doi:10.1016/j.mcat.2022.112172.

Ahrari, F., Yousefi, M., Habibi, Z., & Mohammadi, M. (2023). Cross-linked lipase particles with improved activity; application of a non-toxic linker for cross-linking. *LWT, 173*, 114371. doi:10.1016/j.lwt.2022.114371.

Ahrari, F., Yousefi, M., & Mohammadi, M. (2024). The effect of carbon chain length of cross-linking agent on the functionality of carrier- free immobilized *Thermomyces lanuginosa* lipase particles. *International Journal of Biological Macromolecules, 270*(Pt 1), 132076. doi:10.1016/j.ijbiomac.2024.132076.

Akyüz, A., Tekin, İ., Aksoy, Z., & Ersus, S. (2024). Plant protein resources, novel extraction and precipitation methods: A review. *Journal of Food Process Engineering, 47*(10), e14758. http://onlinelibrary.wiley.com/journal/10.1111/(ISSN)1745-4530. doi:10.1111/jfpe.14758.

Ashjari, M., Garmroodi, M., Ahrari, F., Yousefi, M., & Mohammadi, M. (2020). Soluble enzyme cross-linking: Via multi-component reactions: A new generation of cross-linked enzymes. *Chemical Communications, 56*(67), 9683–9686. http://pubs.rsc.org/en/journals/journal/cc. doi:10.1039/d0cc03429k.

Atha, D. H., & Ingham, K. C. (1981). Mechanism of precipitation of proteins by polyethylene glycols. Analysis in terms of excluded volume. *Journal of Biological Chemistry, 256*(23), 12108–12117. doi:10.1016/s0021-9258(18)43240-1.

Barbosa, O., Ortiz, C., Berenguer-Murcia, A., Torres, R., Rodrigues, R. C., & Fernandez-Lafuente, R. (2014). Glutaraldehyde in bio-catalysts design: A useful crosslinker and a versatile tool in enzyme immobilization. *RSC Advances, 4*(4), 1583–1600. doi:10.1039/c3ra45991h.

Burgess, R. R. (2009). Protein precipitation techniques. *Methods in Enzymology, 463*, 331–342. doi:10.1016/S0076-6879(09)63020-2.

Butler, M. F., Ng, Y. F., & Pudney, P. D. A. (2003). Mechanism and kinetics of the crosslinking reaction between biopolymers containing primary amine groups and genipin. *Journal of Polymer Science, Part A: Polymer Chemistry, 41*(24), 3941–3953. doi:10.1002/pola.10960.

Cao, L., van Langen, L., & Sheldon, R. A. (2003). Immobilised enzymes: Carrier-bound or carrier-free? *Current Opinion in Biotechnology, 14*(4), 387–394. http://www.elsevier.com/locate/copbio. doi:10.1016/S0958-1669(03)00096-X.

Chayen, N. E. (1998). Comparative studies of protein crystallization by vapour-diffusion and microbatch techniques. *Acta Crystallographica Section D: Biological Crystallography, 54*(1), 8–15. doi:10.1107/S0907444997005374.

Cheung, D. T., & Nimni, M. E. (1982). Mechanism of crosslinking of proteins by glutaraldehyde i: Reaction with model compounds. *Connective Tissue Research, 10*(2), 187–199. doi:10.3109/03008208209034418.

Dömling, A. (1998). Isocyanide based multi component reactions in combinatorial chemistry. *Combinatorial Chemistry and High Throughput Screening, 1*(1), 1–22. http://www.eurekaselect.com/585/journal/combinatorial-chemistry-amp-high-throughput-screening. doi:10.2174/1386207301012201181431111.

Duong-Ly, K. C., & Gabelli, S. B. (2014). Salting out of proteins using ammonium sulfate precipitation. *Methods in Enzymology, 541*, 85–94. doi:10.1016/B978-0-12-420119-4.00007-0.

Gao, K., Rao, J., & Chen, B. (2024). Plant protein solubility: A challenge or insurmountable obstacle. *Advances in Colloid and Interface Science, 324*, 103074. doi:10.1016/j.cis.2023.103074.

Habeeb, A. F. S. A. (1967). Preparation of enzymically active, water-insoluble derivatives of trypsin. *Archives of Biochemistry and Biophysics, 119*(1), 264–268. doi:10.1016/0003-9861(67)90453-5.

Havlásek, M., Marques, S. M., Szotkowská, V., Kunka, A., Babková, P., Damborský, J., Prokop, Z., & Bednář, D. (2025). Decoding protein stabilization: Impact on aggregation, solubility, and unfolding mechanisms. *Journal of Chemical Information and Modeling, 65*(16), 8688–8701. http://pubs.acs.org/journal/jcisd8. doi:10.1021/acs.jcim.5c00611.

Ingham, K. C. (1984). Protein precipitation with polyethylene glycol. *Methods in Enzymology, 104*, 351–356.

Jansen, E. F., & Olson, A. C. (1969). Properties and enzymatic activities of papain insolubilized with glutaraldehyde. *Archives of Biochemistry and Biophysics, 129*(1), 221–227. doi:10.1016/0003-9861(69)90169-6.

Khare, A., & Gupta, M. N. (1990). An active insoluble aggregate of E. coli $\beta$-galactosidase. *Biotechnology and Bioengineering, 35*(1), 94–98. doi:10.1002/bit.260350113.

Klibanov, A. M. (2001). Improving enzymes by using them in organic solvents. *Nature, 409*(6817), 241–246. doi:10.1038/35051719.

Liu, Z., & Smith, S. R. (2023). Cross-linked enzyme aggregate (CLEA) preparation from waste activated sludge. *Microorganisms, 11*(8), 1902. www.mdpi.com/journal/microorganisms. doi:10.3390/microorganisms11081902.

Lo Nostro, P., & Ninham, B. W. (2012). Hofmeister phenomena: An update on ion specificity in biology. *Chemical Reviews, 112*(4), 2286–2322. doi:10.1021/cr200271j.

Mateo, C., Palomo, J. M., Van Langen, L. M., Van Rantwijk, F., Sheldon, R. A., & New, A. (2004). Mild cross-linking methodology to prepare cross-linked enzyme aggregates. *Biotechnology and Bioengineering, 86*(3), 273–276. doi:10.1002/bit.20033.

Mateo, C., Grazu, V., & Guisan, J. M. (2013). Immobilization of enzymes on monofunctional and heterofunctional epoxy-activated supports. *Methods in Molecular Biology, 1051*, 43–57. http://www.springer.com/series/7651. doi:10.1007/978-1-62703-550-7_4.

Migneault, I., Dartiguenave, C., Bertrand, M. J., & Waldron, K. C. (2004). Glutaraldehyde: Behavior in aqueous solution, reaction with proteins, and application to enzyme crosslinking. *BioTechniques, 37*(5), 790–802. https://www.future-science.com/journal/btn. doi:10.2144/04375rv01.

Mohammadi, M., Shahedi, M., Ahrari, F., Mostafavi, M., Habibi, Z., & Yousefi, M. (2023). Isocyanide-based multi-component reactions for carrier-free and carrier-bound covalent immobilization of enzymes. *Nature Protocols, 18*(5), 1641–1657. https://www.springer.com/journal/41596. doi:10.1038/s41596-023-00812-z.

Mohammadi, M., Gandomkar, S., Habibi, Z., & Yousefi, M. (2016). One pot three-component reaction for covalent immobilization of enzymes: Application of immobilized lipases for kinetic resolution of: Rac-ibuprofen. *RSC Advances, 6*(58), 52838–52849. http://pubs.rsc.org/en/journals/journalissues. doi:10.1039/c6ra11284f.

Mohammadi, M., Ashjari, M., Dezvarei, S., Yousefi, M., Babaki, M., & Mohammadi, J. (2015). Rapid and high-density covalent immobilization of *Rhizomucor miehei* lipase using a multi component reaction: Application in biodiesel production. *RSC Advances, 5*(41), 32698–32705. http://pubs.rsc.org/en/journals/journalissues. doi:10.1039/c5ra03299g.

Nickerson, J. L., Baghalabadi, V., Dang, Z., Miller, V. A., Little, S. L., & Doucette, A. A. (2022). Organic solvent-based protein precipitation for robust proteome purification ahead of mass spectrometry. *Journal of Visualized Experiments, 180*, e63503. https://www.jove.com/t/63398/determination-photoisomerization-quantum-yield-hydrazone. doi:10.3791/63503.

Niwa, T., Ying, B.-W., Saito, K., Jin, W. Z., Takada, S., Ueda, T., & Taguchi, H. (2009). Bimodal protein solubility distribution revealed by an aggregation analysis of the entire ensemble of *Escherichia coli* proteins. *Proceedings of the National Academy of Sciences, 106*(11), 4201–4206. doi:10.1073/pnas.0811922106.

Novák, P., & Havlíček, V. (2016). Protein extraction and precipitation. In P Ciborowski, & J Silberring (Eds.), *Proteomic profiling and analytical chemistry* (pp. 51–62). Elsevier BV. doi:10.1016/b978-0-444-63688-1.00004-5.

Pimtawong, T., Ren, J., Lee, J., Lee, H.-M., & Na, D. (2025). A review on computational models for predicting protein solubility. *Journal of Microbiology, 63*(1), e2408001. doi:10.71150/jm.2408001.

Qing, R., Hao, S., Smorodina, E., Jin, D., Zalevsky, A., & Zhang, S. (2022). Protein design: From the aspect of water solubility and stability. *Chemical Reviews, 122*(18), 14085–14179. http://pubs.acs.org/journal/chreay. doi:10.1021/acs.chemrev.1c00757.

Quiocho, F. A., & Richards, F. M. (1964). Intermolecular cross linking of a protein in the crystalline state: Carboxypeptidase A. *Proceedings of the National Academy of Sciences, 52*(3), 833–839. doi:10.1073/pnas.52.3.833.

Roberts, C. J. (2014). Protein aggregation and its impact on product quality. *Current Opinion in Biotechnology, 30*, 211–217. http://www.elsevier.com/locate/copbio. doi:10.1016/j.copbio.2014.08.001.

Schein, C. H. (1990). Solubility as a function of protein structure and solvent components. Presse Dienstleistungsgesellschaft mbH und Co. *Nature Biotechnology, 8*(4), 308–317. http://www.nature.com/nbt/index.html. doi:10.1038/nbt0490-308.

Schmit, J. D., & Dill, K. A. (2010). The stabilities of protein crystals. *Journal of Physical Chemistry B, 114*(11), 4020–4027. http://pubs.acs.org/journal/jpcbfk. doi:10.1021/jp9107188.

Shahedi, M., Ahrari, F., & Mohammadi, M. (2025). Multi-component reactions in protein modification and immobilization. *ChemBioChem, 26*(9), e202401010. doi:10.1002/cbic.202401010.

Sheldon, R. A. (2011). Cross-linked enzyme aggregates as industrial biocatalysts. *Organic Process Research and Development, 15*(1), 213–223. doi:10.1021/op100289f.

Spicer, C. D., & Davis, B. G. (2014). Selective chemical protein modification. *Nature Communications, 5*, 4740. http://www.nature.com/ncomms/index.html. doi:10.1038/ncomms5740.

Tacias-Pascacio, V. G., García-Parra, E., Vela-Gutiérrez, G., Virgen-Ortiz, J. J., Berenguer-Murcia, Á., Alcántara, A. R., & Fernandez-Lafuente, R. (2019). Genipin as an emergent tool in the design of biocatalysts: Mechanism of reaction and applications. *Catalysts, 9*(12), 1035. https://www.mdpi.com/2073-4344/9/12/1035/pdf. doi:10.3390/catal9121035.

Tadeo, X., López-Méndez, B., Castaño, D., Trigueros, T., & Millet, O. (2009). Protein stabilization and the hofmeister effect: The role of hydrophobic solvation. *Biophysical Journal, 97*(9), 2595–2603. doi:10.1016/j.bpj.2009.08.029.

Trevino, S. R., Scholtz, J. M., & Pace, C. N. (2007). Amino acid contribution to protein solubility: Asp, Glu, and Ser contribute more favorably than the other hydrophilic amino acids in RNase Sa. *Journal of Molecular Biology, 366*(2), 449–460. https://www.journals.elsevier.com/journal-of-molecular-biology. doi:10.1016/j.jmb.2006.10.026.

Trevino, S. R., Scholtz, J. M., & Pace, C. N. (2008). Measuring and increasing protein solubility. *Journal of Pharmaceutical Sciences, 97*(10), 4155–4166. http://www3.interscience.wiley.com/cgi-bin/fulltext/117905147/PDFSTART. doi:10.1002/jps.21327.

Vihinen, M. (2020). Solubility of proteins. *ADMET and DMPK, 8*(4), 391–399. https://hrcak.srce.hr/en/clanak/354793. doi:10.5599/admet.831.

Wahba, M. I. (2024). A comprehensive review on genipin: An efficient natural cross-linker for biopolymers. *Polymer Bulletin, 81*(16), 14251–14305. https://www.springer.com/journal/289. doi:10.1007/s00289-024-05406-7.

Weiser, D., Varga, A., Kovács, K., Nagy, F., Szilágyi, A., Vértessy, B. G., Paizs, C., & Poppe, L. (2014). Bisepoxide cross-linked enzyme aggregates—New immobilized biocatalysts for selective biotransformations. *ChemCatChem, 6*(5), 1463–1469. http://onlinelibrary.wiley.com/journal/10.1002/(ISSN)1867-3899. doi:10.1002/cctc.201300806.

Wingfield, P. (1998). Protein precipitation using ammonium sulfate. *Current Protocols in Protein Science, 13*(1). doi:10.1002/0471140864.psa03fs13.

Wong, F. W. F., Ariff, A. B., & Stuckey, D. C. (2018). Downstream protein separation by surfactant precipitation: A review. *Critical Reviews in Biotechnology, 38*(1), 31–46. doi:10.1080/07388551.2017.1312266.

Zelinski, T., & Waldmann, H. (1997). Cross-linked enzyme crystals (CLECs): Efficient and stable biocatalysts for preparative organic chemistry. *Angewandte Chemie (International Edition in English), 36*(7), 722–724. http://onlinelibrary.wiley.com/journal/10.1002/(ISSN)1521-3773. doi:10.1002/anie.199707221.

SECTION

# Techniques for carrier-free immobilization

# II

CHAPTER

# Types of cross-linkers for carrier-free immobilization

# 3

**Nashriq Jailani[a], Roshanida A. Rahman[a,b], Nardiah Rizwana Jaafar[a] and Rosli Md. Illias[a,b]**

[a] *Department of Bioprocess & Polymer Engineering, Faculty of Chemical & Energy Engineering, Universiti Teknologi Malaysia, Skudai, Johor, Malaysia,* [b] *The Innovation Centre in Agritechnology for Advanced Bioprocessing, Hub Pendidikan Tinggi Pagoh, Universiti Teknologi Malaysia, Pagoh, Johor, Malaysia*

## 3.1 Introduction

The choice of cross-linker is a significant factor influencing the immobilized activity and characteristics during carrier-free immobilization preparation. Cross-linking transpired between the functional group of the cross-linker and the surface amino acid residues of the enzyme. Since every enzyme is a unique molecule, the establishment of robust contacts and stable cross-linked enzyme immobilization is correlated with the relationship between the types of cross-linker chemistries and the composition of amino acid residues on the enzyme surface. This chapter will discuss the molecular interaction between enzyme surface residues and the cross-linkers that affect the cross-linking nomenclature, the different types of cross-linkers used in carrier-free immobilization preparation from the last 5 years, and its effect on immobilized functional and structural properties.

## 3.2 Understanding molecular interaction between enzyme surface residue with different types of cross-linkers

Throughout the years, numerous chemical cross-linking reagents have been created. The main aim of the cross-linking process is to establish a covalent bond between two adjacent residues within a single polypeptide chain or between two separate polypeptide chains of an enzyme (Mateo et al., 2020). Chemical cross-linking reagents are often composed of two reactive sites connected by a spacer or linker region, usually an alkyl chain (Fass & Thorpe, 2018). Generally, these reactive sites can be categorized into various groups based on their reactivity (Chu et al., 2018).

The common amino acid residues that reacted with different functional groups of the cross-linkers were reported to be lysine, tryptophan, cysteine, aspartate, or glutamate. Fig. 3.1 summarizes the cross-linker with different reactive sites covalently

Carrier-Free Enzyme Immobilization: Fundamentals and Innovations in Biocatalysis.
DOI: https://doi.org/10.1016/B978-0-443-44139-4.00010-1

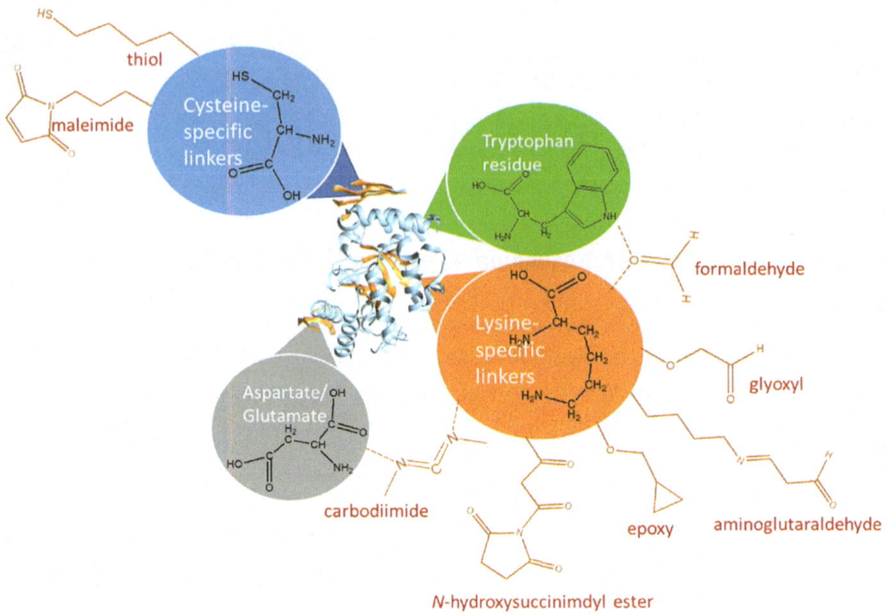


**FIGURE 3.1**

Covalent interaction between different reactive sites of cross-linkers with specific amino acid residues of enzymes.

Examples of amino acids commonly involved in the cross-linking were lysine, tryptophan, cysteine, aspartate, and glutamate.

bonded with these specific amino acid residues. Among all, the primary amino group of lysine and the N-termini of proteins are typically the main targets of the reactive groups in cross-linkers. Ester functional groups, such as N-hydroxysuccinimidyl esters, are predominantly utilized for this purpose (Fig. 3.1). Although these active esters demonstrate high reaction rates, they are also rapidly hydrolyzed in aqueous solutions (Kalkhof & Sinz, 2008). Therefore, obtaining favorable cross-linking yields at low protein concentrations is difficult because of the competing hydrolysis reaction, which also limits the duration of the reaction that can be employed. In addition, the other functional groups, such as epoxy, glyoxyl, aldehyde, and imide, were reported to be reactive against the lysine residue.

Lysine cross-linking offers several advantages, such as the significant presence of lysine residues (around 6% on enzyme surfaces) and a commendable level of reaction specificity. Moreover, when reaction conditions are carefully controlled, including factors like pH, reaction times, and reagent excess, it was shown that the side reactions of N-hydroxysuccinimide (NHS) esters with other amino acids typically do not take place to a significant extent (Kalkhof & Sinz, 2008; Mädler et al., 2009).

Comparable targeted cross-linking, including those utilizing maleimides (Fig. 3.1), can be executed on cysteine residues; however, the limited prevalence of cysteine, often constituting less than 2% of the enzyme surface, diminishes its appeal (Zhang et al., 2008). Alternative cross-linking chemistries are infrequently utilized due to their inability to be carried out under optimal native conditions or because the resultant products are unstable or heterogeneous (Novák & Kruppa, 2008).

Alongside homobifunctional cross-linkers, numerous heterobifunctional linkers have been documented. These may incorporate two distinct reactive groups of two residues such as lysine-, and aspartate/glutamate-reactive, as shown by carbodiimide (Fig. 3.1). Formaldehyde, however, is a notable exception to conventional cross-linkers, as it possesses a single aldehyde group yet may connect two amino acid side chains (lysine and tryptophan) through a two-step synthesis (Fig. 3.1). While lysine and tryptophan residues are generally the focus, formaldehyde is noted to be a less selective reagent (Sutherland et al., 2008; Toews et al., 2008).

## 3.3 Effect of length and type of interaction of different cross-linkers on the cross-linking nomenclature

In preparation for carrier-free immobilization that generally involves cross-linking between enzymes with the cross-linkers, numerous parameters can affect biocatalyst activity at the molecular level, including linker length and the type of covalent cross-linking employed, producing different nomenclature products.

In covalent cross-linking, hydrophobic surfaces frequently induce protein denaturation by facilitating interactions between the hydrophobic side chains of the protein and the surface. A molecular spacer or linker is utilized to mitigate surface effects and address these undesirable interactions (Talbert & Goddard, 2012). A crucial aspect to take into account is the spacer's length. As shown in Fig. 3.2A, short linkers are generally perceived to enhance the rigidity of the enzyme, while longer linkers are considered to reduce steric hindrance and provide more efficient multipoint covalent attachment due to the increased number of residues available to the linker (Santos et al., 2015).

In addition, the different types of covalent interactions between the cross-linkers and the surface of enzyme residues also affect the overall performance. The flexibility as a result of using the longer cross-linker, for instance, does not always necessitate better enzyme immobilization performance. The interaction between the cross-linkers and the residues can be categorized into single-point covalent interaction (SCI) or multipoint covalent interaction (MCI), as depicted in Fig. 3.2B.

MCI of enzymes to cross-linkers has demonstrated stabilizing effects across a broad range of enzymes. One possible advantage of MCI is the rigidification of the enzyme, which is expected to enhance structural stability compared to SCI (Fig. 3.2B) (Rodrigues et al., 2013). However, the accomplishment of MCI is somewhat complex, as the formation of many covalent connections between an enzyme and its surface

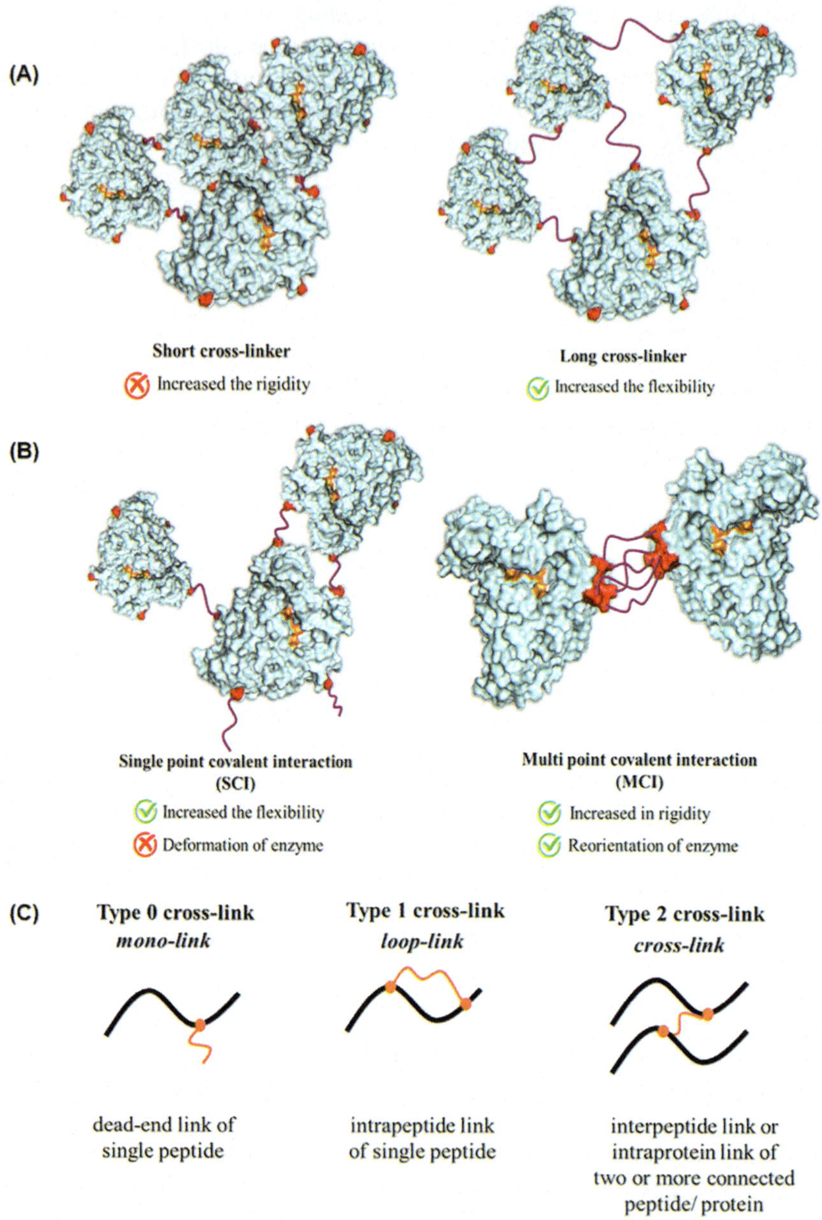


**FIGURE 3.2**

Impact of the length and different types of interaction towards nomenclature products of cross-linking in carrier-free immobilization.

(A) length of the cross-linker, (B) different types of interactions, and (C) cross-linking nomenclature.

may lead to structural alterations in the enzyme if the bonds are not appropriately matched. Thus, the technique requires meticulous adjustment of the linkers, reaction conditions, and protein modifications (Mateo et al., 2006).

On the other hand, SCI often incorporates a singular type of reactive functional group, predominantly the primary amino groups of lysine residues, whose reactivity may be meticulously adjusted through precise pH regulation (Hoarau et al., 2017). The initial residue interacts to attach the protein to the surface, whereas subsequent interactions with adjacent residues enhance the protein's stability. The inability to regulate the orientation of enzymes relative to the surface is a limitation of this approach (Hoarau et al., 2017).

The overall synthesis and distribution of cross-linking are governed by intricate reaction kinetics that entail cross-linking competition among many pathways. This competition will determine the ultimate ratio of the specified nomenclature of the cross-linking, which can be categorized into type 0, type 1, and type 2 cross-links, as shown in Fig. 3.2C. Type 0 cross-link, which is a mono-link, commonly known as a dead-end product, occurs when a single reactive arm of a homo- or bifunctional cross-linker successfully binds to a residue on a protein, such as a lysine, while the second reactive arm does not identify a compatible partner and is subsequently quenched, usually via hydrolysis. An example of this type of linker is bis-(sulfosuccinimidyl) suberate (BS3), where this (NHS) ester is a water-soluble, membrane-impermeable cross-linker widely used for cell-surface cross-linking (Mateo et al., 2006).

A type 1 cross-link, a loop-link, often classified as an intra-protein cross-link, is defined by the effective connection of two reactive sites via a bifunctional linker, with both linked residues originating from the same protease-generated peptide fragment. Type 2 cross-links, on the other hand, refer to interpeptide or intraprotein links that connect two or more peptides or proteins. Distinguishing a loop-link from a standard intra-protein cross-link is essential. Both link two residues on the same polypeptide chain; however, a loop-link occurs within a single peptide, while a canonical intra-protein cross-link connects two distinct peptides ($\alpha$ and $\beta$) derived from the same protein. Loop-links are essential for establishing localized distance constraints that limit the flexibility of short, contiguous regions. In contrast, the intra-cross-link linkages serve as the primary source of information for ascertaining the quaternary structure of multisubunit complexes, identifying protein-protein interactions, and delineating the interfaces of those interactions (Santos et al., 2015; Talbert & Goddard, 2012).

## 3.4 Different types of cross-linkers used in carrier-free immobilization

The development of the carrier-free enzyme immobilization approach has led to the reporting of various types of cross-linkers over time. Among all the reported cross-linkers, macromolecular synthetic or natural cross-linkers are often used for enzyme cross-linking. Apart from that, tailor-functionalized cross-linkers started to gain attention for use in cross-linking. These cross-linkers were functionalized, whether by substitution or addition of certain functional groups or a combination of synthetic and natural biopolymers, to facilitate the cross-linking between enzymes in immobilization.

Table 3.1 summarizes the types of cross-linkers used in the development of cross-linked enzyme aggregates (CLEAs), cross-linked enzymes (CLEs), and cross-linked enzyme crystals (CLECs) immobilization from the last 5 years. From the table, these types of cross-linkers were categorized into natural molecules and biopolymers, synthetic, combinations of synthetic and natural, and functionalized cross-linkers, which are further discussed.

### 3.4.1 Natural molecules and biopolymer-based cross-linkers

A natural molecule or biopolymer cross-linker is a chemical agent sourced from natural origins, such as plants or microbes, that establishes strong interactions, either covalent or ionic, between adjacent enzyme molecules. Either small molecules or biopolymers are becoming recognized as a safer, more sustainable option than several synthetic cross-linkers. The primary benefits of utilizing natural cross-linkers are their biocompatibility and nontoxicity. Employing cross-linkers sourced from renewable resources enhances ecological and sustainable manufacturing processes, frequently eliminating the necessity for harmful solvents or high-energy conditions. The cross-linker sources from natural sources used in carrier-free immobilization over the last 5 years are listed in Table 3.1.

According to the literature, genipin (Fig. 3.3A) is used for the development of *Trametes versicolor* laccase CLEAs (Hong et al., 2021). Genipin is a colorless, water-soluble iridoid that is part of the extensive group of monoterpenoids characterized by a cyclopentan-[C]-pyran structure framework. It is synthesized from geniposide, a glucoside derived from the fruits of *Genipa americana* and *Gardenia jasminoides* (Hong et al., 2021), which is enzymatically degraded to genipin in the presence of the $\beta$-glucosidase enzyme, as illustrated in Fig. 3.4A.

On the other hand, another small natural molecule, such as cysteamine and carboxylic acids (Fig. 3.3A), was reported from the same group, involving the development of CLEs for lipase immobilization. Endogenously, cysteamine is derived from coenzyme A degradation, in which this particular coenzyme A is synthesized from the combination of cysteine and vitamin B5 (pantothenate) as illustrated in Fig. 3.4B (Besouw et al., 2013). Long-chain dicarboxylic acids such as dodecanedioic acid (DDDA), azelaic acid, and adipic acid are derived from vegetable oils and can be synthesized through biotransformation via microbiological methods, which have been extensively reviewed previously (Lee et al., 2019).

Based on the study, Ahrari et al. (2022) were the first to report the use of 1,11-undecanedicarboxylic acid, also known as DDDA, as a cross-linker for CLEs immobilization of *Rhizomucor miehei* lipase. In the following years, the same group explored cysteamine as an alternative to synthetic cross-linkers for the development

**Table 3.1** Examples of different cross-linkers used in carrier-free immobilization for the last five years.

| Cross-linker | Type of cross-linker | Enzyme | Enzyme source | Class of enzyme | References |
|---|---|---|---|---|---|
| **Cross-linked enzyme aggregates (CLEAs) immobilization** | | | | | |
| Dialdehyde amylopectin | Functionalized macromolecule | Polyethylene terephthalate hydrolase (PETase) | *Ideonella sakaiensis* | Hydrolase | Lee et al. (2025) |
| Hyperbranched polymer | Synthetic | Laccase | NA[a] | Oxidoreductase | Peng et al. (2025) |
| Glutaraldehyde | Synthetic | Commercial cellulolytic cocktails (Celluclast and Cellic CTec2) | NA | Hydrolase | Morais et al. (2025) |
| Dextran | Natural macromolecule | Levanase | *Shouchella lehensis* G1 | Hydrolase | Makki et al. (2024) |
| Amylopectin | Natural macromolecule | PETase | *I. sakaiensis* | Hydrolase | Lin et al. (2024) |
| O-carboxymethyl chitosan | Functionalized macromolecule | Cyclodextrin gluconotransferase (CGTase) and maltogenic amylase (MAG1) | *S. lehensis* G1 | Transferase and hydrolase | Yip et al. (2024) |
| Glutaraldehyde | Synthetic | Dehydrogenase/reductase 1 | *Lactobacillus fermentum* | Oxidoreductase | Guo et al. (2024) |
| Glutaraldehyde | Synthetic | Engineered xylanase R8 (N58D) | *Neocallimastix patriciarum* | Hydrolase | Cheng et al. (2024) |
| Fructose-glutaraldehyde | Combi-synthetic-natural | Laccase | *Trametes versicolor* | Oxidoreductase | Dey et al. (2023) |
| Glutaraldehyde-putrescine | Combi-synthetic-natural | Lipase | *Candida rugosa* | Hydrolase | Khiari et al. (2023) |

(continued on next page)

**Table 3.1** Examples of different cross-linkers used in carrier-free immobilization for the last five years—cont'd

| Cross-linker | Type of cross-linker | Enzyme | Enzyme source | Class of enzyme | References |
|---|---|---|---|---|---|
| Chitosan-glutaraldehyde | Combi-synthetic-natural | CGTase | *S. lehensis* G1 | Transferase | Jailani et al. (2023) |
| Glutaraldehyde | Synthetic | Phytase Quantum Blue | NA | Hydrolase | Henninger et al. (2023) |
| Dextran polyaldehyde | Natural macromolecule | Sucrose isomerase | *Pantoea dispersa* UQ68J | Isomerase | Chen et al. (2023) |
| Chitosan | Natural macromolecule | CGTase | *S. lehensis* G1 | Transferase | Jailani et al. (2022) |
| Glutaraldehyde | Synthetic | Cellulase, xylanase and pectinase | *Bacillus* sp. | Hydrolase | Amadi et al. (2022) |
| Glutaraldehyde | Synthetic | $\beta$-Mannanase | NA | Hydrolase | Behera et al. (2022) |
| Glutaraldehyde | Synthetic | $\alpha$-amylase | *Aspergillus fumigatus* | Hydrolase | Ullah et al. (2021) |
| Genipin | Natural | Laccase | *Trametes versicolor* | Oxidoreductase | Hong et al. (2021) |
| Chitosan | Natural macromolecule | MAG1 | *S. lehensis* G1 | Hydrolase | Nawawi et al. (2020) |

*(continued on next page)*

| | | | | | |
|---|---|---|---|---|---|
| **Cross-linked enzymes (CLEs) immobilization** | | | | | |
| 1,11-undecanedicarboxylic acid, azelaic acid, and adipic acid | Natural | Lipase | *Thermomyces lanuginosa* | Hydrolase | Ahrari et al. (2024) |
| Glutaraldehyde | Synthetic | Co-lipases | *Rhizomucor miehei* (sn-1,3 specific) and *Candida antarctica* (nonspecific) | Hydrolase | Ahrari and Mohammadi (2024) |
| Cysteamine | Natural | Lipase | *Rhizomucor miehei* and *Thermomyces lanuginosa* | Hydrolase | Ahrari et al. (2023) |
| 1,11-undecanedicarboxylic acid | Natural | Lipase | *Rhizomucor miehei* | Hydrolase | Ahrari et al. (2022) |
| **Cross-linked enzyme crystals (CLECs) immobilization** | | | | | |
| Glutaraldehyde | Synthetic | Papain | *Carica papaya* | Hydrolase | Hao et al. (2024) |
| Bis-maleimidoethane | Synthetic | Halohydrin dehalogenase | *Iumatobacter coccineus* | Lyase | Staar et al. (2022) |
| Glutaraldehyde | Synthetic | Lipase | *Aspergillus* sp. | Hydrolase | Fernández-Penas et al. (2021) |
| Glutaraldehyde | Synthetic | Artificial metalloenzyme iron@NikA | NA | NA | Lopez et al. (2020) |

[a] *Not available.*

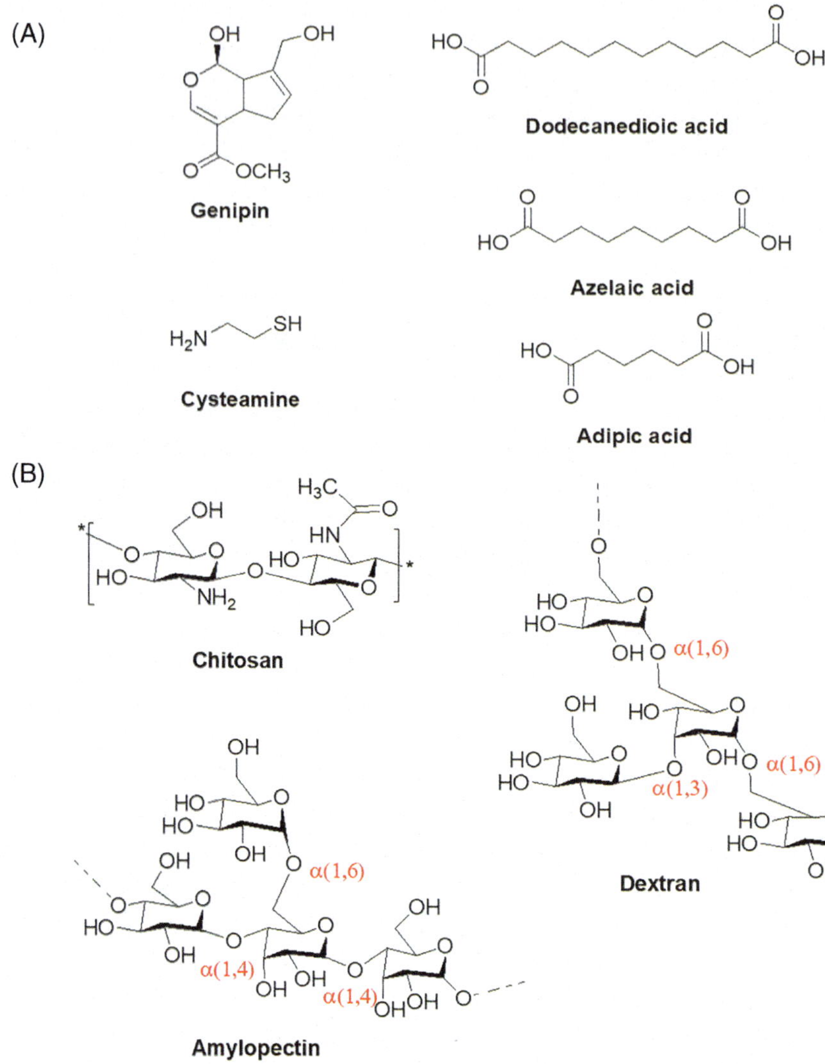


**FIGURE 3.3**

The chemical structures of natural small molecules and macromolecular biopolymer cross-linkers used in carrier-free immobilization from the reported studies from the last five years.

(A) natural small molecule and (B) biopolymer cross-linker.

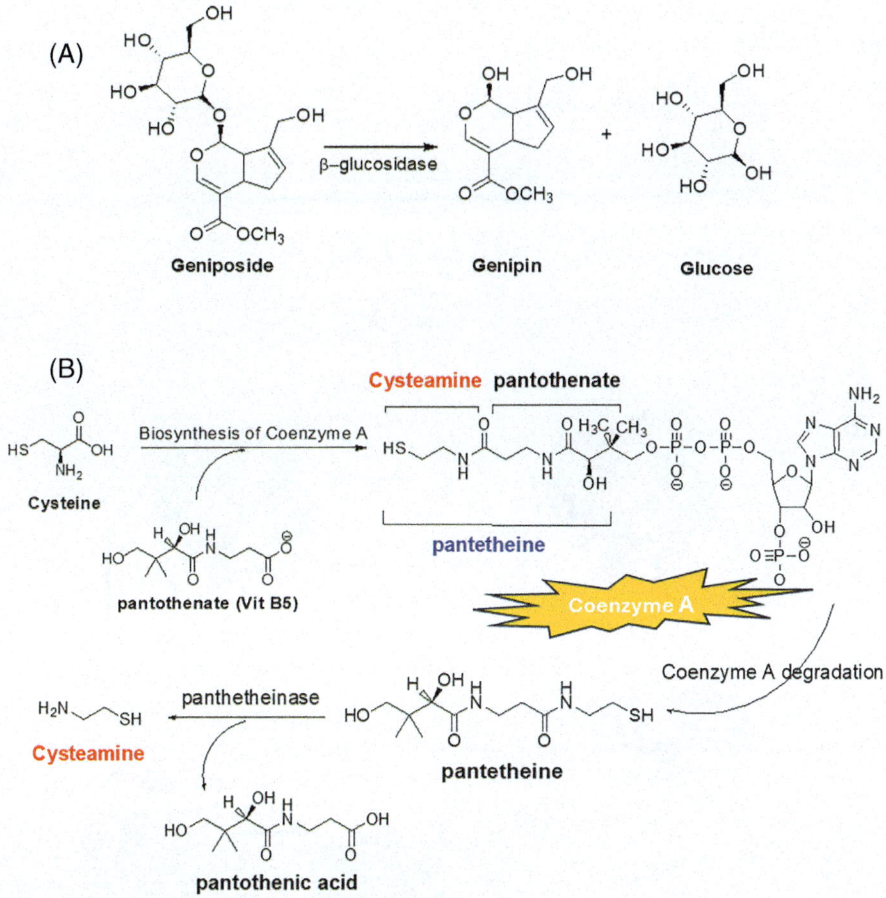


**FIGURE 3.4**

Biosynthesis of genipin and cysteamine cross-linker.

(A) Genipin synthesis from geniposide source and (B) cysteamine produced from the combination of cysteine and vitamin B5.

of lipases from *R. miehei* and *Thermomyces lanuginosus* CLEs immobilization (Ahrari et al., 2023). Subsequently, the effect of different carbon chain lengths of carboxylic acids utilizing DDDA, azelaic acid, and adipic acid was studied by the same group to develop CLEs of lipase from *T. lanuginose* (Ahrari et al., 2024). According to the report, the type of cross-linking involved using cysteamine and carboxylic acids following isocyanide-based multicomponent reactions (ICMRs). The ICMRs have garnered significant attention in the covalent immobilization of enzymes, which is a recognized reaction in organic chemistry wherein three or more reactants, often comprising an aldehyde, an amine, a carboxylic acid, and an isocyanide, are

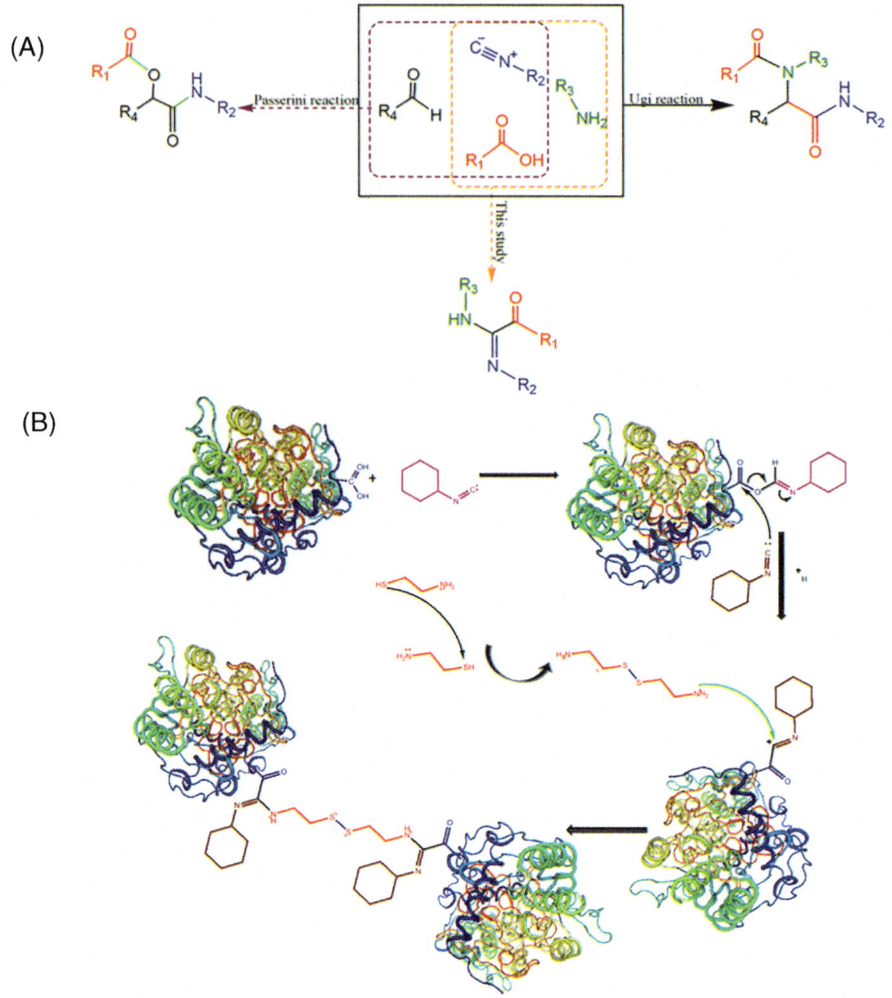


**FIGURE 3.5**

Isocyanide-based multicomponent reactions (ICMRs).

(A) Three types of ICMRs and (B) suggested mechanism for the cross-linking reaction using cysteamine cross-linker.

*From Ahrari, F., Yousefi, M., Habibi, Z., & Mohammadi, M. (2023). Cross-linked lipase particles with improved activity: Application of a non-toxic linker for cross-linking.* LWT, 173, *114371.*

concurrently introduced into a reaction mixture to yield a novel product in a one-pot procedure as illustrated in Fig. 3.5A. Following this, the mechanism of cysteamine for the cross-linking mechanism was suggested, as depicted in Fig. 3.5B.

In the meantime, macromolecule chitosan (CS) was reported to be the best cross-linker for the development of maltogenic amylase and cyclodextrin glucosyltransferase from *Shouchella lehensis* G1 (Jailani et al., 2022; Nawawi et al., 2020). The integration of computational with experimental screening of the cross-linker in developing CLEAs was first reported by Jailani et al. (2022). Out of seven cross-linkers tested, including polyethylene glycol, pectin, benzoquinone, glutaraldehyde (GA), and dialdehyde starch, CS was found to show the highest binding affinity and stability towards cyclodextrin gluconotransferase (CGTase) G1 through docking and molecular dynamics simulation. In corroboration with the computational analysis, the experimental results present that CS recovered the highest CGTase G1 activity (84%) compared to other cross-linkers tested. In conjunction with this work, the development of polyethylene terephthalate hydrolase (PETase)-CLEAs reported by Lin et al. (2024) used the computational work to cut the laborious step in cross-linker screening and found the amylopectin recovered more than 80% of its activity recovery for plastic degradation analysis. While dextran biopolymer was reported as the optimal cross-linker for the CLEAs of levanase from *S. lehensis* G1 (Makki et al., 2024).

Chitosan, dextran, and amylopectin (Fig. 3.3B) are examples of natural polysaccharides or biopolymer macromolecules. Chitosan is primarily derived from chitin, for which shrimp and crabs are the predominant sources referenced in the literature for CS manufacture, while other species such as lobster, crayfish, and oyster have also been employed (Kou et al., 2021). Dextran is an exopolysaccharide synthesized by certain species of lactic acid bacteria, for example, belonging to the genera of *Lactococcus, Lactobacillus, Streptococcus, Aerococcus*, and *Carnobacterium* through fermentation, usually in the presence of sucrose (Díaz-Montes, 2021). Amylopectin is one of the two main components of starch and is derived from plants, particularly in staple foods such as wheat and maize or root vegetables like potatoes and cassava (Jaiswal & Chibbar, 2017).

Other than the mentioned cross-linkers in Table 3.1, several natural-based cross-linkers have also been reported. For example, Ayhan et al. (2012) presented a robust and biocompatible glucose oxidase, peroxidase, and urease CLEAs, employing the biocompatible L-lysine biomolecule as the cross-linker. L-lysine is an indispensable amino acid naturally occurring in high-protein foods. In the study, the hydroxyl groups of threonine and serine residues of enzymes were first oxidized with periodate to produce aldehydes, which subsequently cross-linked with L-lysine (Ayhan et al., 2012). In addition, CLEAs of glucoamylase were synthesized, as reported by Talekar et al. (2014), leveraging a pectin cross-linker. Pectin is a group of polysaccharides abundant in galacturonic acid. It is a major part of the cell walls in all higher plants, and it is especially common in fruits like apples, oranges, and lemons (Roman-Benn et al., 2023). It was reported that from the study, CLEAs of glucoamylase employing pectin as a cross-linker resulted in enhanced thermal stability and excellent reusability compared to the free counterpart. These unmodified natural and biopolymers were reported to primarily stabilize the enzyme aggregates through noncovalent interactions such as hydrogen bonding between the polymer's abundant hydroxyl groups and the enzyme's surface residues or ionic interactions, such as a positively charged amine group with a negatively charged enzyme residue. The

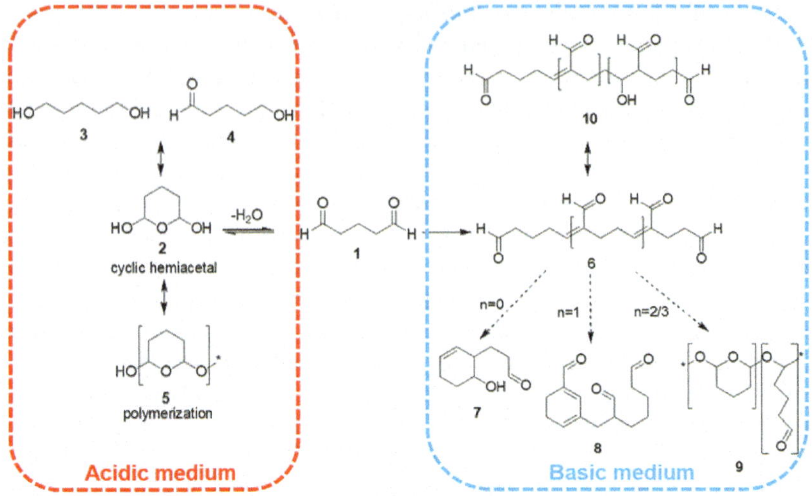


**FIGURE 3.6**

The possible glutaraldehyde structures coexist based on different pH levels.

The red circle indicates the acidic medium, while the blue circle indicates the basic medium.

ionic interaction, however, has some disadvantages, where this approach is very mild but often provides lower operational stability compared to covalent methods, as the linkages can be disrupted by changes in salt concentration.

### 3.4.2 Synthetic micro/macromolecules cross-linkers

From Table 3.1, GA is still the predominant synthetic cross-linker employed in carrier-free immobilization techniques. Although its toxicity may pose a concern, GA has been utilized for decades for protein cross-linking due to its easy handling and ready availability. GA is a bifunctional aldehyde (structure 1, Fig. 3.6) characterized by a low molecular weight, appears as a colorless liquid with a pungent odor, and is soluble in water, alcohol, and organic solvents (Barbosa et al., 2014). It is typically available as an aqueous solution with a pH of approximately 3 to 4. However, the structure of GA in aqueous solution has been a topic of debate and controversy, as GA can exist in forms such as monomeric dialdehyde, dimer, trimer, and polymer depending on pH conditions (Fig. 3.6). Therefore, the mechanism of the cross-linking is still not fully understood and remains explored by many researchers.

From Fig. 3.6, it was demonstrated that structures 1, 2, 3, and 4 coexist in equilibrium in the GA solution as a result of the hydration of structure 1 following its dissolution in water (Hardy et al., 1969). Conversely, a study reported by Whipple and Ruta (1974) indicated that aqueous GA primarily exists as a cyclic hemiacetal (2), which is distributed between two geometric isomeric forms (cis and trans). On the other hand, Korn et al. (1972) identified an equilibrium among free GA (1), cyclic hemiacetal (2), and their oligomers (5) through proton-nucleus magnetic resonance ($^1$H-NMR) analysis. Consequently, in a commercial aqueous GA solution, structures 1, 2, 3, 4, and 5 may coexist under acidic conditions. In the meantime, an aldol condensation of GA chains may spontaneously occur in an aqueous solution under basic conditions, favoring structures 6 and 8 (Fig. 3.6). It was also documented that the substantial quantities of $\alpha$, $\beta$-unsaturated aldehydes (6) are capable of ring formation (8) (Richards & Knowles, 1968). Structures 7, 9, and 10 can also exist in the aqueous solution of GA in alkaline media, resulting from polymerization via an intramolecular-intermolecular mechanism (9) and dimolecular aldol condensation (7 and 10) (Aso & Aito, 1962).

The interactions of GA with amino groups in acidic environments are seen in Fig. 3.7. Fig. 3.7A illustrates a nucleophilic attack occurring by the amino group of the proteins on GA. The primary amine remains unprotonated as it serves as a nucleophile attacking the electrophilic carbon. Despite the protonated amine groups' loss of capacity to attack the carbonyl carbon, dehydration occurs rapidly under acidic conditions. Conversely, the last stage of C–O bond cleavage and dehydration is disfavored in basic circumstances. The Schiff's base exhibits instability in highly acidic environments, where the imine bond may undergo hydrolysis, resulting in the regeneration of GA and amine groups (Barbosa et al., 2014).

Two mechanisms for the linkage of GA with amino groups under basic conditions have been proposed from $\alpha$, $\beta$ unsaturated aldehyde (structure 6): Schiff's base formation and Michael's addition, as illustrated in Fig. 3.7B. The Schiff base reaction involves the interaction between internal aldehyde groups and primary amino groups, resulting in the formation of imine linkages. The second mechanism entails Michael's addition to a double carbon-carbon bond. In addition, Wine et al. (2007) conducted the initial investigation into GA cross-linking using high-resolution X-ray diffraction. Their findings align with other reviews, suggesting that the product generated in alkaline media results from the concurrent formation of Schiff's base and Michael addition. The proposal includes an additional mechanism for cross-linking under acidic conditions, specifically the formation of Schiff's bases and the C–N bond through Anti-Markovnikov addition at the adjacent carbon of C–N. The low stability of the end product obtained through Michael's addition can be attributed to the disruption of the double C=C bond, which induces structural instability (Barbosa et al., 2014). Conversely, the stabilization of adducts during Michael's addition does not necessitate a reduction agent (Migneault et al., 2004).

Interestingly, a recently published work performed by Peng et al. (2025) studied an alternative synthetic cross-linker other than commonly used GA. This study employs hyperbranched polymers (HPBs) as an innovative, in situ-generated cross-linker for the preparation of CLEAs of laccase, leading to the formation of hyperbranched polymer-cross-linked laccase aggregates (HPCLEAs). This method overcomes a significant limitation of conventional CLEAs preparation, specifically the challenge of regulating the cross-linking process caused by the irregular voids that develop within enzyme aggregates when utilizing traditional cross-linkers such as GA. Fig. 3.8

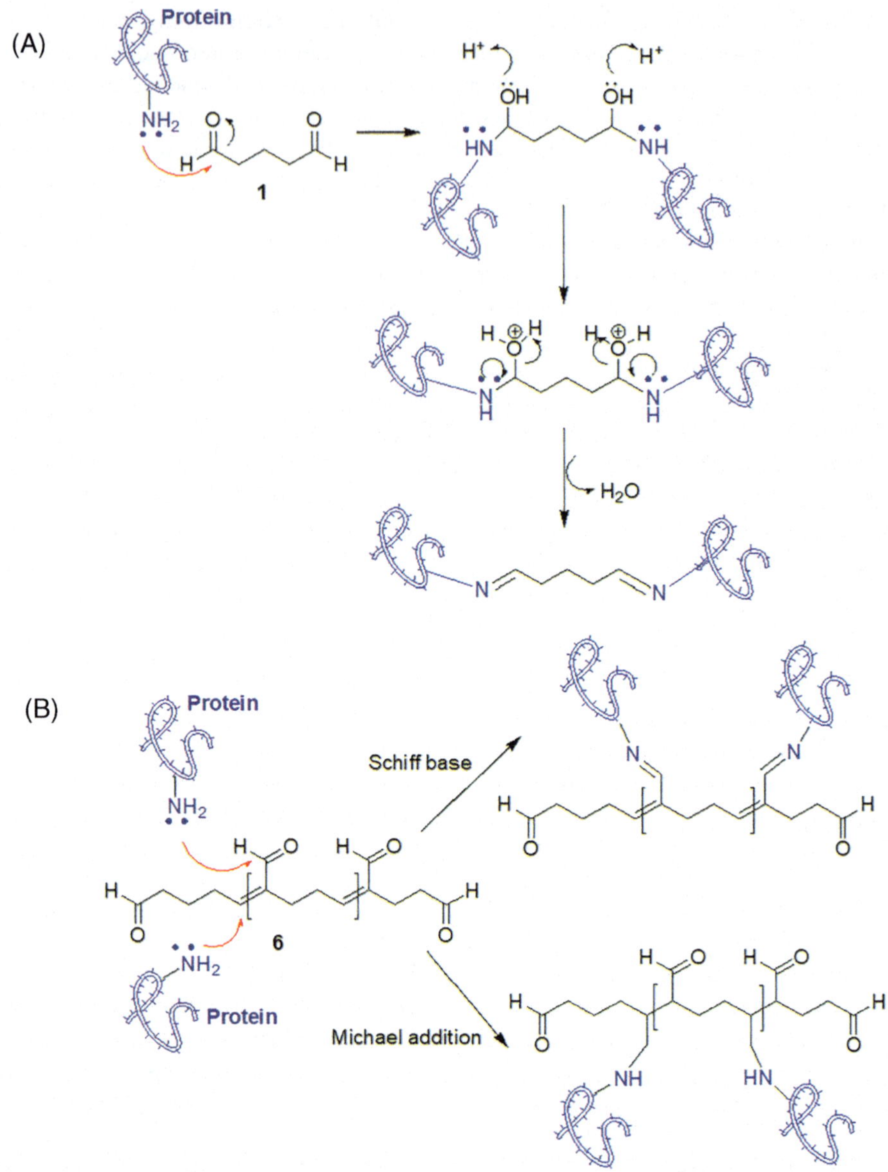


**FIGURE 3.7**

The reaction mechanism of glutaraldehyde cross-linker under different conditions.

(A) acidic and (B) basic conditions.

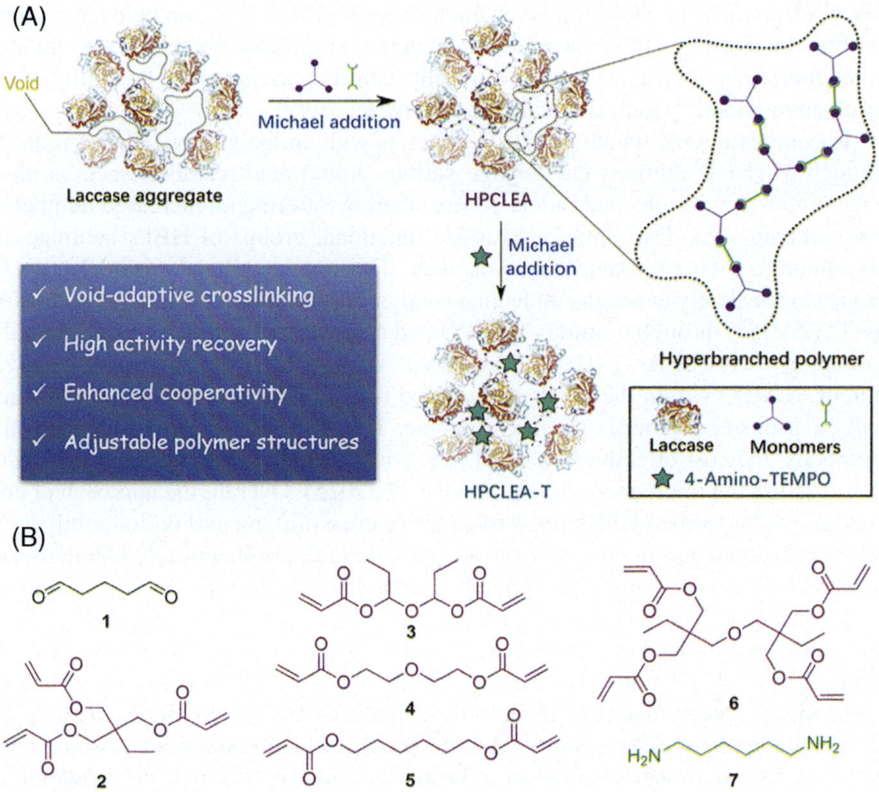


**FIGURE 3.8**

Synthesis of hyperbranched polymer-cross-linked laccase aggregates (HPLCEA).

(A) Schematic representation of HPCLEAs formation and HPCLEAs with covalently immobilized TEMPO (HPCLEAs-T) via Michael addition reactions and (B) chemical structures of glutaraldehyde (1) and the monomers (2–7) utilized in the synthesis of HPCLEAs.

*With permission from Peng, Y., Xiong, X., Qiu, M., Lu, Y., Chen, T., & Xu, Z. (2025). Hyperbranched polymer-crosslinked laccase aggregates for efficient aerobic oxidation of alcohols.* Enzyme and Microbial Technology, 189, *110673.*

depicts the schematic representation of HPCLEAs and the HPBs utilized in the reported study.

According to the study, hyperbranched polymers present numerous advantages, establishing them as a noteworthy technology in carrier-free enzyme immobilization. As seen in Fig. 3.8A, the primary characteristic is the void-adaptive cross-linking process. HPBs are synthesized in situ within the irregular voids of laccase aggregates. The internal generation enables the polymer structure to conform to and fill irregular spaces effectively. In addition, the characteristics of HPCLEAs can be controlled by altering the structures of the acrylate monomers (compounds 2-6, Fig. 3.8B) utilized in the formation of HPBs. The structural adjustability pertains to the flexibility of the resultant polymers, which is crucial for activity recovery.

In contrast to GA, which primarily interacts with amine groups, HBPs produced through Michael additions can engage various amino acid residues, such as thiol, guanidinium, imidazole, and indole groups, thereby offering an increased number of cross-linking sites. The numerous surface functional groups of HBPs facilitate the co-immobilization of functional molecules. This study utilized residual acrylate groups to covalently attach the molecular catalyst 2,2,6,6-tetramethyl-1-piperidine-*N*-oxyl (TEMPO) through 4-amino-TEMPO and aza-Michael addition, resulting in the formation of HPCLEAs-T (HPCLEAs-TEMPO) (Fig. 3.8A). The compartmentalization of TEMPO within the HBPs is proposed to inhibit laccase denaturation, which will be discussed further in the next subtopic. The HPCLEAs method integrates the simplicity and cost-effectiveness of CLEAs with controlled structural parameters of the cross-linking microenvironment (Peng et al., 2025). Overall, the approach of employing in situ-formed HPBs for void-adaptive cross-linking and co-immobilization presents a significant opportunity for the systematic development and facilitation of efficient cooperative catalysis in a durable, carrier-free material.

Nevertheless, from the list in Table 3.1, there are several other reported synthetic cross-linkers employed in carrier-free immobilization to replace commonly used GA. Wang et al. (2011) documented the application of *p*-benzoquinone as a cross-linking agent for the development of recombinant lipase CLEAs from *Geobacillus* sp. *p*-Benzoquinone may exist as both a natural substance and a synthesized compound, where *p*-benzoquinone is produced industrially, primarily through the oxidation of hydroquinone, aniline, or other benzene derivatives. It was exhibited that lipase-CLEAs from the study achieved 5-fold enhanced thermostability compared to lipase-CLEAs produced with GA as a cross-linker (Wang et al., 2011).

In addition, another study reported that the CLEAs of *Candida rugosa* lipase were created by activating the enzyme's free carboxyl groups using the carbodiimide-succinimide mechanism, followed by cross-linking with various sizes of polyethyleneimines (PEIs) (Velasco-Lozano et al., 2014). In the study, the synthesis of carboxyl-lipase-CLEAs was conducted in two phases utilizing the established chemistry of carbodiimide. Initially, 3-[3-dimethylaminopropyl] carbodiimide (EDC) is utilized, followed by N-hydroxysuccinimide (NHS) to facilitate their reaction with primary amines. The activated carboxyl groups were subsequently covalently linked to the main amino groups of the PEIs. It was observed that the optimal lipase-CLEAs were achieved using 1.3 kDa PEI as a cross-linker, exhibiting 1.3-fold higher thermal stability compared to CLEAs produced via the conventional amino cross-linking approach with GA (Velasco-Lozano et al., 2014).

Other than that, an epoxy-based cross-linker has been reported in developing CLEAs of two enzyme classes by Weiser et al. (2014). In the study, lipases from *Pseudomonas fluorescens* (AK), *Burkholderia cepacia* (PS), and *Candida antarctica* (CaL B), as well as phenylalanine ammonia-lyase (PAL) from *Petroselinum crispum*, were studied using glycerol diglycidyl ether (GDE) bisepoxide cross-linker in the preparation of their respective CLEAs. According to the study, poly(glycidyl ethers), including GDE, are extensively utilized as additives for cross-linking polymers containing amine, hydroxyl, or carboxyl groups. Their application as cross-linking agents in the preparation of CLEAs has not been documented, but GDE may serve as effective cross-linkers in the formation of CLEAs due to their capacity to establish stable bonds under mild circumstances with the amine groups of lysine, as well as with the sulfur- and oxygen-containing residues of cysteine, tyrosine, aspartate, or glutamate. It was observed that the GDE-CLEAs of lipase PS and PAL exhibited greater mechanical stability compared to the corresponding CLEAs cross-linked with GA. Concurrently, the catalytic activity of the lipases (AK, CaL B, and PS) cross-linked with GDE in CLEAs exceeded that of the respective lipases-CLEAs cross-linked with GA, suggesting that the utilization of GDE and analogous bisepoxide cross-linkers expands the potential of cross-linking technology for enzyme immobilization.

### 3.4.3 Combi-synthetic-natural cross-linkers

Despite utilizing a natural molecule or biopolymer as a green alternative to a synthetic cross-linker, researchers begin to apply both synthetic and natural cross-linkers, which constitutes a sophisticated approach to address the challenges typically linked to single-component cross-linking. For example, the limitation of using synthetic GA mentioned in Subsection 3.4.2 is that cross-linking is ineffective when the enzyme does not possess an adequate number of properly oriented surface amino groups, specifically lysine residues. Therefore, the combination with natural molecules or biopolymers may enhance the reactive sites or improve physical and mechanical properties of the final immobilized product, addressing issues of low coupling efficiency and inadequate physical stability. On the other hand, as mentioned in Subsection 3.4.1, utilizing natural cross-linkers alone frequently exhibits inadequate mechanical stability or elevated solubility due to weak hydrogen bonding or ionic interactions occurring in cross-linking, resulting in diminished reusability and operational longevity. Thus, the synthetic component is added, which may chemically enhance the natural material or establish a more durable connection between enzyme molecules.

From Table 3.1, lipase CLEAs-like copolymers (CLEAs-LCs) were developed using a combination of GA and putrescine (PTR), a natural polyamine (1,4-butanediamine), which serves as a co-feeder and cross-linker to enhance the number of reactive sites and facilitate cross-linking (Khiari et al., 2023). The preparation involves sequential, one-pot cross-linking steps with a significant pH alteration (pH jump) to facilitate amine activation, as seen in Fig. 3.9A. From the study, the pH is suddenly elevated, generally from about 5.5–6.5 to roughly pH 8.0, with the gradual addition of NaOH (Fig. 3.9B). The increase in pH deprotonates the amine groups of PTR, converting $-NH_3^+$ to $-NH_2$, thereby activating them. The newly deprotonated, highly nucleophilic amines of PTR swiftly react with the aldehyde groups found on GA and the GA-activated enzyme surfaces. This abrupt and intense chemical

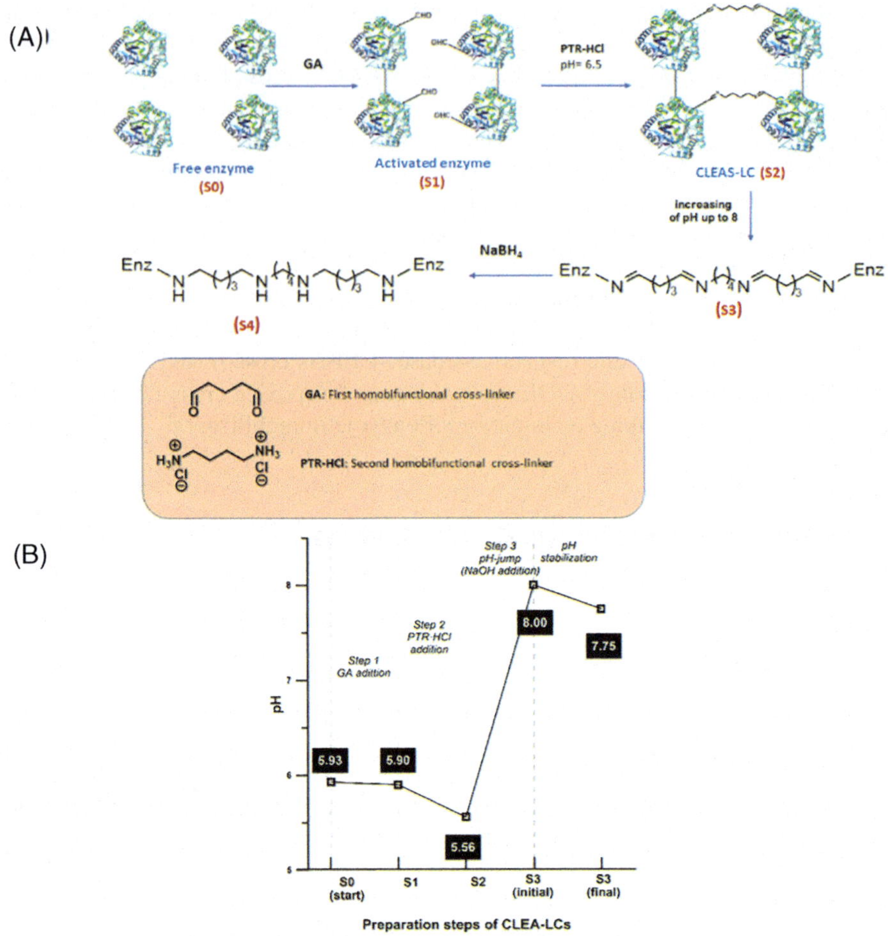


**FIGURE 3.9**

The CLEAs-like copolymer (CLEAs-LC) of lipase formation.

(A) Schematic representation of the stages involved in the manufacture of CLEAs-like copolymer of lipase. S1 shows a GA-activated enzyme, while S2 shows a copolymer in liquid form prior to pH transition and (B) pH assessment during the formulation of CLEAs-LC.

*From Khiari, O., Bouzemi, N., Sánchez-Montero, J. M., & Alcántara, A. R. (2023). Easy and versatile technique for the preparation of stable and active lipase-based CLEA-like copolymers by using two homofunctional cross-linking agents: Application to the preparation of enantiopure ibuprofen.* International Journal of Molecular Sciences, 24 *(17), 13664.*

reaction amplifies cross-linking by generating multiple Schiff bases (−N=C−), which eventually leads to the development of the solid CLEAs-LC. In the report, it was indicated that PTR readily interacts with the aldehyde groups of GA, indicating its role as a natural spacer or linker to connect enzyme molecules that may be excessively

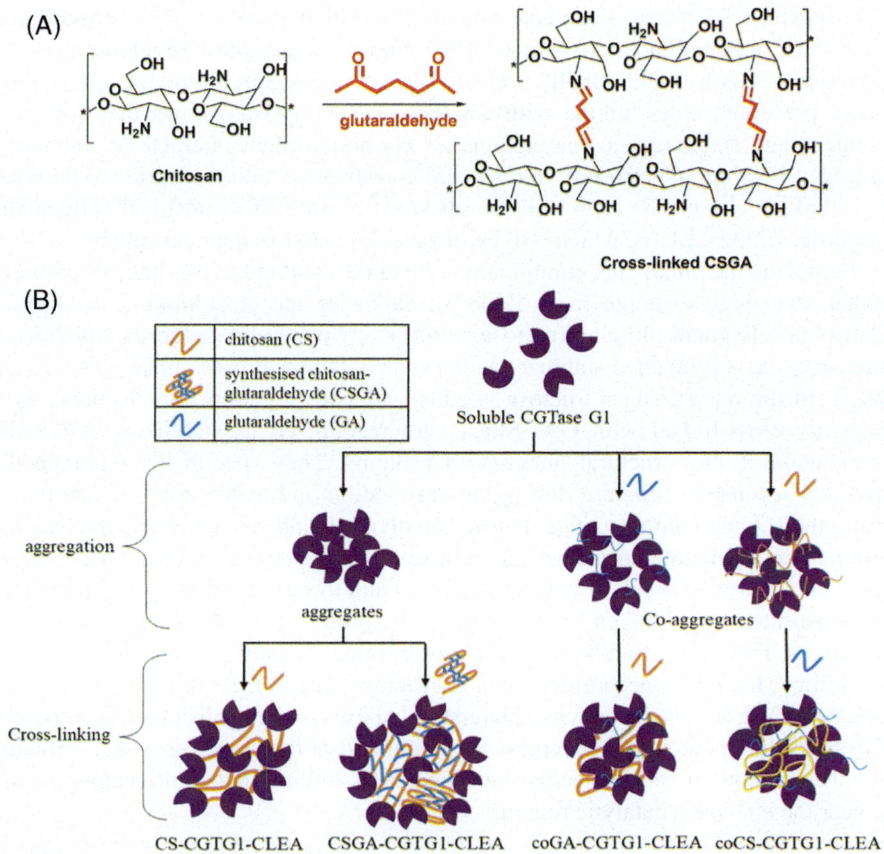


**FIGURE 3.10**

Schematic diagram of the synthesis of chitosan-glutaraldehyde (CSGA) and the formation of CGTase G1-CLEAs using a different strategy.

(A) Chemical cross-linking product (CSGA) of GA with chitosan and (B) schematic representation of the formation of CGTase G1-CLEAs.

*With permission from Jailani, N., Jaafar, N. R., Rahman, R. A., & Illias, R. M. (2023). Robust cross-linked cyclodextrin glucanotransferase from* Bacillus lehensis *G1 aggregates using an improved cross-linker and a new co-aggregant for the production of cyclodextrins.* Enzyme and Microbial Technology, 169, *110283.*

distant or deficient in surface lysine residues for direct GA-GA or GA-enzyme cross-linking (Khiari et al., 2023).

Another example of a combi natural-synthetic cross-linker employed in carrier-free immobilization is reported by Jailani et al. (2023), who developed a CLEAs of CGTase G1. The natural polymer CS was initially chemically cross-linked with the synthetic agent GA by a Schiff base approach to produce the novel chitosan-glutaraldehyde (CSGA) cross-linker (Fig. 3.10A). This pretreatment was employed to enhance the physical, chemical, and mechanical properties of the cross-linker prior to its application in cross-linking the enzyme aggregates. In addition, in the same report, GA was additionally evaluated as a co-aggregant during the aggregation phase, preceding cross-linking with CS (Fig. 3.10B). It was shown that GA, as a co-aggregant, stabilizes enzyme aggregates via noncovalent interactions, including hydrogen bonding, electrostatic forces, and coordination bonding, prior to the final cross-linking phase. The chemically synthesized combi CSGA positively affects the properties of the CLEAs, which will be discussed further in the next subtopic.

Following the interesting combination of natural-synthetic cross-linker aforementioned, one more example from Table 3.1 indicates the cross-linking in laccase-CLEAs development utilizing fructose involves a two-step procedure in which fructose serves as a protective stabilizer rather than a chemical cross-linker (Dey et al., 2023). In the report, sugar fructose is added after aggregation prior to being synthetically cross-linked with GA. Sugars are recognized as stabilizers that assist in maintaining the structural integrity of proteins. They specifically maintain the protein's secondary structure during the desolvation and aggregation phases by altering the solvent characteristics. During desolvation and cross-linking, the enzyme experiences significant structural and conformational alterations. In the presence of sugar, the hydrogen bonds between the protein and the surrounding water molecules are substituted by hydrogen bonds between the protein and the hydroxyl groups of the sugars. This interaction is essential for preserving the natural shape of the protein. In addition, the monosaccharides, such as fructose, due to their diminutive size, can rapidly diffuse across the enzyme's interstitial spaces and establish hydrogen bonds. This procedure is thought to decrease the likelihood of the GA cross-linker diffusing into the enzyme, thereby reducing unwanted cross-linking that could happen within or near the enzyme's catalytic region.

### 3.4.4 Functionalized cross-linkers

In light of finding a new green natural source of cross-linkers, utilizing a hyperbranched synthetic polymer cross-linker, or employing an intriguing combination of natural and synthetic molecules used in cross-linking, as explained in Subsections 3.4.1, 3.4.2, and 3.4.3 above, other approaches, such as designing and improving the cross-linker's properties by functionalization, have been explored. Employing functionalized cross-linkers is an excellent approach to address the constraints of traditional carrier-free immobilization by creating a more efficient, rigid, and customized covalent network. Functionalization entails the chemical modification of a precursor material to introduce or enhance the density of highly reactive groups essential for the cross-linking reaction.

According to the report by Jailani et al. (2022), who initially integrated the computational method to screen the cross-linker in CLEAs development, Lee et al. (2025) and Yip et al. (2024) from the same group subsequently utilized the information of the molecular interaction between the cross-linker and enzyme surface for the

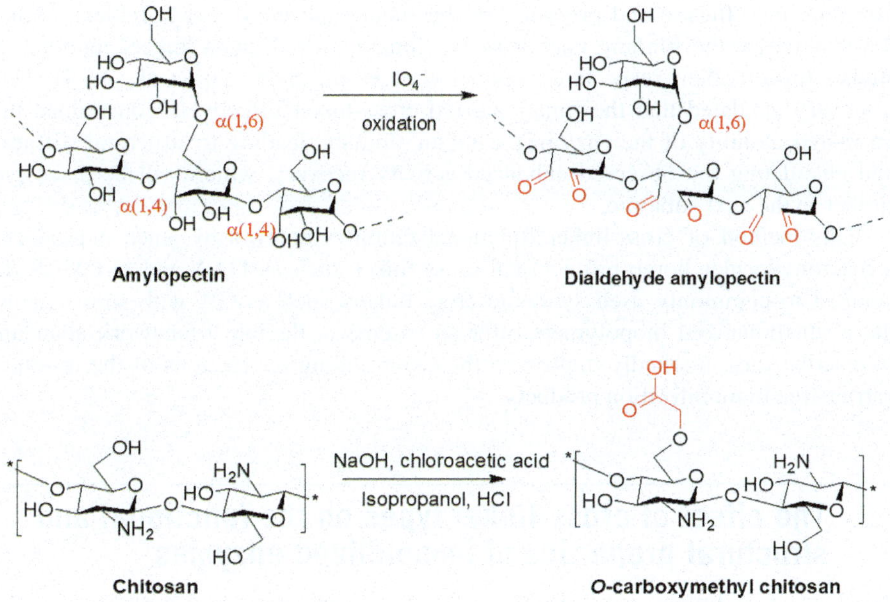


**FIGURE 3.11**

The chemical structure of the functionalization of macromolecular biopolymers.

The functionalization of amylopectin into dialdehyde amylopectin and chitosan into *O*-carboxymethyl chitosan.

design and functionalization of the cross-linker, enhancing the CLEA of polyethylene terephthalate hydrolase (PETase) and combi cyclodextrin glucanotransferase-maltogenic amylase G1 (CGTase-MAG1), respectively (Table 3.1). In their study, amylopectin has been oxidized into dialdehyde amylopectin (DAA) for PETase-CLEAs development, while CS has been modified into *O*-carboxymethyl chitosan (*O*-CMCS) and used as the cross-linker for combi CGTase-MAG1-CLEAs formation. The chemical modification of the cross-linkers is shown in Fig. 3.11. Raw polysaccharides such as amylopectin and CS are appealing due to their affordability and availability; nevertheless, they exhibit insufficient chemical reactivity to establish stable Schiff base C=N bonds with the amine $-NH_2$ residues on the enzyme's surface. The functionalization stage is crucial for introducing a high density of reactive aldehyde groups, which facilitates the creation of a stiff, multipoint covalent bond network.

Based on the studies, the main objective of the docking investigation was not only to design the cross-linker but also to confirm that the active site remained unobstructed by the cross-linking procedure. By verifying that the substrate maintained proper binding to the immobilized enzyme active pocket, the minimal activity loss to negligible conformational alterations can be confidently ascribed, thereby affirming the efficacy and precision of the functionalized strategy. Tables 3.2 and 3.3 summarize the binding energy of the functionalized cross-linkers reported in studies towards their respective enzymes (Lee et al., 2025; Yip et al., 2024). This discovery validated that the functionalized cross-linker effectively constrained the undesired mobility of the enzyme's catalytic domain, leading to structural stiffness and elucidating the observed enhanced activity recovery, which will be discussed further in the next subtopic.

This method of cross-linker functionalization is noteworthy since it produces a macromolecular homo-bifunctional cross-linker such as DAA and *O*-CMCS. In contrast to commonly used synthetic cross-linkers such as GA with short chains, these functionalized biopolymers offer an extensive, flexible framework abundant in reactive sites, markedly improving the cross-linking interactions of the resultant carrier-free immobilization products.

## 3.5 The effect of cross-linker types on the functional and structural properties of immobilized enzymes

Types of cross-linkers may affect the functional and structural properties of the immobilized enzyme. Since each enzyme is unique, in this subtopic, it's worth discussing and comparing how the effect of the different types of cross-linkers and the different approaches to immobilization may affect the properties of the same enzyme. Table 3.4 classified the enzymes of PETase, Combi CGTase-MAG1, CGTase, Lipase, and Laccase as either immobilized using CLEAs, CLECs, or CLEs utilizing different types of cross-linkers. The characteristics in terms of activity recovery, selectivity, stability, morphology, and particle sizes were compared and discussed within the same enzyme to examine the impact of types of cross-linkers on the functional and structural properties of the immobilized products.

### 3.5.1 Biocatalyst functional properties (activity, selectivity, and stability)

The discourse on elucidating how the selection of cross-linker and immobilization methodology substantially influences the characteristics of immobilized enzymes can be discussed using lipases from the reported study. As listed in Table 3.4, carrier-free CLEs and co-CLEs of lipase are produced from an ICMRs framework utilizing a natural cross-linker; CLECs of lipase utilizing a synthetic cross-linker; and CLEAs-like copolymers (CLEAs-LCs) of lipase reported using the natural-synthetic cross-linker combination (Table 3.4).

First, the investigation into the impact of cross-linker carbon chain length uncovers a significant correlation between the size of the linker and the characteristics of the biocatalyst for *Thermomyces lanuginosa* lipase (TLL) CLEs. The carbon chain length of di-acidic cross-linkers, utilized in ICMRs to produce CLE of lipase, is

**Table 3.2** Binding energy of dialdehyde amylopectin with ***Is***PETase. Binding site of the cross-linker on the surface of ***Is***PETase where the catalytic site was represented in red, while the cross-linkers were depicted as dots and color-coded by element.

| Cross-linker | Binding energy (kcal/mol) | Binding position of cross-linker | Hydrogen bonds interacting residues | Hydrogen bond length (Å) | Hydrophobic interacting residues |
|---|---|---|---|---|---|
| Dialdehyde amylopectin (DAA) | −4.6 | 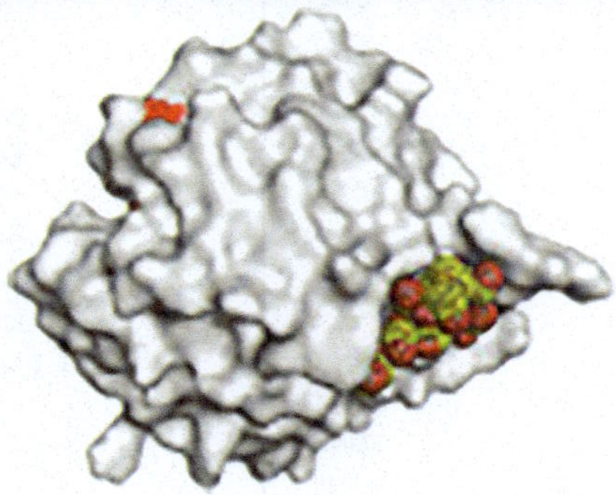 | Thr195<br>Pro174<br>Lys177<br>Phe261<br>Arg224<br>Arg260<br>Asn264 | 2.80, 2.85<br>3.00<br>2.77<br>3.17<br>3.15<br>3.04, 2.96, 3.30<br>2.80, 2.98, 3.00 | Leu176,<br>Ser175,<br>Pro197, Thr266,<br>Ser290, His291,<br>Asn288,<br>Cys289,<br>Ala226, Asn225 |

*With permission from Lee, Y. L., Jaafar, N. R., Huyop, F., Bakar, F. D. A., Rahman, R. A., & Md Illias, R. (2025). Functionalization of amylopectin as a strategy to improve polyethylene terephthalate hydrolase-cross-linked enzyme aggregate (IsPETase-CLEA) in plastic degradation. International Journal of Biological Macromolecules, 306, 141492.*

**Table 3.3** Binding energy and the interaction of CGTase and MAG1 with various functional groups on chitosan and functionalized ***O***-carboxymethyl chitosan cross-linkers in docking study.

| Cross-linker | CGTase | | MAG1 | |
|---|---|---|---|---|
| | **Swissdock (ΔG, kcal/mol)** | **AutodockVina (ΔG, kcal/mol)** | **Swissdock (ΔG, kcal/mol)** | **AutodockVina (ΔG, kcal/mol)** |
| chitosan | -6.66 | −6.9 | −8.93 | −6.2 |
| *O*-CMCS[a] | −8.92 | −7.8 | −9.68 | −6.6 |
| | **H-bonds interacting residues (length, Å)** | **Hydrophobic interacting residues** | **H-bonds interacting residues (length, Å)** | **Hydrophobic interacting residues** |
| chitosan | Asn267 (3.18)<br>Glu268 (2.62)<br>Asn308 (2.94, 2.91)<br>Glu567 (2.97, 3.11)<br>Glu569 (3.17)<br>Glu601 (2.81, 2.86)<br>Ser630 (2.85) | Glu229<br>Gly230<br>Thr233<br>Ser600<br>Lys609<br>Ile611 | Leu61 (2.89)<br>Thr401 (3.26)<br>Phe510 (2.83 | Lys60<br>Leu62<br>Lys63<br>Thr64<br>His394<br>Gln397<br>Phe509<br>His511 |
| *O*-CMCS | His264 (2.86)<br>Glu268 (2.97)<br>Glu601 (2.86)<br>Asn604 (3.07)<br>Ala610 (2.87)<br>Tyr627 (2.98, 2.96) | Glu229<br>Gln261<br>Gly548<br>Tyr549<br>Glu569<br>Leu571<br>Tyr594<br>Ser600<br>Gly603<br>Lys609<br>Ile611<br>Gly612 | Gln10 (3.30)<br>Leu15 (2.91)<br>Leu17 (2.79)<br>Phe125 (2.97)<br>Arg339 (3.31, 3.13)<br>Gln367 (3.12, 3.18)<br>Gln370 (2.73) | Gln11<br>Tyr16<br>Leu116<br>Glu123<br>Val124<br>His335<br>Pro364<br>Asp369<br>Pro408 |

*With permission from Yip, Y. S., Jaafar, N. R., Rahman, R. A., Puspaningsih, N. N. T., Jailani, N., & Illias, R. M. (2024). Improvement of combined cross-linked enzyme aggregates of cyclodextrin glucanotransferase and maltogenic amylase by functionalization of cross-linker for maltooligosaccharides synthesis. International Journal of Biological Macromolecules, 273, 133241.*

[a] *O-carboxymethyl chitosan.*

**Table 3.4** Effect of different cross-linkers towards functional and structural properties of the immobilized enzymes.

| Enzyme | Type of immo-bilization | Cross-linker | Cross-linking condition | Activity recovery | Catalytic efficiency/ specific activity | Operational stability | Morphology | Size of particles | References |
|---|---|---|---|---|---|---|---|---|---|
| PETase | CLEAs | Amylo-pectin | 0.5% amylopectin at 4°C for 1 h | 81.9% | 0.001 $mM^{-1}s^{-1}$ | Maintained 43.1% of its initial activity after 5th cycle and completely lost its activity at 6th cycle | Between type 1 (ball/uniform structure) and type 2 (less defined structure) | 160 nm | Lin et al. (2024) |
| | | DAA[a] | 0.5% DAA at 4°C for 30 min | 75% | 0.007 $mM^{-1}s^{-1}$ | Retained 38% activity after 7th cycle | Exhibited type 1 aggregates | 144 μm | Lee et al. (2025) |
| Combi CGTase-MAG1 | CLEAs | Chitosan | 0.15% chitosan at 4°C for 1.5 h | 54.1% | 3.68 $mLs^{-1}g^{-1}$ | Retained 36.19% of its initial activity after 5th cycle | Combination of type 1 (MAG1) and type 2 (CGTase) | 121 μm | Yip et al. (2023) |
| | | O-CMCS[b] | 0.65% O-CMCS at 4°C for 3.5 h | 74% | 13.34 $mLs^{-1}g^{-1}$ | Retained 50.0% of its initial activity after 5th cycle | Resemble combi CGTase-MAG1 using chitosan however possessed higher cavities | 823 μm | Yip et al. (2024) |
| CGTase | CLEAs | Chitosan | 0.15% chitosan at 4°C for 1 h | 88.36% | 41.11 $mM^{-1}s^{-1}$ | Maintained 17.9% of initial total CDs after 5th cycle | Exhibited a type 2 structure with compact structure | 177 μm | Jailani et al. (2022) |
| | | CSGA[c] | 0.15% CSGA at 4°C for 1 h | 94.67% | 70.31 $mM^{-1}s^{-1}$ | Retained 34.3% of initial total CDs after 5th cycle | Resembles as type 2 aggregates with increased crystallinity and porosity | 411 μm | Jailani et al. (2023) |

(*continued on next page*)

**Table 3.4** Effect of different cross-linkers towards functional and structural properties of the immobilized enzymes—cont'd

| Enzyme | Type of immo-bilization | Cross-linker | Cross-linking condition | Activity recovery | Catalytic efficiency/ specific activity | Operational stability | Morphology | Size of particles | References |
|---|---|---|---|---|---|---|---|---|---|
| Lipase | CLEs | Cystea-mine | 1244.3 mM of cysteamine at 25°C in the presence of 130 mM of cyclohexyl isocyanide | 97% | 33.55 U/mg | Retained 79% of its initial activity after 4th cycle | A structure-less and accumulated form | NR[d] | Ahrari et al. (2023) |
| | | DADD[e] | 131 mM of DADD with 260.8 mM of cyclohexyl isocyanide | 86.3% | 147.8 U/mg | Retained 60% of its initial activity after 5th cycle | A structure-less and accumulated form | 297.78 nm | Ahrari et al. (2024) |
| | | Azelaic acid | 283.4 mM of azelaic acid with 353 mM of cyclohexyl isocyanide | 82.8% | 88.4 U/mg | Losing 80%–90% of their activities in the 5th cycle | Exhibited an accumulated form | 205.71 nm | |
| | | Adipic acid | 110.6 mM of adipic acid with 219 mM of cyclohexyl isocyanide | 70.6% | 46.5 U/mg | Losing 80%–90% of their activities in the 5th cycle | Exhibited flower-like structures | 236.32 nm | |
| | CLECs | GA | 2.5% (v/v) GA for 48 h at 20°C | NR | NR | Retained 60% of its initial activity after 10th cycle | NR | 8 ± 2.52 μm | Fernández-Penas et al. (2021) |

*(continued on next page)*

| | | | | | | | | | |
|---|---|---|---|---|---|---|---|---|---|
| | CLEAs | GA-putrescine | 5% of GA and molar coefficient of GA: putrescine (2000:1000) shaking for 1 h at 120 rpm | 99% | 600 U/mg | Maintained 93% of its initial enantiomeric excess values | Amorphous surface with the presence of randomly distributed pores | 100–200 µm | Khiari et al. (2023) |
| Laccase | CLEAs | Hyper-branched polymer | Combination of 20 mM of trifunctional and 50 mM of diamine monomer at 30°C for 19 h | 99% | 35.3 $mM^{-1}min^{-1}$ | Maintained 50% of its initial activity after 4th cycle | Irregular shapes | 2–10 µm | Peng et al. (2025) |
| | | Fructose-GA | 10 mM of GA added with 10 mg/mL of fructose cross-link for 6 h at 20°C | 45% | 0.541 $\mu M^{-1}min^{-1}$ | Retained 80% of their initial activity even after 10th cycle | Small spherical aggregates of enzymes clustered together | 1 µm | Dey et al. (2023) |
| | | Genipin | 1.8 mM genipin for 12 h | 15% | 145 $M^{-1}s^{-1}$ | Retained 60% of their initial activity after 7th cycle | Ball-like shape | 0.4 µm | Hong et al. (2021) |

[a] *Dialdehyde Amylopectin*
[b] *O-carboxymethyl chitosan*
[c] *Chitosan-glutaraldehyde*
[d] *Not reported*
[e] *dodecanedioic acid*

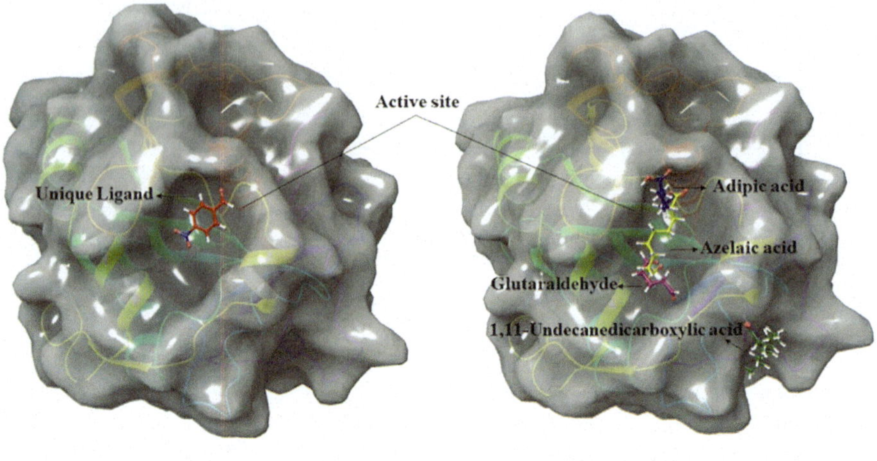


**FIGURE 3.12**

Molecular docking of lipase with different types of cross-linkers.

Cross-linkers used were glutaraldehyde, DADD, azelaic acid, and adipic acid.

*With permission from Ahrari, F., Yousefi, M., & Mohammadi, M. (2024). The effect of carbon chain length of cross-linking agent on the functionality of carrier- free immobilized* Thermomyces lanuginosa *lipase particles.* International Journal of Biological Macromolecules, 270 *(Pt 1), 132076.*

directly associated with the resultant enzyme activity. The longest linker, 1,11-undecanedicarboxylic acid (DADD) (11 carbons), produced the highest specific activity (147.8 U/mg for TLL), exceeding the activities achieved with lesser linkers. In silico analysis, as shown in Fig. 3.12, indicates that the shortest linkers, such as adipic acid (4 carbons) and azelaic acid (7 carbons), exert a significant inhibitory effect by penetrating and obstructing the enzyme's active site, an issue circumvented by the longer C11 DADD (Ahrari et al., 2024).

Conversely, the small-molecule linker cysteamine exhibited outstanding efficacy, attaining an immobilization yield of up to 97%. The elevated activity, along with its nontoxic characteristics, renders it a superior substitute for conventional cross-linkers such as GA (Ahrari & Mohammadi, 2024). However, the development of lipase-CLEAs using the GA-PTR natural-synthetic cross-linker resulted in the recovery of almost all lipase activity (99%). Furthermore, the reported studies reveal that lipase-CLEAs exhibit a superior specific activity of 600 U/mg in comparison to other cross-linkers used in CLEs immobilization (Khiari et al., 2023).

From Table 3.4, it appears that the use of the nontoxic linker cysteamine had the best operational stability, attaining 79% of their original activity after four reaction cycles. While the longest carboxylic acid linker, DADD, also worked well for stability, keeping 60% of its initial activity after five cycles. On the other hand, the shorter cross-linkers of azelaic acid (C7) and adipic acid (C4) showed much less operational stability, losing 80%–90% of their activity by the fifth cycle. This loss is probably caused by both enzyme leaching, since the shorter linkers may have occurred with fewer or weaker cross-links, and strong inhibition from linkers getting into the active site, which causes irreversible deactivation over time (Ahrari et al., 2024).

On the other hand, lipase-CLECs, which yielded the most structured variant of immobilized enzymes, demonstrated a robust operational stability in both aqueous solutions and organic solvents (Fernández-Penas et al., 2021). The crystal lattice, interconnected with a linker such as GA in the report, secures the enzyme in its native, active conformation, rendering it exceptionally resistant to denaturation and mechanical stress experienced during repeated recycling and utilization in harsh conditions; therefore, it appeared to maintain 60% of its initial activity after ten cycles. This organized, extremely stable configuration is widely regarded as the benchmark for prolonged operational use.

In terms of selectivity, the impact on selection for the enrichment of omega-3 fatty acids frequently entails a compromise with activity. The C7 azelaic acid linker, albeit producing lower activity (88.4 U/mg), exhibited the best selectivity of 22 in the catalytic liberation of eicosapentaenoic acid (EPA) and docosahexaenoic acid (DHA) from fish oil at 25°C in pH 7 for 8 h. This indicates that the intermediate length of the C7 linker is ideal for establishing a microenvironment that promotes selective substrate binding rather than optimizing the reaction rate. The cysteamine-CLEs had a selectivity of 0.3, whereas the C11 DADD and C4 adipic acid CLEs demonstrated a selectivity of 6, respectively (Ahrari et al., 2023, 2024). The overall stability of the resultant biocatalysts exhibited consistent enhancement across many linkers (cysteamine, DADD, azelaic acid, and adipic acid), with all derivatives displaying improved thermal and co-solvent stability relative to their free enzyme equivalents. The dual-linker method for CLEAs-like copolymers (CLEAs-LCs) employing GA and PTR yielded exceptional retention of catalytic characteristics and significant enhancements in thermal and storage stability, especially for *Candida rugosa* lipase (Khiari et al., 2023).

From Table 3.4, Lin et al. (2024, 2025) evaluate the efficacy of two distinct cross-linking agents, which use a native natural amylopectin (Amy) and functionalized DAA, on the biocatalytic characteristics of *Is*PETase-CLEAs. The study examines the influence of macromolecular cross-linker functionalization, in which the primary distinction resides in the chemical groups available for cross-linking (Fig. 3.11). It was reported that a native Amy cross-linker depends on weak ionic interactions and hydrogen bonding, while DAA resulting from periodate oxidation offers highly reactive aldehyde groups that establish robust, covalent Schiff base linkages with the amino group residues on the enzyme surface.

It was interesting to note that using the same optimized cross-linking, the activity recovery of *Ideonella sakaiensis* polyethylene terephthalate hydrolase (*Is*PETase)-CLEAs using DAA (75%) is slightly lower than that of using Amy (82%) as the cross-linker. Nevertheless, utilizing macromolecular DAA improves the catalytic efficiency of *Is*PETase-CLEAs by 7 times compared to an unfunctionalized cross-linker. In addition, DAA- *Is*PETase-CLEAs had a greater operational stability, where it can be used for more than seven cycles, whereas Amy-

*Is*PETase-CLEAs were reported to lose their activity after five cycles of reuse. The distinction is directly ascribed to the nature of the bonding of the macromolecular functionalized DAA. The covalent, multipoint attachment facilitated by DAA establishes a stiff, durable structure that enhances resistance to mechanical shear and denaturation throughout repeated usage and cycles. Native Amy cross-linker is predominantly influenced by ionic or hydrogen bonds, rendering them more vulnerable to disruption, resulting in enzyme leakage and diminished activity over time. Reducing leaching and maintaining the conformation of the enzyme's active site by using functionalized DAA in the study enhances specific activity towards the PET substrate (Lee et al., 2025; Lin et al., 2024).

In the meantime, four distinct enzyme aggregates, comprising two Combi-CLEAs of CGTase with MAG1 and two individual CGTase-CLEAs, emphasizing the influence of various cross-linkers on their characteristics, were reported from the same group (Table 3.4) (Jailani et al., 2022, 2023; Yip et al., 2023, 2024). The four enzyme preparations showed differing performances across the assessed parameters. The used of combi natural-synthetic cross-linker CSGA, enhanced single-enzyme CGTase-CLEAs, exhibited the best activity recovery (94.67%) and selectivity (70.31 $mM^{-1}s^{-1}$), succeeded by the baseline CGTase-CLEAs using CS cross-linker (88.36% activity recovery, 41.11 $mM^{-1}s^{-1}$ selectivity). The CGTase-CLEAs preparation using CSGA shows higher activity recovery and substrate binding compared to the basic CGTase-CS CLEAs, indicating improved interaction due to the superior cross-linker.

On the other hand, using the functionalized *O*-CMCS (*O*-carboxymethyl CS) in developing combi CGTase-MAG1-CLEAs demonstrated an enhanced activity recovery of 74% and a selectivity of 13.34 $mLs^{-1}g^{-1}$, representing a significant advancement compared to the original Combi CGTase-MAG1 aggregate using CS as a cross-linker, which exhibited 54.1% activity recovery and 3.68 $mLs^{-1}g^{-1}$ selectivity. The increase in efficacy is ascribed to the functionalization of the cross-linker, since *O*-CMCS exhibited a greater binding affinity for both enzymes, resulting in improved cross-linking efficiency. The enhanced catalytic efficiency in combi-CLEAs using *O*-CMCS cross-linker was attributed to the elongated side chain of the carboxymethyl group, which promoted a more flexible conformation, hence substantially decreasing the Michaelis constant.

The adjustments to the cross-linker were essential for operational stability, as shown in Table 3.4. The combi CGTase-MAG1-CLEAs using *O*-CMCS cross-linker exhibited superior performance among the combined aggregates, maintaining 50% of its original activity after five cycles, in contrast to merely 36.19% for the Combi CGTase-MAG1 aggregate using unfunctionalized CS as a cross-linker. The improved reusability is associated with the heightened stiffness of the protein backbone cross-linked with the *O*-CMCS structure. Besides, the recyclability of CGTase-CLEAs using CSGA cross-linker was also significantly enhanced by 56.6% relative to the fundamental CGTase-Chitosan CLEAs after five cycles. The limited reusability of the basic CGTase-CS CLEAs (retaining merely 17.9% of the total product) resulted

**Table 3.5** Synthesis of laccase-CLEAs using different hyperbranched polymers as a cross-linker.

| CLEAs | Crosslinkers | Relative activity (%) |
|---|---|---|
| 1CLEA-1 | 1 (20 mM) | 82 ± 10 |
| 2CLEA-1 | 2 (20 mM) | 79 ± 5 |
| HPCLEA-1 | 2 (20 mM) and 7 (10 mM) | 22 ± 7 |
| HPCLEA-2 | 2 (20 mM) and 7 (20 mM) | 42 ± 5 |
| HPCLEA-3 | 2 (20 mM) and 7 (30 mM) | 75 ± 14 |
| HPCLEA-4 | 2 (20 mM) and 7 (50 mM) | 99 ± 2 |
| HPCLEA-5 | 2 (20 mM) and 7 (100 mM) | 75 ± 10 |
| HPCLEA-6 | 3 (20 mM) and 7 (30 mM) | 54 ± 4 |
| HPCLEA-7 | 4 (20 mM) and 7 (30 mM) | 61 ± 15 |
| HPCLEA-8 | 5 (20 mM) and 7 (30 mM) | 82 ± 5 |
| HPCLEA-9 | 6 (20 mM) and 7 (30 mM) | 64 ± 12 |

*Data with permission from Peng, Y., Xiong, X., Qiu, M., Lu, Y., Chen, T., & Xu, Z. (2025). Hyperbranched polymer-crosslinked laccase aggregates for efficient aerobic oxidation of alcohols. Enzyme and Microbial Technology, 189, 110673.*

directly from their inadequate mechanical stability, which the novel CSGA cross-linker aimed to address in the reported study.

As for laccase enzyme in CLEA development, different types of cross-linkers categorized as synthetic cross-linkers (hyperbranched polymers), combinations of synthetic-natural cross-linkers (fructose-GA), and natural molecules (genipin) were compared (Table 3.4). The selection of a cross-linker establishes a distinct trade-off between optimizing initial enzyme activity and attaining long-term stability and reusability. The hyper-branched polymer (HPB) system is engineered to introduce the void-adaptive cross-linking process, which generates the polymer in situ within the enzyme aggregates, thereby reducing the conformational alterations that usually cause activity loss during cross-linking (Peng et al., 2025). Peng et al. (2025) studied the different combinations of monomers as cross-linkers for laccase-CLEAs development. The activity recovery of each HPB used is summarized in Table 3.5, while the structure of the monomers employed in the study is depicted in Fig. 3.8B (Subsection 3.4.2).

From Table 3.5, a diamine monomer (7, Fig. 3.8B) was amalgamated with a trifunctional monomer (2, Fig. 3.8B) to facilitate the in situ synthesis of HPBs within the irregular voids of laccase aggregates in CLEAs. Elevating the concentration of 7 enhanced the relative activity of HPCLEAs, signifying an augmented degree of cross-linking, which is crucial for activity restoration. Instead of 2, other acrylate monomers (3–6, Fig. 3.8B) were also utilized in conjunction with 7 to synthesize HPCLEAs (Table 3.5) exhibiting favorable activities, indicating that the modulation of activity can be achieved by structural variations in HPBs. The flexibility of HPBs, in addition to the degree of cross-linking, is influenced by the structures of acrylate monomers and likely plays a crucial role in activity recovery. As a result, the

HPCLEAs using the combination of diamine and trifunctional monomer in laccase-CLEAs demonstrate exceptional initial performance, achieving an impressive 99% activity recovery (Peng et al., 2025).

Nevertheless, the elevated initial activity of the aforementioned HPBs compromises their operational stability, as these CLEAs only maintained 50% activity after four cycles. In contrast, utilizing the fructose-aided GA cross-linker, which utilizes a protective sugar to facilitate cross-linking, emphasizes stability and reusability in laccase-CLEAs developed by Dey et al. (2023). It compromises some initial functionality, achieving a moderate 45% recovery of activity, yet offers superior operational stability, maintaining 80% of its starting activity after 10 cycles, signifying a highly resilient and durable structure. In the meantime, the genipin cross-linker, a natural and nontoxic substance, constitutes the most rigid system, where its vigorous cross-linking leads to the lowest activity recovery at 15%, yet provides high operational stability (60% activity retained after 7 cycles) and enhanced thermal and pH stability relative to GA-based CLEAs. In addition, the elevated internal cross-linking density enhances the selectivity profile, demonstrated by its superior substrate affinity relative to the other CLEAs developed using HPBs or fructose-GA cross-linkers, yielding a catalytic efficiency of 145 $M^{-1}s^{-1}$ (Hong et al., 2021).

### 3.5.2 Biocatalyst structural properties (size and shape)

The example of using a functionalized macromolecular biopolymer cross-linker has shown an impact on the structural properties of *Is*PETase-CLEAs (Table 3.4). Interestingly, DAA was reported to have a smaller size compared to native Amy analyzed by dynamic light scattering analysis (Fig. 3.13), however, utilizing native Amy cross-linker significantly produced smaller, nano-sized CLEAs particles (0.16 μm), indicating that the macromolecular structure facilitated homogeneous, confined aggregation. An increase in particle size to 144 μm was reported when using the DAA cross-linker, resulting in the formation of larger, micro-sized particles. This may be due to less constrained cross-linking, which facilitated greater aggregation among protein molecules.

In addition, Amy-*Is*PETase-CLEAs displays aggregates that tend to possess a rougher, less homogeneous structure, perhaps resulting in considerable mass transfer constraints (Lin et al., 2024). However, DAA-*Is*PETase-CLEAs aggregates exhibit a more homogeneous, with higher cavities, and globular or spherical morphology. Scanning electron microscopy images (Fig. 3.13), as shown by standard CLEAs studies, would demonstrate that the strong, covalent cross-links established by DAA-*Is*PETase-CLEAs result in a more distinct and compact outer structure. This regulated shape is advantageous for manipulation and separation. Moreover, employing a macromolecular cross-linker such as DAA has been proposed to enhance substrate accessibility relative to tiny cross-linkers like GA, which may form excessively dense, inaccessible aggregates.

When comparing the uses of functionalized biopolymer (*O*-CMCS) with a combination of natural-synthetic biopolymer CSGA in the development of combi

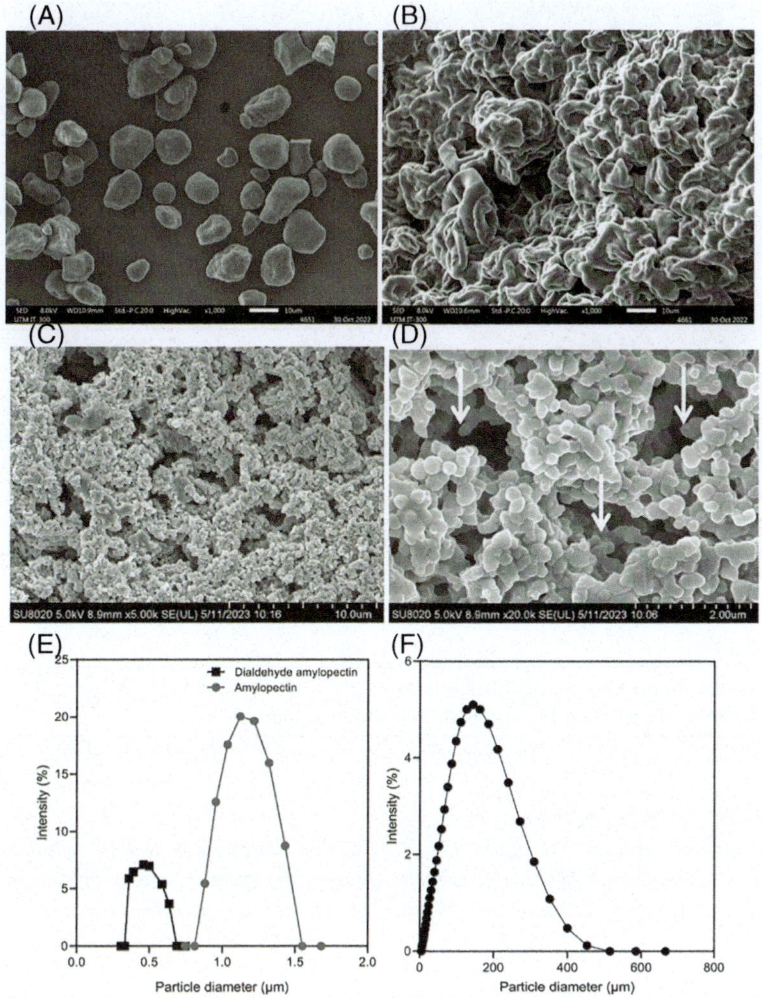


**FIGURE 3.13**

The SEM pictures and particle distributions of the functionalized cross-linker and IsPETase-CLEAs

FESEM of (A) amylopectin and (B) dialdehyde amylopectin (DAA) at 1000× magnification; FESEM images of (C) DAA-*Is*PETase-CLEAs at 5000× and (D) DAA-*Is*PETase-CLEAs at 20,000×; dynamic light scattering (DLS) curves for (E) amylopectin and DAA and (F) DAA-*Is*PETase-CLEAs.

*Data with permission from Lee, Y. L., Jaafar, N. R., Huyop, F., Bakar, F. D. A., Rahman, R. A., & Md Illias, R. (2025). Functionalization of amylopectin as a strategy to improve polyethylene terephthalate hydrolase-cross-linked enzyme aggregate (IsPETase-CLEA) in plastic degradation.* International Journal of Biological Macromolecules, 306, *141492.*

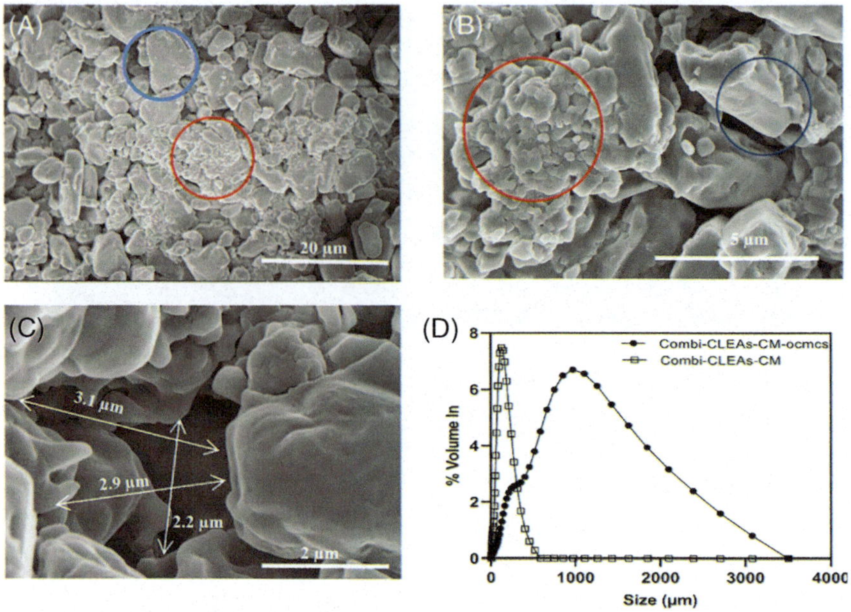


**FIGURE 3.14**

FESEM pictures and particle distributions of combi CGTase-MAG1-CLEAs were developed using a functionalized *O*-carboxymethyl chitosan cross-linker.

The FESEM magnifications are shown at (A) 2000×, (B) 10,000×, (C) 20,000×, and (D) denotes the distribution of particle sizes. The *red* circle indicates CGTase, whilst the *blue* circle indicates MAG1 enzymes.

*Data with permission from Yip, Y. S., Jaafar, N. R., Rahman, R. A., Puspaningsih, N. N. T., Jailani, N., & Illias, R. M. (2024). Improvement of combined cross-linked enzyme aggregates of cyclodextrin glucanotransferase and maltogenic amylase by functionalization of cross-linker for maltooligosaccharides synthesis.* International Journal of Biological Macromolecules, 273, *133241.*

CGTase-MAG1 and single CGTase-CLEAs, a difference in morphology and particle size is shown to influence the performance outcomes (Table 3.4). As seen in Fig. 3.14, both combi-CLEAs (using CS and *O*-CMCS cross-linker) comprised a blend of Type 1, indicating a MAG1, and Type 2, indicating a CGTase structure; however, the combi-CLEAs using *O*-CMCS displayed a more pliable architecture with more cavities, facilitating enzyme conformation and stability (Yip et al., 2024). The CGTase-CS CLEAs exhibited a compact Type 2 structure, whereas the enhanced CSGA preparation was characterized as a Type 2 aggregate with augmented crystallinity and porosity, the increased porosity facilitating the mitigation of substrate diffusional limitations (Fig. 3.15) (Jailani et al., 2023). The smallest particle was the Combi CGTase-MAG1 aggregate at 121 µm, whereas the largest was when using the *O*-CMCS cross-linker at 823 µm. The employed CSGA (411 µm) and *O*-CMCS in

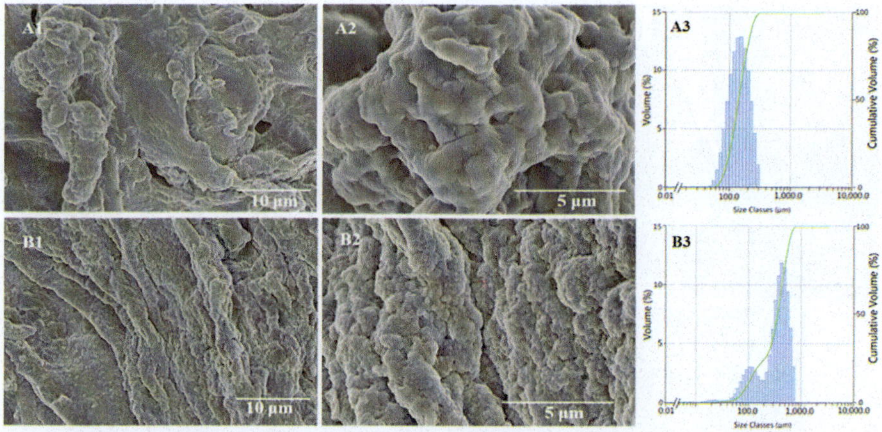


**FIGURE 3.15**

FESEM pictures and particle dimensions of CGTase-CLEAs.

(A) CLEAs developed using chitosan (CS) and (B) CLEAs developed using a synthetic chitosan-glutaraldehyde (CSGA). The FESEM magnifications employed are (1) 3000× and (2) 10,000×, while (3) denotes their corresponding particle sizes.

*Data with permission from Jailani, N., Jaafar, N. R., Rahman, R. A., & Illias, R. M. (2023). Robust cross-linked cyclodextrin glucanotransferase from* Bacillus lehensis *G1 aggregates using an improved cross-linker and a new co-aggregant for the production of cyclodextrins.* Enzyme and Microbial Technology, 169, *110283.*

CLEAs aggregates, being bigger than the baseline CLEAs, are more appropriate for continuous operation in a column reactor due to their capacity to diminish pressure loss (Figs. 3.14 and 3.15).

The morphology and dimensions of immobilized lipase particles are significantly influenced by the preparation methods (CLEs, CLEAs, CLECs) and the particular cross-linker employed, resulting in unique structural attributes and particle sizes. CLEs, which are developed by direct cross-linking of the soluble enzyme, generally yield structures that are less defined than conventional crystalline or aggregated forms. From Fig. 3.16, CLEs formulated with linkers such as DADD and azelaic acid are shown to possess an accumulated or random configuration. Nonetheless, variances are present within the CLEs developed using adipic acid, where the flower-like structures are shown as depicted in Fig. 3.16C.

In contrast, the CLEs of lipase developed using small molecule cysteamine, as depicted in Fig. 3.16D, show a different morphology compared to that of lipase-CLEs using different chain lengths of carboxylic acid, where it appears as an aggregate form. In the meanwhile, the CLEAs-like copolymers exemplified by those synthesized by the dual-linker system of GA and PTR exhibit a distinct morphological route (Fig. 3.16E). The GA-PTR CLEAs-LCs do not expressly specify a particular morphology, such as flower-like or rod-like forms; however, CLEAs appeared to commonly yield amorphous aggregates (Khiari et al., 2023).

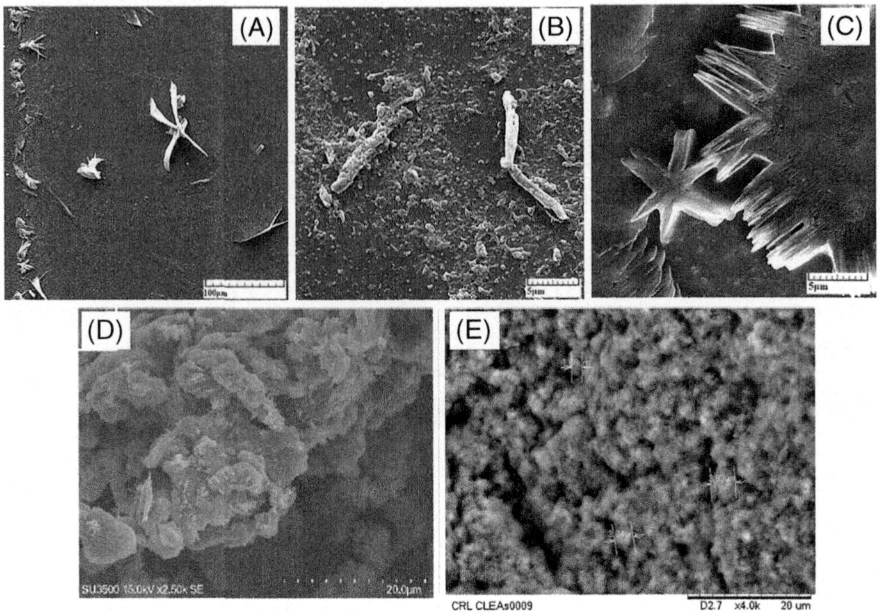


**FIGURE 3.16**

SEM images of lipase-CLEs and lipase-CLEAs developed using different types of cross-linkers.

Lipase-CLEs using (A) DAA, (B) azelaic acid, (C) adipic acid, (D) cysteamine cross-linkers, and (E) CLEAs-lipase using GA-putrescine cross-linker.

*Figures (A), (B), and (C) with permission from Ahrari, F., Yousefi, M., & Mohammadi, M. (2024). The effect of carbon chain length of cross-linking agent on the functionality of carrier- free immobilized* Thermomyces lanuginosa *lipase particles.* International Journal of Biological Macromolecules, 270 *(P1), 132076. Figure (D) from Ahrari, F., Yousefi, M., Habibi, Z., & Mohammadi, M. (2023). Cross-linked lipase particles with improved activity: Application of a non-toxic linker for cross-linking.* LWT, 173 *(December 2022), 114371. Figure (E) from Khiari, O., Bouzemi, N., Sánchez-Montero, J. M., & Alcántara, A. R. (2023). Easy and versatile technique for the preparation of stable and active lipase-based CLEA-like copolymers by using two homofunctional cross-linking agents: Application to the preparation of enantiopure ibuprofen.* International Journal of Molecular Sciences, 24 *(17), 13664.*

The dimensions of these lipase-CLEs range from nano to micro scale (Table 3.4). Interestingly, the lipase-CLEs developed using different carbon chain lengths of 11, 7, and 4 (DADD, azelaic acid, and adipic acid, respectively) do not possess a particle size in order. DADD-CLEs appeared as the biggest particles at 297.78 nm, followed by adipic acid (C4) CLEs with a particle size of 236.32 nm, while lipase-CLEs developed using azelaic acid (C7) produced the smallest particles, measuring roughly 205.71 nm. As intriguing as this result, the C7 azelaic acid linker, albeit producing lower activity as aforementioned, exhibited the best selectivity in the catalytic liberation of EPA and DHA from fish oil, which indicates that the intermediate length of the C7 linker is ideal for establishing a microenvironment that promotes selective substrate binding rather than optimizing the reaction rate.

In the meantime, lipase-CLECs constitute the most structurally specified variant of immobilized enzymes. According to Fernández-Penas et al. (2021), the benefit of lipase-CLECs is that the ordered structure preserves the native enzyme conformation, resulting in elevated activity, while the crystal matrix produces mechanically resilient particles. As a result, the lipase-CLECs were reported to have a bigger size of 8 μm as compared to those lipase-CLEs. The dimensions of these crystals are determined by the crystallization conditions, but can attain the scale necessary for industrial applications (Fernández-Penas et al., 2021). However, lipase developed using the CLEAs technique, which combines GA with PTR, exhibited the largest particle size of 100–200 μm (Khiari et al., 2023). In contrast to CLEs and CLECs, CLEAs are created through the initial aggregation of enzymes, followed by cross-linking, justifying the large size of particles.

From Table 3.4, it was indicated that the chemical composition of the cross-linkers directly influences the ultimate morphology and particle size of the laccase-CLEAs, which subsequently impacts mass transfer and mechanical strength. The HPB cross-linker generates the biggest particles, distinguished by irregular forms and a wide size range of 2–10 μm, depicted in Fig. 3.17. This broader, less uniform structure correlates with diminished mechanical robustness, elucidating its inadequate operational stability (Peng et al., 2025). The Fructose-GA system, on the other hand, produces intermediate-sized particles, generally tiny spherical aggregates, with an average diameter of 1 μm (Fig. 3.17). This spherical structure indicates an amalgamated form that equilibrates an adequate surface area with the requisite mechanical strength for repeated utilization, hence enhancing its operational stability (Dey et al., 2023).

Conversely, genipin used as a cross-linker yields the tiniest and most homogeneous particles of laccase-CLEAs, characterized by a spherical morphology with a minimal diameter of merely 0.4 μm. The nano-sized, precisely defined morphology is particularly beneficial as it markedly diminishes external diffusion constraints for the substrate; nevertheless, the compactness necessary for this diminutive size also constrains enzyme mobility, resulting in the lowest recorded initial activity compared to other cross-linkers used (Hong et al., 2021). In summary, a longer or bigger cross-linker influences its larger particle size, and different types of cross-linkers utilized affect its morphological form.

## 3.6 Conclusions

As a conclusion from this chapter, the selection of a cross-linker is the primary factor influencing the structural and functional characteristics of carrier-free immobilization. Synthetic cross-linkers, primarily GA, continue to be the primary agents for cross-linking in carrier-free immobilization, nevertheless, their inconsistent cross-linking patterns have prompted the exploration of novel methodologies. This procedure

**FIGURE 3.17**

SEM images of laccase-CLEAs developed using different cross-linkers.

Cross-linker (A) trifunctional monomer, (B) a combination of trifunctional with diamine monomer, (C) fructose-GA, (D) GA, and (E) genipin.

*Figures (A) and (B) with permission from Peng, Y., Xiong, X., Qiu, M., Lu, Y., Chen, T., & Xu, Z. (2025). Hyperbranched polymer-crosslinked laccase aggregates for efficient aerobic oxidation of alcohols.* Enzyme and Microbial Technology, 189, *0673. Figures (C) and (D) with permission from Dey, B., Panwar, V., & Dutta, T. (2023). Fructose-aided cross-linked enzyme aggregates of laccase: An insight on its chemical and physical properties.* Process Biochemistry, 133, *220–227. Figure (E) with permission from Hong, J., Jung, D., Park, S., Oh, Y., Oh, K. K., & Lee, S. H. (2021). Immobilization of laccase via cross-linked enzyme aggregates prepared using genipin as a natural cross-linker.* International Journal of Biological Macromolecules, 169, *541–550.*

encompasses the synthesis of hyperbranched polymers, employing a void-adaptive cross-linking to accommodate aggregate voids and interact with a broader spectrum of amino acid residues, thereby enhancing structural and functional control. In addition, screening a greener cross-linker from a natural source is still being explored, where the natural macromolecules, biopolymers, including CS, amylopectin, and dextran, and small molecules like genipin, cysteamine, and carboxylic acids, show a promising alternative to toxic synthetic cross-linkers. In the meantime, combi-synthetic-natural systems gain interesting attention, whereby employing the natural component in the synthetic linkers may act as a spacer to augment reaction sites, thus enhancing stability and reusability of the immobilized enzyme. The most sophisticated strategy employed for the development of a suitable cross-linker to date is through functionalized cross-linkers, typically designed through computational methods, to establish a tailored covalent network that effectively restricts the mobility of the catalytic domain, thereby enhancing structural stiffness and activity recovery. Ultimately, the diverse methodologies concentrate on attaining MCIs and refining the cross-linker's length and chemistry to enhance enzyme rigidification while maintaining structural integrity and superior catalytic performance.

## References


Ahrari, F., Yousefi, M., Habibi, Z., & Mohammadi, M. (2022). Application of undecanedicarboxylic acid to prepare cross-linked enzymes (CLEs) of *Rhizomucor miehei* lipase (RML): Selective enrichment of polyunsaturated fatty acids. *Molecular Catalysis, 520*, 112172. doi:10.1016/j.mcat.2022.112172.

Ahrari, F., Yousefi, M., Habibi, Z., & Mohammadi, M. (2023). Cross-linked lipase particles with improved activity: Application of a non-toxic linker for cross-linking. *LWT, 173*, 114371. doi:10.1016/j.lwt.2022.114371.

Ahrari, F., & Mohammadi, M. (2024). Combined cross-linking of *Rhizomucor miehei* lipase and *Candida antarctica* lipase B for the effective enrichment of omega-3 fatty acids in fish oil. *International Journal of Biological Macromolecules, 260*(Pt 2), 129362. doi:10.1016/j.ijbiomac.2024.129362.

Ahrari, F., Yousefi, M., & Mohammadi, M. (2024). The effect of carbon chain length of cross-linking agent on the functionality of carrier- free immobilized *Thermomyces lanuginosa* lipase particles. *International Journal of Biological Macromolecules, 270*(Pt 1), 132076. doi:10.1016/j.ijbiomac.2024.132076.

Amadi, O. C., Awodiran, I. P., Moneke, A. N., Nwagu, T. N., Egong, J. E., & Chukwu, G. C. (2022). Concurrent production of cellulase, xylanase, pectinase and immobilization by combined cross-linked enzyme aggregate strategy: Advancing tri-enzyme biocatalysis. *Bioresource Technology Reports, 18*, 101019. doi:10.1016/j.biteb.2022.101019.

Aso, C., & Aito, Y. (1962). Intramolecular-intermolecular polymerization of glutaraldehyde. *Bulletin of the Chemical Society of Japan, 35*(8), 1426. doi:10.1246/bcsj.35.1426.

Ayhan, H., Ayhan, F., & Gülsu, A. (2012). Highly biocompatible enzyme aggregates crosslinked by L-lysine. *Turkish Journal of Biochemistry, 37*(1), 14–20. doi:10.5505/tjb.2012.00719.

Barbosa, O., Ortiz, C., Berenguer-Murcia, A., Torres, R., Rodrigues, R. C., & Fernandez-Lafuente, R. (2014). Glutaraldehyde in bio-catalysts design: A useful crosslinker and a versatile tool in enzyme immobilization. *RSC Advances, 4*(4), 1583–1600. doi:10.1039/c3ra45991h.

Behera, S., Dev, M. J., & Singhal, R. S. (2022). Cross-linked $\beta$-mannanase aggregates: Preparation, characterization, and application for producing partially hydrolyzed guar gum. *Applied Biochemistry and Biotechnology, 194*(5), 1981–2004. doi:10.1007/s12010-022-03807-w.

Besouw, M., Masereeuw, R., Van Den Heuvel, L., & Levtchenko, E. (2013). Cysteamine: An old drug with new potential. *Drug Discovery Today, 18*(15–16), 785–792. doi:10.1016/j.drudis.2013.02.003.

Chen, N., Chang, B., Shi, N., Lu, F., & Liu, F. (2023). Robust and recyclable cross-linked enzyme aggregates of sucrose isomerase for isomaltulose production. *Food Chemistry, 399*, 134000. doi:10.1016/j.foodchem.2022.134000.

Cheng, H. L., Hsu, W. T., Su, Y. H., Lee, Y. J., Tangthi, H. N., Wu, C. T., & Lee, L. L. (2024). Cross-linked enzyme aggregates of xylanase, XynR8(N58D), for effective degradation of untreated lignocellulosic biomass. *Biocatalysis and Biotransformation, 42*(5), 629–643. doi:10.1080/10242422.2024.2305969.

Chu, F., Thornton, D. T., & Nguyen, H. T. (2018). Chemical cross-linking in the structural analysis of protein assemblies. *Methods, 144*, 53–63. doi:10.1016/j.ymeth.2018.05.023.

Dey, B., Panwar, V., & Dutta, T. (2023). Fructose-aided cross-linked enzyme aggregates of laccase: An insight on its chemical and physical properties. *Process Biochemistry, 133*, 220–227. doi:10.1016/j.procbio.2023.09.006.

Díaz-Montes, E. (2021). Dextran: Sources, structures, and properties. *Polysaccharides, 2*(3), 554–565. doi:10.3390/polysaccharides2030033.

Fass, D., & Thorpe, C. (2018). Chemistry and enzymology of disulfide cross-linking in proteins. *Chemical Reviews, 118*(3), 1169–1198. doi:10.1021/acs.chemrev.7b00123.

Fernández-Penas, R., Verdugo-Escamilla, C., Martínez-Rodríguez, S., & Gavira, J. A. (2021). Production of cross-linked lipase crystals at a preparative scale. *Crystal Growth and Design, 21*(3), 1698–1707. doi:10.1021/acs.cgd.0c01608.

Guo, J., Yue, X., Hou, Y., Wang, Y., Liu, Y., Liu, G., He, Y., Ma, L., Zhou, L., & Jiang, Y. (2024). Immobilization of cross-linked enzymes aggregates on hierarchical covalent organic frameworks: Highly stable chemoenzymatic nanoreactor for asymmetric synthesis of optically active halohydrins. *International Journal of Biological Macromolecules, 278*(Pt 2), 134641. doi:10.1016/j.ijbiomac.2024.134641.

Hao, Q., Guo, P., Wang, H., Su, M., Li, C., Hu, Y., & Su, W. (2024). Synergistic precipitant powered self-assembly of papain for cross-linked enzyme crystals preparation. *Particuology, 85*, 102–112. doi:10.1016/j.partic.2023.03.023.

Hardy, P. M., Nicholls, A. C., & Rydon, H. N. (1969). The nature of glutaraldehyde in aqueous solution. *Journal of the Chemical Society D: Chemical Communications*, (10), 565. doi:10.1039/c29690000565.

Henninger, C., Hoferer, M., Ochsenreither, K., & Eisele, T. (2023). Cross-linked phytase aggregates for improved phytate degradation at low pH in animal feed. *European Food Research and Technology, 249*(9), 2377–2386. doi:10.1007/s00217-023-04299-6.

Hoarau, M., Badieyan, S., & Marsh, E. N. G. (2017). Immobilized enzymes: Understanding enzyme-surface interactions at the molecular level. *Organic and Biomolecular Chemistry, 15*(45), 9539–9551. doi:10.1039/c7ob01880k.

Hong, J., Jung, D., Park, S., Oh, Y., Oh, K. K., & Lee, S. H. (2021). Immobilization of laccase via cross-linked enzyme aggregates prepared using genipin as a natural cross-linker. *International Journal of Biological Macromolecules, 169*, 541–550. doi:10.1016/j.ijbiomac.2020.12.136.

Jailani, N., Jaafar, N. R., Suhaimi, S., Mackeen, M. M., Bakar, F. D. A., & Illias, R. M. (2022). Cross-linked cyclodextrin glucanotransferase aggregates from *Bacillus lehensis* G1 for cyclodextrin production: Molecular modeling, developmental, physicochemical, kinetic and thermodynamic properties. *International Journal of Biological Macromolecules, 213*, 516–533. doi:10.1016/j.ijbiomac.2022.05.170.

Jailani, N., Jaafar, N. R., Rahman, R. A., & Illias, R. M. (2023). Robust cross-linked cyclodextrin glucanotransferase from *Bacillus lehensis* G1 aggregates using an improved cross-linker and a new co-aggregant for the production of cyclodextrins. *Enzyme and Microbial Technology, 169*, 110283. doi:10.1016/j.enzmictec.2023.110283.

Jaiswal, S., & Chibbar, R. N. (2017). Amylopectin small chain glucans form structure fingerprint that determines botanical origin of starch. *Carbohydrate Polymers, 158*, 112–123. doi:10.1016/j.carbpol.2016.11.059.

Kalkhof, S., & Sinz, A. (2008). Chances and pitfalls of chemical cross-linking with amine-reactive N-hydroxysuccinimide esters. *Analytical and Bioanalytical Chemistry, 392*(1–2), 1–8. doi:10.1007/s00216-008-2231-5.

Khiari, O., Bouzemi, N., Sánchez-Montero, J. M., & Alcántara, A. R. (2023). Easy and versatile technique for the preparation of stable and active lipase-based CLEA-like copolymers by using two homofunctional cross-linking agents: Application to the preparation of enantiopure ibuprofen. *International Journal of Molecular Sciences, 24*(17), 13664. doi:10.3390/ijms241713664.

Korn, A. H., Feairheller, S. H., & Filachoine, E. M. (1972). Glutaraldehyde: Nature of the reagent. *Journal of Molecular Biology, 65*(3), 525–529. doi:10.1016/0022-2836(72)90206-9.

Kou, S., Peters, L. M., & Mucalo, M. R. (2021). Chitosan: A review of sources and preparation methods. *International Journal of Biological Macromolecules, 169*, 85–94. doi:10.1016/j.ijbiomac.2020.12.005.

Lee, H., Sugiharto, Y. E. C., Lee, H., Jeon, W., Ahn, J., & Lee, H. (2019). Biotransformation of dicarboxylic acids from vegetable oil–derived sources: Current methods and suggestions for improvement. *Applied Microbiology and Biotechnology, 103*(4), 1545–1555. doi:10.1007/s00253-018-9571-7.

Lee, Y. L., Jaafar, N. R., Huyop, F., Bakar, F. D. A., Rahman, R. A., & Illias, R. M. D. (2025). Functionalization of amylopectin as a strategy to improve polyethylene terephthalate hydrolase-cross-linked enzyme aggregate (IsPETase-CLEA) in plastic degradation. *International Journal of Biological Macromolecules, 306*, 141492. doi:10.1016/j.ijbiomac.2025.141492.

Lin, L. Y., Jaafar, N. R., Ling, J. G., Huyop, F., Abu Bakar, F. D., Rahman, R. A., & Illias, R. M. (2024). Cross-linked enzyme aggregates of polyethylene terephthalate hydrolyse (PETase) from *Ideonella sakaiensis* for the improvement of plastic degradation. *International Journal of Biological Macromolecules, 263*, 130284. doi:10.1016/j.ijbiomac.2024.130284.

Lopez, S., Marchi-Delapierre, C., Cavazza, C., & Ménage, S. (2020). A selective sulfide oxidation catalyzed by heterogeneous artificial metalloenzymes Iron@NikA. *Chemistry – A European Journal, 26*(70), 16633–16638. doi:10.1002/chem.202003746.

Mädler, S., Bich, C., Touboul, D., & Zenobi, R. (2009). Chemical cross-linking with NHS esters: A systematic study on amino acid reactivities. *Journal of Mass Spectrometry, 44*(5), 694–706. doi:10.1002/jms.1544.

Makki, H. H., Jaafar, N. R., Jailani, N., Alqasem, A. A., Rahmat, Z., & Illias, R. M. (2024). Strategy in synthesizing longer-chain levan-type fructooligosaccharides by selective dextran macromolecular cross-linked *Bacillus lehensis* G1 endolevanase aggregate immobilization. *Catalysts, 14*(9), 584. doi:10.3390/catal14090584.

Mateo, C., Palomo, J. M., Fuentes, M., Betancor, L., Grazu, V., López-Gallego, F., Pessela, B. C. C., Hidalgo, A., Fernández-Lorente, G., Fernández-Lafuente, R., & Guisán, J. M. (2006). Glyoxyl agarose: A fully inert and hydrophilic support for immobilization and high stabilization of proteins. *Enzyme and Microbial Technology, 39*(2), 274–280. doi:10.1016/j.enzmictec.2005.10.014.

Mateo, C., Pessela, B. C. C., Fuentes, M., Torres, R., Betancor, L., Hidalgo, A., Fernandez-Lorente, G., Fernandez-Lafuente, R., & Guisan, J. M. (2020). Stabilization of multimeric enzymes via immobilization and further cross-linking with aldehyde-dextran. *Methods in Molecular Biology, 2100*, 175–187. doi:10.1007/978-1-0716-0215-7_11.

Migneault, I., Dartiguenave, C., Bertrand, M. J., & Waldron, K. C. (2004). Glutaraldehyde: Behavior in aqueous solution, reaction with proteins, and application to enzyme crosslinking. *Biotechniques, 37*(5), 790–802. doi:10.2144/04375rv01.

Morais, J. R. F., Costa, I. O., Padilha, C. E. A., Rios, N. S., & Santos, E. S. D. (2025). Improving reusability of biocatalysts by exploiting cross-linked enzyme aggregates (CLEAs) with commercial cellulolytic cocktails for hydrolysis of green coconut waste. *Sustainability (Switzerland), 17*(9), 4221. doi:10.3390/su17094221.

Nawawi, N. N., Hashim, Z., Manas, N. H. A., Azelee, N. I. W., & Illias, R. M. (2020). A porous-cross linked enzyme aggregates of maltogenic amylase from *Bacillus lehensis* G1: Robust biocatalyst with improved stability and substrate diffusion. *International Journal of Biological Macromolecules, 148*, 1222–1231. doi:10.1016/j.ijbiomac.2019.10.101.

Novák, P., & Kruppa, G. H. (2008). Intra-molecular cross-linking of acidic residues for protein structure studies. *European Journal of Mass Spectrometry, 14*(6), 355–365. doi:10.1255/ejms.963.

Peng, Y., Xiong, X., Qiu, M., Lu, Y., Chen, T., & Xu, Z. (2025). Hyperbranched polymer-crosslinked laccase aggregates for efficient aerobic oxidation of alcohols. *Enzyme and Microbial Technology, 189*, 110673. doi:10.1016/j.enzmictec.2025.110673.

Richards, F. M., & Knowles, J. R. (1968). Glutaraldehyde as a protein cross-linking reagent. *Journal of Molecular Biology, 37*(1), 231–233. doi:10.1016/0022-2836(68)90086-7.

Rodrigues, R. C., Ortiz, C., Berenguer-Murcia, Á., Torres, R., & Fernández-Lafuente, R. (2013). Modifying enzyme activity and selectivity by immobilization. *Chemical Society Reviews, 42*(15), 6290–6307. doi:10.1039/c2cs35231a.

Roman-Benn, A., Contador, C. A., Li, M. W., Lam, H. M., Ah-Hen, K., Ulloa, P. E., & Ravanal, M. C. (2023). Pectin: An overview of sources, extraction and applications in food products, biomedical, pharmaceutical and environmental issues. *Food Chemistry Advances, 2*, 100192. doi:10.1016/j.focha.2023.100192.

Santos, J. C. S. D., Barbosa, O., Ortiz, C., Berenguer-Murcia, A., Rodrigues, R. C., & Fernandez-Lafuente, R. (2015). Importance of the support properties for immobilization or purification of enzymes. *ChemCatChem, 7*(16), 2413–2432. doi:10.1002/cctc.201500310.

Staar, M., Henke, S., Blankenfeldt, W., & Schallmey, A. (2022). Biocatalytically active and stable cross-linked enzyme crystals of halohydrin dehalogenase HheG by protein engineering. *ChemCatChem., 14*(9), e202200145. doi:10.1002/cctc.202200145.

Sutherland, B. W., Toews, J., & Kast, J. (2008). Utility of formaldehyde cross-linking and mass spectrometry in the study of protein-protein interactions. *Journal of Mass Spectrometry, 43*(6), 699–715. doi:10.1002/jms.1415.

Talbert, J. N., & Goddard, J. M. (2012). Enzymes on material surfaces. *Colloids and Surfaces B: Biointerfaces, 93*, 8–19. doi:10.1016/j.colsurfb.2012.01.003.

Talekar, S., Nadar, S., Joshi, A., & Joshi, G. (2014). Pectin cross-linked enzyme aggregates (pectin-CLEAs) of glucoamylase. *RSC Advances, 4*(103), 59444–59453. doi:10.1039/C4RA09552A.

Toews, J., Rogalski, J. C., Clark, T. J., & Kast, J. (2008). Mass spectrometric identification of formaldehyde-induced peptide modifications under in vivo protein cross-linking conditions. *Analytica Chimica Acta, 618*(2), 168–183. doi:10.1016/j.aca.2008.04.049.

Ullah, H., Pervez, S., Ahmed, S., Haleem, K. S., Qayyum, S., Niaz, Z., Nawaz, M. A., Nawaz, F., Subhan, F., & Tauseef, I. (2021). Preparation, characterization and stability studies of cross-linked α-amylase aggregates (CLAAs) for continuous liquefaction of starch. *International Journal of Biological Macromolecules, 173*, 267–276. doi:10.1016/j.ijbiomac.2021.01.057.

Velasco-Lozano, S., López-Gallego, F., Vázquez-Duhalt, R., Mateos-Díaz, J. C., Guisán, J. M., & Favela-Torres, E. (2014). Carrier-free immobilization of lipase from candida rugosa with polyethyleneimines by carboxyl-activated cross-linking. *Biomacromolecules, 15*(5), 1896–1903. doi:10.1021/bm500333v.

Wang, A., Zhang, F., Chen, F., Wang, M., Li, H., Zeng, Z., Xie, T., & Chen, Z. (2011). A facile technique to prepare cross-linked enzyme aggregates using p-benzoquinone as cross-linking agent. *Korean Journal of Chemical Engineering, 28*(4), 1090–1095. doi:10.1007/s11814-010-0476-0.

Weiser, D., Varga, A., Kovács, K., Nagy, F., Szilágyi, A., Vértessy, B. G., Paizs, C., & Poppe, L. (2014). Bisepoxide cross-linked enzyme aggregates: New immobilized biocatalysts for selective biotransformations. *ChemCatChem., 6*(5), 1463–1469. doi:10.1002/cctc.201300806.

Whipple, E. B., & Ruta, M. (1974). Structure of aqueous glutaraldehyde. *Journal of Organic Chemistry, 39*(12), 1666–1668. doi:10.1021/jo00925a015.

Wine, Y., Cohen-Hadar, N., Freeman, A., & Frolow, F. (2007). Elucidation of the mechanism and end products of glutaraldehyde crosslinking reaction by X-ray structure analysis. *Biotechnology and Bioengineering, 98*(3), 711–718. doi:10.1002/bit.21459.

Yip, Y. S., Manas, N. H. A., Jaafar, N. R., Rahman, R. A., Puspaningsih, N. N. T., & Illias, R. M. (2023). Combined cross-linked enzyme aggregates of cyclodextrin glucanotransferase and maltogenic amylase from *Bacillus lehensis* G1 for maltooligosaccharides synthesis. *International Journal of Biological Macromolecules, 242*, 124675. doi:10.1016/j.ijbiomac.2023.124675.

Yip, Y. S., Jaafar, N. R., Rahman, R. A., Puspaningsih, N. N. T., Jailani, N., & Illias, R. M. (2024). Improvement of combined cross-linked enzyme aggregates of cyclodextrin glucanotransferase and maltogenic amylase by functionalization of cross-linker for maltooligosaccharides synthesis. *International Journal of Biological Macromolecules, 273*, 133241. doi:10.1016/j.ijbiomac.2024.133241.

Zhang, Q., Crosland, E., & Fabris, D. (2008). Nested Arg-specific bifunctional crosslinkers for MS-based structural analysis of proteins and protein assemblies. *Analytica Chimica Acta, 627*(1), 117–128. doi:10.1016/j.aca.2008.05.074.

CHAPTER

# Cross-linked enzyme crystals (CLECs)

4

**Lina Ahlborn and Anett Schallmey**

*Institute for Biochemistry, Biotechnology and Bioinformatics, Technische Universität Braunschweig, Braunschweig, Germany*

## 4.1 Introduction

Cross-linked enzyme crystals (CLECs), represent one possibility to immobilize an enzyme in a carrier-free fashion. As the name indicates, CLECs are prepared through crystallization of the enzyme of interest and subsequent chemical cross-linking of individual enzyme molecules within the crystal. This way, very stable enzyme particles are obtained that do not dissolve in aqueous or nonaqueous solutions, such as organic solvents. This concept had been introduced already in Quiocho and Richards (1964) for crystals of carboxypeptidase A (Quiocho & Richards, 1964, 1966; Quiocho et al., 1967), and was later adapted for the efficient immobilization of a wide range of enzymes (see Section 4.4. *Application of cross-linked enzyme crystals*). In the 1990s, Altus Biologics even commercialized CLECs as immobilized enzyme preparations, e.g., of *Candida rugosa* lipase (ChiroCLEC-CR), and *Pseudomonas cepacia* lipase (ChiroCLEC-PC). Despite this great development and the observed impressive stabilities of CLECs, the CLEC technology has not experienced such a widespread application as, e.g., carrier-based enzyme immobilization techniques, especially on an industrial scale. This is mainly attributed to the requirement for (partially) purified enzymes, the fact that not all enzymes crystallize easily, and challenges in efficient protein crystallization on a technical scale. With the recent development of protein engineering methods to improve the crystallizability and cross-linking of proteins, the application of the CLEC technology to virtually any enzyme of interest should be possible today. Moreover, protein crystallization in batch has also advanced to a mature method for protein purification on an industrial scale. This chapter aims to give an overview of available methods in CLEC preparation, the corresponding CLEC characteristics, as well as possible applications of CLECs in biocatalysis and beyond.

**Carrier-Free Enzyme Immobilization: Fundamentals and Innovations in Biocatalysis.**
**DOI: https://doi.org/10.1016/B978-0-443-44139-4.00001-0**

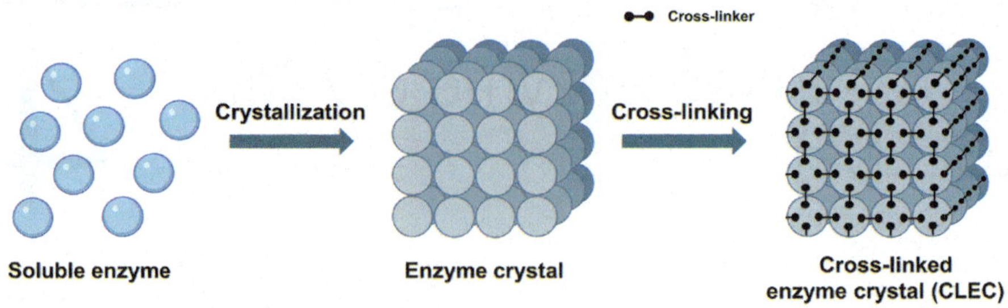


**FIGURE 4.1**

Preparation of cross-linked enzyme crystals.

CLECs are prepared by crystallization of purified, soluble enzyme, followed by chemical cross-linking using a bifunctional cross-linker.

## 4.2 Cross-linked enzyme crystal preparation

The preparation of CLECs is a two-step process (Fig. 4.1). First, the purified and soluble enzyme is precipitated in a controlled manner to form microcrystals. These microcrystals are subsequently cross-linked using a chemical cross-linker to form inter- and intramolecular covalent bonds between the enzyme molecules in the crystal lattice. Thus, stable enzyme particles are obtained that do not dissolve in changing environments (Roy & Abraham, 2004). The shape of the resulting CLECs is thereby dependent on the packing of the individual enzyme molecules within the crystal.

### 4.2.1 Enzyme crystallization

The first step in CLEC preparation is enzyme crystallization. The crystallization of a protein consists of two successive phases, starting with nucleation and followed by crystal growth, as represented in a phase diagram (Fig. 4.2). Nucleation requires a supersaturated enzyme solution. Under these conditions, the high enzyme concentration promotes frequent molecular encounters, leading to the formation of nuclei that can subsequently grow and associate with each other (Nanev, 2007). Once a nucleus reaches the critical size, it can either fall apart or grow into a crystal. However, if the enzyme concentration is too high, uncontrolled enzyme aggregation takes place. On the other hand, no nucleation occurs if the enzyme concentration is too low (Bijelic & Rompel, 2018).

To reach the state of supersaturation, precipitation agents can be added to the enzyme solution to decrease the overall enzyme solubility. The precipitation agents decrease enzyme solubility by withdrawing water from the hydration shell of the proteins. In this state, the enzyme molecules are more likely to associate with each other and form small crystallization nuclei (Rupp, 2015). Precipitation agents can be inorganic salts, nonionic macromolecular organic compounds (i.e., polymers) such

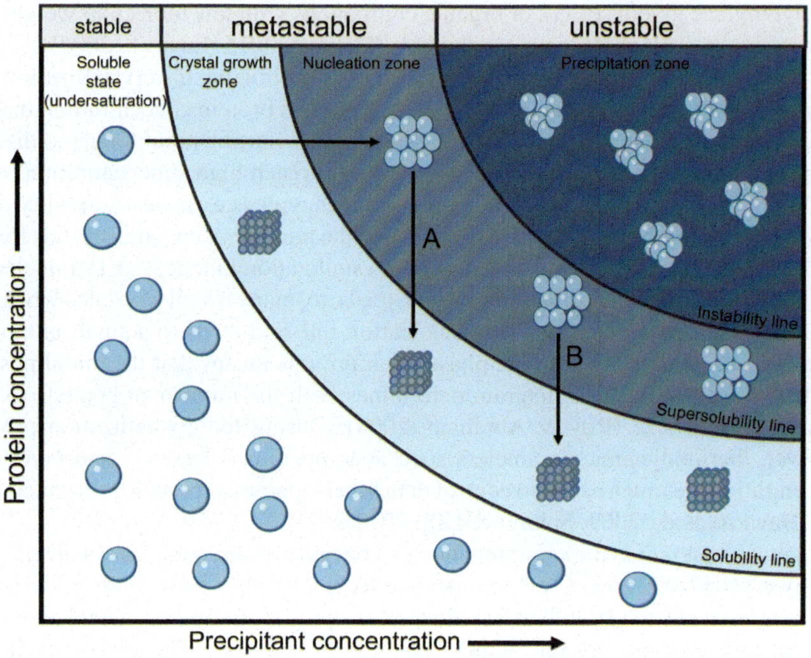


**FIGURE 4.2**

Phase diagram of protein crystallization.

Protein crystallization is commonly described as a function of protein and precipitant concentration, two key parameters that are varied during crystallization experiments. The solubility line separates the soluble state and the crystal growth zone. The stable state corresponds to a single-phase, thermodynamically stable solution in which the protein molecules (shown as blue circles) remain completely dissolved. When the solubility line is crossed, the solution becomes supersaturated. The supersaturated region is subdivided by the instability line into a metastable and an unstable zone. Within the metastable zone, the supersolubility line further distinguishes the nucleation zone from the crystal growth zone. The nucleation zone is characterized by a higher supersaturation, enabling the formation of stable crystal nuclei. As nuclei grow and crystals start to appear, the protein concentration decreases and the solution shifts into the crystal growth zone, where the enzyme crystals increase in size. At very high supersaturation, the unstable zone is reached, resulting in proteins precipitation and the formation of unstructured aggregates. The two main crystallization methods are indicated by *arrows A* and *B*. *Arrow A* represents vapor-diffusion techniques. In these methods, the protein solution initially resides in the stable state and is gradually brought into the supersaturated zone by vapor diffusion. Once the supersaturation is achieved, enzyme nuclei form, followed by crystal growth. In contrast, *arrow B* represents batch crystallization, in which the enzyme solution is brought into the supersaturation state at the beginning of the crystallization process. Under these conditions, nucleation occurs, followed by crystal growth over time.

as polyethylene glycol (PEG), or organic compounds with low molecular weight, like organic solvents, alcohols, or sugar alcohols (Hekmat, 2015; Lu et al., 2002). Another possibility to promote enzyme crystallization is by optimizing the crystallization conditions. Usually, the temperature, pH, buffer type, and protein concentration, but also precipitant type and concentration are varied. A good starting point for crystallization is represented by a high enzyme concentration (to reach high supersaturation) and a pH that is adjusted to the isoelectric point of the enzyme (i.e., lowest solubility of the enzyme) (Roy & Abraham, 2004). The size and the number of enzyme crystals formed are determined by the nucleation rate (i.e., crystallization kinetics). A fast nucleation rate occurs at higher supersaturation and leads to many small crystals, whereas a lower supersaturation causes a slow nucleation rate and leads to a small number of large crystals. With respect to the phase diagram, this means that the initial position of the system in the phase diagram determines both the number of crystals formed as well as their size (Roy & Abraham, 2004). During the crystallization process, however, thermodynamic parameters such as temperature, enzyme, and precipitant concentration are much easier to control than kinetic parameters such as the nucleation rate (Navarro et al., 2009; Schmit & Dill, 2012).

In many cases, enzyme crystallization can be quite challenging, as not all enzymes readily crystallize. Thus, it is recommended to first identify parameters under which the enzyme is able to crystallize and afterwards optimize these parameters to produce good-quality enzyme crystals with a desired crystal size. The crystal quality is very important and directly influences the stability of the resulting CLECs, as the crystal lattice itself serves as a structural scaffold during enzyme immobilization (Margolin, 1996).

### 4.2.2 Crystallization methods

To identify suitable crystallization parameters in the seemingly endless possibilities of combinations, a practical approach is to start with an initial crystallization screening using, e.g., a sparse matrix. In a sparse matrix screen, random combinations of a complete matrix are tested to reduce the number of parameters for subsequent optimization experiments (Jancarik & Kim, 1991). In the optimization experiments, crystallization parameters are systematically varied around the initial successful conditions, creating a multidimensional matrix that yields enzyme crystals. This approach allows for screening a large number of variables with minimal laboratory effort and consumption of materials and time (Jancarik & Kim, 1991). However, it may miss promising crystallization conditions if only a rather small number of conditions is tested in the first screening round. An alternative and simple possibility for initial screenings is commercially available, ready-to-use screening kits (available from, e.g., Jena Bioscience, Molecular Dimension, or Creative Biostructure). These kits usually offer a wide range of crystallization conditions, and there are even specialized kits, e.g., for difficult-to-crystalize enzymes like membrane proteins (e.g., from MiTeGen or Hampton Research), available.

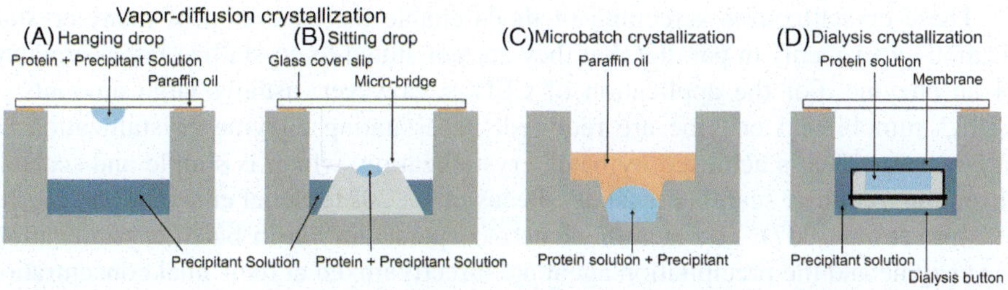


**FIGURE 4.3**

Overview of small-scale crystallization techniques.

Vapor-diffusion methods are divided into hanging, (A) and sitting drop, (B) methods. In both approaches, the wells are covered with a glass cover slip and sealed with paraffin oil. The crystallization drop contains a mixture of enzyme and precipitant, while the reservoir contains only the precipitant at high concentration. Equilibrium between the drop and the reservoir occurs via vapor diffusion, gradually increasing supersaturation in the drop, which promotes nucleation. In microbatch crystallization, (C) enzyme and precipitant are mixed to generate a supersaturated solution from the beginning. The resulting crystallization cocktail is placed in a well and covered with paraffin oil to prevent evaporation. In dialysis crystallization, (D) the enzyme solution is placed in a dialysis button covered with a semi-permeable membrane, which is then immersed in a well containing the precipitant solution. Small precipitants can cross the dialysis membrane, slowly brining the enzyme solution into the supersaturated state and promoting enzyme crystallization.

Crystallization screening experiments are usually performed in microscale using vapor diffusion (by hanging or sitting drop) (Bijelic & Rompel, 2018; Chayen, 1998), micro batch (Chayen, 1998), or micro dialysis crystallization (Thomas et al., 1989; Zeppezauer et al., 1968) (Fig. 4.3). Among these, hanging and sitting drop crystallization are the most commonly used methods for screening. In this setup, the reservoir contains the buffer solution and a highly concentrated precipitation agent, while the drop consists of the buffer solution, a lower-concentrated precipitation agent, and the enzyme. Via vapor diffusion, the system tries to reach an equilibrium, and water vapor diffuses from the drop into the reservoir. This leads to a further concentration of the enzyme and the precipitation agent in the drop, resulting in the formation of a supersaturated solution, which is the starting point for enzyme crystallization (corresponding to *arrow A* in Fig. 4.2) (Bijelic & Rompel, 2018; Chayen, 1998). In dialysis crystallization (Fig. 4.3D), the enzyme solution is separated from the solution containing the precipitation agent via a dialysis membrane. Small precipitation agents can cross the dialysis membrane and slowly increase their concentration in the enzyme solution (Kwan et al., 2023). This finally results in a supersaturated solution that promotes enzyme crystallization. A different mechanism is used in microbatch crystallization (Fig. 4.3C). Here, the enzyme and the precipitation agent are mixed in concentrations that yield a supersaturated solution from the beginning (corresponding to *arrow B* in Fig. 4.2) (Chayen, 1998).

These crystallization screening methods enable the execution of many crystallization experiments in parallel, but they are not suited to crystallize larger amounts of an enzyme. For the application of CLECs, however, usually large amounts of CLEC-immobilized enzyme are required, necessitating enzyme crystallization on larger scale. This is achieved by batch crystallization, which is simple and scalable (from micro to liter scale) (Fernández-Penas et al., 2021; Hebel et al., 2013a, 2013b; Hekmat et al., 2007). To reach the state of supersaturation in batch crystallization, the enzyme and the precipitation agent are directly mixed at their final concentration at the beginning of the experiment (Lu et al., 2002). In the milliliter and liter scale, batch crystallization methods change from a nonagitated to an agitated process called stirred batch crystallization (Hebel et al., 2013a). In this process, the most common precipitation agents are halide salts or PEG (Roy & Abraham, 2004). When using an agitated process for enzyme crystallization, several considerations should be taken into account. (1) The crystallization kinetics can change depending on the stirring speed. For example, an increased impeller tip speed led to an earlier onset of lysozyme crystallization (Smejkal et al., 2013). (2) The stirring speed also influences the size of the enzyme crystals. Increasing the agitation rate generally leads to the formation of smaller crystals (Schmidt et al., 2005; Smejkal et al., 2013). This effect is primarily attributed to enhanced shear forces within the system, which can promote crystal breakage and thus limit overall crystal growth (Filip et al., 2024). Today, preparative and large-scale protein crystallization are industrially relevant processes, serving either as a formulation step for protein therapeutics or as an enzyme/protein purification strategy (Hekmat, 2015). Large-scale enzyme purification via crystallization is an inexpensive method that has been shown to be directly applicable to crude protein mixtures, as shown, e.g., for recombinant L-methionine c-lyase or glucose isomerase (Takakura et al., 2006; Visuri, 1985). However, scaling up a protein crystallization process can be challenging, as also factors such as hydrodynamics and shear sensitivity of the protein to be crystallized have to be considered (Hekmat, 2015). Traditional stirred batch crystallization is limited by the shear forces generated during the process, which can adversely affect enzyme crystal quality (Schmidt et al., 2005; Smejkal et al., 2013). As an alternative, continuous crystallization in flow reactors has emerged in recent years. The advantage of such reactors can be a reduced shear stress that is applied to the enzyme crystals (Neugebauer & Khinast, 2015). This can even enable the formation of protein crystal forms that are not possible under standard conditions (Thomas et al., 2021). On the other hand, enzyme crystallization on a microchip has also been reported. The McCLEC reactor enabled the *in situ* enzyme crystallization, cross-linking, and CLEC application on a single micro-scale device (Conejero-Muriel et al., 2015, 2016).

### 4.2.3 Limitations of crystallization and opportunities via protein engineering

Some enzymes, however, are extremely difficult, if not impossible, to crystallize. Enzyme crystallization is not only influenced by the crystallization conditions but also highly depends on the properties of the enzyme. Enzymes are inherently flexible and dynamic structures, and are thus unlikely to form rigid constructs like an enzyme crystal (Durbin & Feher, 1996). The interactions between enzyme molecules in an enzyme crystal are usually weak and noncovalent, such as ionic, dipole, van der Waals interactions, and hydrogen bonds, as well as a few in number (Rupp, 2009). Moreover, they are highly specific and depend on the surface characteristics of the enzyme to be crystallized (Bijelic & Rompel, 2018; Roy & Abraham, 2004). Thus, if an enzyme of interest is very difficult to crystallize, protein engineering can be used to improve its crystallizability by adjusting the surface characteristics of the enzyme. Reported methods in this respect include surface entropy reduction (SER), crystal contact engineering, or ionic interaction engineering, amongst others (Grob et al., 2020; Hermann et al., 2021; Nowotny et al., 2019; Walla et al., 2021). In SER, bulky, solvent-exposed amino acid residues, which have a high conformational entropy, are exchanged by small residues such as alanine (Cooper et al., 2007). This reduction in surface entropy lowers the entropic penalty associated with forming an ordered crystal lattice. As a result, the free energy barrier for nucleation and crystal growth is reduced, making crystallization thermodynamically more favorable. However, the identification of suitable positions on the enzyme surface for mutagenesis is not trivial. To address this challenge, the surface entropy reduction prediction (SERp) server (Goldschmidt et al., 2007) provides a computational tool for identifying surface residues that are suitable candidates for entropy-reducing engineering. Using these techniques, many hard-to-crystallize enzymes have been successfully crystallized until now (Goldschmidt et al., 2007). Moreover, the same methods can also be applied to further improve the crystallization kinetics of an enzyme for which initial crystallization conditions have already been identified (Nowotny et al., 2019; Staar et al., 2022a; Wichmann et al., 2023).

### 4.2.4 Cross-linking of enzyme crystals

Chemical cross-linking of the enzyme molecules within a crystal is a crucial step in the preparation of CLECs, as it provides additional stability to the enzyme crystals through the formation of intermolecular covalent bonds. This in turn immobilizes the individual enzyme molecules within the crystal lattice, resulting in stable particles (Lalonde, 1997). By contrast, without cross-linking, enzyme crystals tend to redissolve when exposed to environments that differ from the immobilization cocktail (Roy & Abraham, 2004).

Cross-linkers used for the preparation of CLECs are divided into two groups, homobifunctional and heterobifunctional cross-linkers (Fig. 4.4) (Margolin & Navia, 2001). Homobifunctional cross-linkers exhibit two identical reactive groups that react with certain functional groups of the enzyme. In contrast, heterobifunctional cross-linkers possess two different reactive groups, and the cross-linking reaction occurs between different functional groups of the enzyme (Mattson et al., 1993). Typical reactive groups of cross-linkers include aldehydes, imido esters or succinimidyl esters, which primarily react with primary amine groups of enzymes such as the ε-amino groups of lysine residues or the N-terminal amino groups. Other functional groups of enzymes, such as carboxyl and sulfhydryl groups, can be targeted by cross-linking as well (Hermanson, 2013). Not all cross-linkers exhibit a strict specificity towards particular functional groups. Instead, cross-linkers such as glutaraldehyde (GA) or *p*-azidobenzoyl hydrazide (ABH) react with multiple reactive groups of enzymes (Mattson et al., 1993; Migneault et al., 2004). Choosing the right type of cross-linker for each individual enzyme is of great importance as the characteristics of the cross-linker influence the properties of the final CLECs. The linker length, for example, influences the flexibility of the enzyme crystal, with longer linkers being associated with increased crystal flexibility (Wong & Wong, 1992). In addition, cross-linking can influence the diffusion of the substrate through the crystal pores. A large cross-linker or a high degree of cross-linking can decrease the permeability of the enzyme crystals (Roy & Abraham, 2004). Some cross-linkers also exhibit additional functions such as photo activation or cleavability (Margolin & Navia, 2001; Mattson et al., 1993; Yan et al., 2015). CLECs that are cross-linked by cleavable cross-linkers are stable in one set of conditions but dissolve in different environments (Hetrick et al., 2014; Margolin & Navia, 2001). This property is of particular interest, e.g., for the application of CLECs in drug delivery (Hetrick et al., 2014).

Although a large variety of chemical cross-linkers exists, many of them are even commercially available, GA has been used by far the most for CLEC preparation so far (Yan et al., 2015). Glutaraldehyde offers several advantages, such as low cost, high efficiency, and easy handling. Despite the extensive use of GA as a cross-linker, its cross-linking mechanism is not fully understood yet (Migneault et al., 2004). Glutaraldehyde is a dialdehyde cross-linker that primarily reacts with the ε-amino groups of lysine residues, forming strong covalent bonds (Yan et al., 2015). However, activity towards the amino acids arginine, tyrosine, tryptophan, and phenylalanine has been shown as well (Migneault et al., 2004). Moreover, GA tends to polymerize in aqueous solutions, resulting in variable linker lengths that enable intra- and intermolecular cross-linking over various distances (Migneault et al., 2004). However, when lysine or other GA-reactive residues are located in the active site of an enzyme, cross-linking via GA can result in enzyme inactivation (Ayala et al., 2002; Staar et al., 2022a, 2022b). Hence, alternative cross-linkers will be preferred in such cases. Examples of the use of alternative cross-linkers include CLEC formation of *Candida rugosa* lipase, halohydrin dehalogenase HheG, and glucose oxidase, which have been cross-linked using 1,4-di-[3′-(2′-pyridyldithio)propion-amido]butane (DPDPB), bis-maleimidoethane (BMOE), and 1,8-diamino-octane, respectively (Margolin & Navia, 2001; Margolin et al., 2000; Staar et al., 2022a). In the latter case, the cross-linker did not react with amino acid residues of the enzyme, but aldehyde groups of the sugar molecules present on the surface of glycosylated glucose oxidase instead (Margolin et al., 2000).

To generate active and stable CLECs, the cross-linking conditions must be carefully optimized. Insufficient cross-linking generates CLECs where the outer layers of the crystal are fixed, but the core of the enzyme crystal is not cross-linked. Such CLECs look normal in the cross-linking cocktail but expand in low ionic strength buffers and shrink in concentrated solutions (i.e., high ionic strength), resulting in

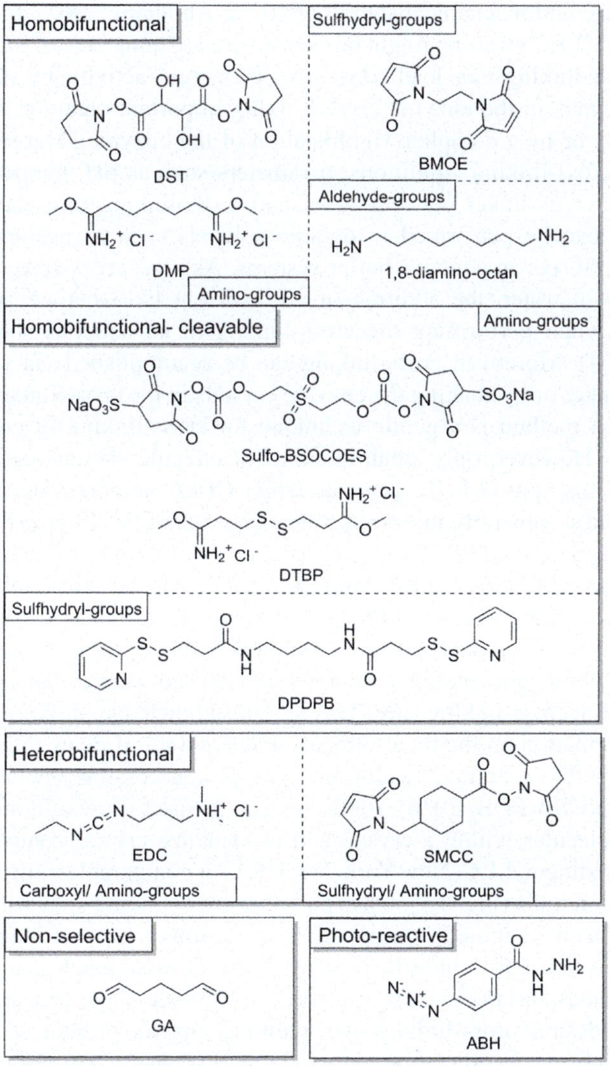


**FIGURE 4.4**

Typical bifunctional reagents used for cross-linking of enzyme crystals and their targeted functional groups on the enzyme surface

[*ABH*, *p*-azidobenzoyl hydrazide; *BMOE*, bis-maleimidoethane; *DMP*, dimethyl pimelimidate; *DPDPB*, 1,4-di-[3´-(2´-pyridyldithio)propion-amido]butane; *DST*, disuccinimidyl tartrate; *DTBP*, dimethyl-3,3´-dithiopropionimidat-dihydrochlorid; *EDC*, 1-ethyl-3-[3-dimethylaminopropyl]carbodiimide hydrochloride; *GA*, glutaraldehyde; *SMCC*, succinimidyl 4-(N-maleimidomethyl)cyclohexan-1-carboxylate; *Sulfo-BSOCOES*, bis[2-(sulfosuccinimidooxycarbonyloxy)
ethyl]sulfone].

a loss of storage and mechanical stability (Roy & Abraham, 2004). In special cases, however, such CLEC swelling might be even desired (Chung et al., 2025). In contrast, extensive cross-linking can lead to a loss of enzyme activity by either blocking the water channels in the enzyme crystal, fixing important catalytic amino acids in the active side, or by a complete rigidification of the enzyme (Margolin, 1996). To optimize the cross-linking conditions, parameters such as pH, temperature, protein concentration, cross-linker concentration, and cross-linking time can be adjusted. Cross-linking can be performed in organic solvents such as acetone or dimethyl sulfoxide (DMSO) or in aqueous buffer systems. As most cross-linkers display only low solubility in water, the addition of a co-solvent like acetone or DMSO may be necessary when performing the cross-linking in an aqueous solution (Roy & Abraham, 2004). Moreover, cross-linking can be accomplished via vapor diffusion of the cross-linker or by soaking the enzyme crystals in the cross-linker solution. The vapor diffusion method is a gentle technique for cross-linking of enzyme crystals (Lusty, 1999). However, only small amounts of enzyme crystals can be processed at once using this approach. To generate larger CLEC amounts for application, the soaking method is generally preferred, since large quantities of enzyme crystals can be processed.

To find a suitable cross-linker, enzyme crystals can be cross-linked with a variety of different chemical cross-linkers, followed by an analysis of the resulting CLEC stability and residual activity (trial-and-error principle). Alternatively, the arrangement of enzyme molecules within the crystal lattice can be analyzed to identify potential cross-linking sites. This requires knowledge of the enzyme's crystal structure. Information on the three-dimensional structure of many different enzymes and their symmetry operations within the crystals is accessible via the Protein Data Bank (PDB) (RCSB PDB, 2026). PDB files can be used to rebuild the arrangement of enzyme molecules within a crystal and to visualize crystal contacts using, e.g., PyMOL (Schrödinger, LLC, New York, NY, USA) or comparable molecular modeling and visualization tools (Fig. 4.5). Once potential cross-linking sites within the crystal contact have been identified, a compatible cross-linker (or combination of cross-linkers) is selected that is able to react with the desired functional groups of the enzyme molecules and fits to their distance in the crystal. A similar approach can be used to install desired cross-linking sites within an enzyme of interest for subsequent CLEC formation, as recently reported for halohydrin dehalogenase HheG (Staar et al., 2022a). Cross-linking of this enzyme using GA resulted in stable but inactive CLECs. Hence, crystal contact engineering was employed to enable the use of alternative cross-linkers (Staar et al., 2022a, 2022b). To this end, the protein binding energy prediction (PRODIGY)-CRYSTAL webserver (Jiménez-García et al., 2019) was used to identify amino acids in HheG that faced each other at the crystal interface (Fig. 4.5). Protein engineering was then used to replace such residues by either cysteine or lysine for later cysteine- or lysine-specific cross-linking. This way, HheG mutants D114C and V46K were obtained, which could be cross-linked efficiently using BMOE or dithiobis(succinimidyl propionate) (DSP), respectively (Staar et al., 2022a, 2022b). Using only low cross-linker concentrations, highly stable and catalytically active

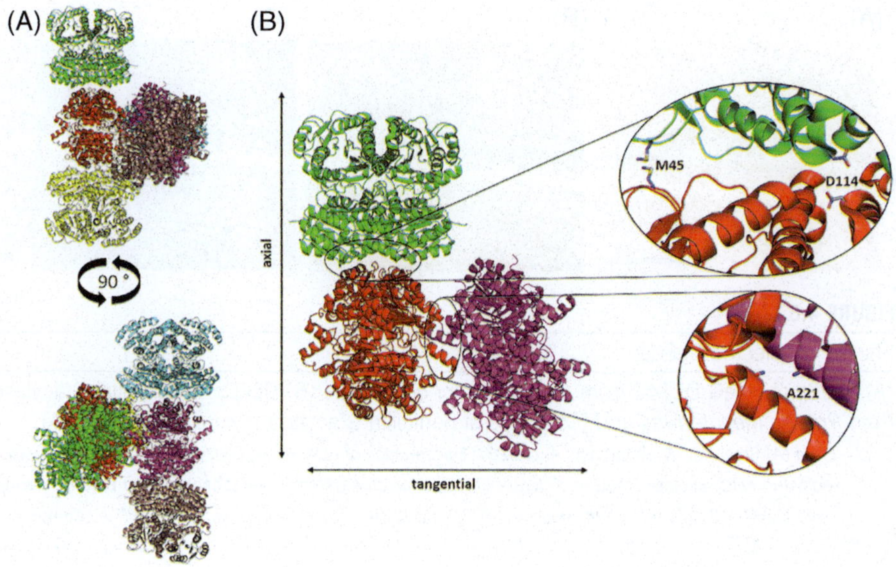


**FIGURE 4.5**

Crystal contact analysis of wild-type HheG based on its crystal structure (PDB ID: 5O30).

The arrangement of HheG tetramers within the unit cell is illustrated in section A, while B highlights suitable amino acid positions for replacement by cysteine in the axial and tangential crystal contacts of HheG.

*From Staar, M., Henke, S., Blankenfeldt, W., & Schallmey, A. (2022a). Biocatalytically active and stable cross-linked enzyme crystals of halohydrin dehalogenase HheG by protein engineering.* ChemCatChem. 14*(9), e202200145. https://doi.org/10.1002/cctc.202200145. licensed under the Creative Commons Attribution license.*

HheG CLECs could be attained. Moreover, HheG D114C even displayed improved crystallization kinetics compared to wild-type HheG.

## 4.3 Cross-linked enzyme crystal characteristics

CLECs of various shapes and sizes have been reported in literature (Fig. 4.6) (Margolin & Navia, 2001). CLEC shape is thereby dependent on the actual packing of the individual enzyme molecules within the crystal. Thus, it is highly enzyme-dependent and cannot be easily adjusted. Changing the crystallization conditions, however, can lead to different crystal morphologies for the same enzyme (Lee et al., 2002). In contrast, the size of a crystal, and thus the resulting CLEC, can be easily adjusted through, e.g., crystallization time. CLEC morphology is usually analyzed by microscopic techniques, such as scanning electron microscopy (Perveen et al., 2022),

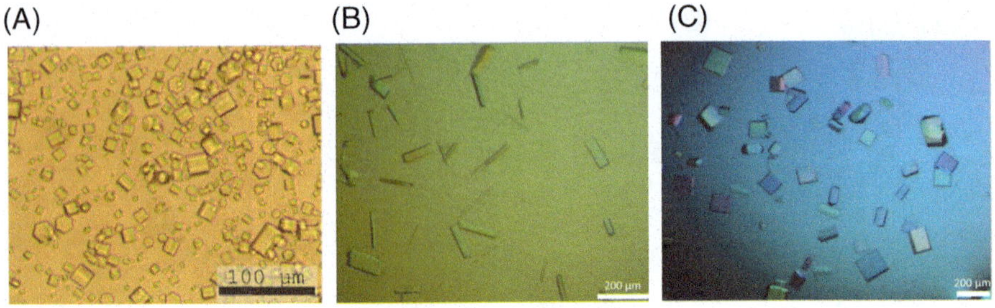


**FIGURE 4.6**

Shapes of selected CLECs.

(A) CLECs of HheG D114C from *Ilumatobacter coccineus*, (B) CLECs of penicillin G acylase from *Priestia megaterium*, and (C) CLECs of penicillin G acylase from *Bacillus* sp. FJAT.

*From (A) Staar, M., & Schallmey, A. (2023). Performance of cross-linked enzyme crystals of engineered halohydrin dehalogenase HheG in different chemical reactor systems.* Biotechnology and Bioengineering. 120*(11), 3210–3223. https://doi.org/10.1002/bit.28528. (B and C) Kubiak, M., Mayer, J., Kampen, I., Schilde, C., & Biedendieck, R. (2021). Structure-properties correlation of cross-linked penicillin G acylase crystals.* Crystals, 11*(4), 451. https://doi.org/10.3390/cryst11040451. licensed under the Creative Commons Attribution license.*

while the CLEC size distribution can be determined via particle size analysis using, e.g., laser diffraction (Staar & Schallmey, 2023).

The great advantage of CLECs compared to soluble enzymes or carrier-based enzyme immobilization is their high activity-to-volume ratio, as protein crystals represent the highest protein density possible (even higher than what can be achieved in highly concentrated solutions of soluble enzymes) (Roy & Abraham, 2004). This characteristic is highly desirable for biocatalytic applications. Thus, a CLEC-based packed-bed reactor (PBR) ideally only comprises catalytically active protein in high packing density, enabling outstanding space-time yields (STYs) in continuous-flow biotransformations. Moreover, CLEC-immobilized enzymes exhibit much higher specific activities in comparison to enzymes that have been immobilized on a noncatalytic carrier or embedded in a polymeric matrix (Roy & Abraham, 2004). In the latter cases, the immobilized material consists mostly of inert carrier, while in CLECs, the enzyme molecules represent the catalytic species and the solid support (in the form of the crystal lattice) simultaneously. Therefore, remarkable volumetric productivities can also be achieved in biocatalytic batch reactions when CLECs are applied.

Even though CLECs consist of rather fragile protein molecules that have been chemically cross-linked with each other, their mechanical and operational stabilities are impressive. CLECs withstand high shear forces, as, e.g., caused by mixing in a reactor, without particle fragmentation and resist elevated pressures, as, e.g., occurring in a PBR (Roy & Abraham, 2004). This makes CLECs highly suitable for repetitive or long-term application, as they also usually exhibit significantly improved half-life times. For example, the half-life time of the halohydrin dehalogenase HheG under reaction conditions increased from 23 h for soluble enzyme to more than 80 days for the respective CLECs (Staar et al., 2022a, 2022b). Moreover, CLECs can be applied in a broad range of reaction conditions as they generally exhibit much higher stability towards pH, temperature, and nonaqueous reaction media (e.g., organic solvents, ionic liquids, and supercritical fluids) compared to their soluble enzymes. This dramatic stability improvement upon CLEC formation is explained by the dense and highly ordered packing of the enzyme molecules within the crystal, which results in additional polar (i.e., ionic or hydrogen bond) and hydrophobic interactions, as well as a significant increase in buried surface area (Margolin & Navia, 2001). Moreover, chemical cross-linking of the enzyme molecules within the crystal lattice introduces additional covalent linkages, which reduce molecular motion further and thus minimize protein unfolding. As such, protein molecules within a CLEC are even resistant towards proteolysis (Clair & Navia, 1992). It should be noted that this considerable stability increase is virtually independent of CLEC size, while their mechanical stability correlates with the degree of chemical cross-linking (Margolin & Navia, 2001). In some cases, an effect of the crystal shape on the mechanical stability of the respective CLECs has also been described.

CLEC size, however, does have an impact on the specific activity of the immobilized enzyme (Margolin & Navia, 2001). While CLEC formation is actually a rather mild immobilization process, the observed specific activities of CLEC-immobilized enzymes are still reduced compared to the specific activities of corresponding soluble enzymes. CLECs represent microporous, heterogeneous catalysts. Thus, substrate molecules have to diffuse through pores and channels present in the crystal in order to reach the active sites of enzyme molecules in the crystal interior. The larger the size of the crystal or the larger the substrate molecule, the more limiting the diffusion will be. Overall, smaller CLECs with a size below 10 μm are generally preferred in terms of activity, as they exhibit a high surface-area-to-volume ratio and are only insignificantly mass-transfer limited, if the substrate is not too big (Margolin & Navia, 2001). On the other hand, larger CLECs offer superior separation and handling properties (Roy & Abraham, 2004). Moreover, the enzyme's conformational flexibility within the crystal also plays a major role in activity, especially if small structural changes have to occur for effective substrate binding or during catalysis (Margolin & Navia, 2001).

## 4.4 Application of cross-linked enzyme crystals

The chemical cross-linking of protein crystals had initially been investigated in the context of X-ray structure determination more than 60 years ago, as a means of stabilizing the crystals in solution (Quiocho & Richards, 1964). At that time, the authors already mentioned the possibility of using this approach to generate active but insoluble enzyme particles, which can be removed from the reaction solution, analogous to the resin-attached enzymes reported earlier (Bar-Eli & Katchalski, 1960). Quickly, the potential of CLEC formation for enzyme immobilization has been recognized, leading to a large number of scientific publications until the early 2000s (Roy & Abraham, 2004). Most of these studies focused on the application of CLECs as heterogeneous catalysts in biotransformation, as the CLECs enabled high catalyst loading, the use of nonaqueous reaction media, as well as catalyst reuse. Moreover, the commercialization of different CLECs by Altus Biologics gave a broader scientific community access to these remarkable biocatalysts. Apart from their application in biocatalysis, CLECs have also been studied in biosensor and medical applications, or for their use as microscaffolds (Margolin & Navia, 2001; Roy & Abraham, 2004).

### 4.4.1 Application in biocatalysis

Many CLEC applications in the field of biocatalysis have focused on hydrolases such as lipases or proteases. This is not surprising as many of these enzymes have been commercially available for over 60 years ago. Moreover, the application of these enzymes in the reverse, i.e., nonhydrolytic, reaction direction (e.g., for ester or peptide synthesis) requires the use of nonaqueous reaction media such as organic solvents. Non-immobilized enzymes, however, display only little stability in these solvents and cannot be reused, which is why enzyme immobilization became important. Altus Biologics, for example, commercialized CLECs based on ChiroCLEC-CR and ChiroCLEC-PC, which have been applied in various esterification and transesterification reactions (Collins et al., 1998; Lalonde et al., 1995). The application of these lipases as CLECs instead of (carrier-immobilized) crude enzyme preparations even had another significant benefit. As crude enzyme preparations often contain mixtures of different hydrolytic enzymes, the outcome of a biocatalytic reaction in terms of enantiopurity of the desired product was often insufficient. In contrast, CLEC preparation is accompanied by a (at least partial) purification of the enzyme, apart from immobilization. Hence, often much higher enantioselectivities could be achieved when using CLECs as a biocatalyst. Thus, using ChiroCLEC-CR in the hydrolysis of $\alpha$-substituted carboxylic acid esters yielded 3 to 50 times higher enantioselectivities compared to reactions with crude enzyme (Fig. 4.7A) (Lalonde et al., 1995). The half-life $\tau_{1/2}$ of these CLECs at 40°C and pH 7 was found to be 13 days compared to less than 5 h for the purified soluble lipase, which also enabled a repetitive batch application of the CLECs over 18 cycles in the kinetic resolution of ketoprofen. Even higher enantioselectivity improvements could be realized in the esterification of the racemic carboxylic acids ibuprofen and *p*-chloro-2-phenoxypropionic acid in nonpolar organic solvents (Fig. 4.7B) (Persichetti et al., 1996). Interestingly, a much faster inactivation of the ChiroCLEC-CR catalyst at 40°C has been reported in 0.2 M PEG 1000/acetate buffer, pH 5, in the presence of 10 mM $CaCl_2$ ($\tau_{1/2} = 72$ h), which might be caused by the significantly lower pH (Brady et al., 2004).

Employing ChiroCLEC-PC, the kinetic resolution of racemic 1-phenylethyl acetate in batch to provide (*R*)-1-phenylethanol in high enantiomeric excess (Fig. 4.7C) was even scaled up to 100 L scale (Collins et al., 1998). Instead of a cosolvent, the

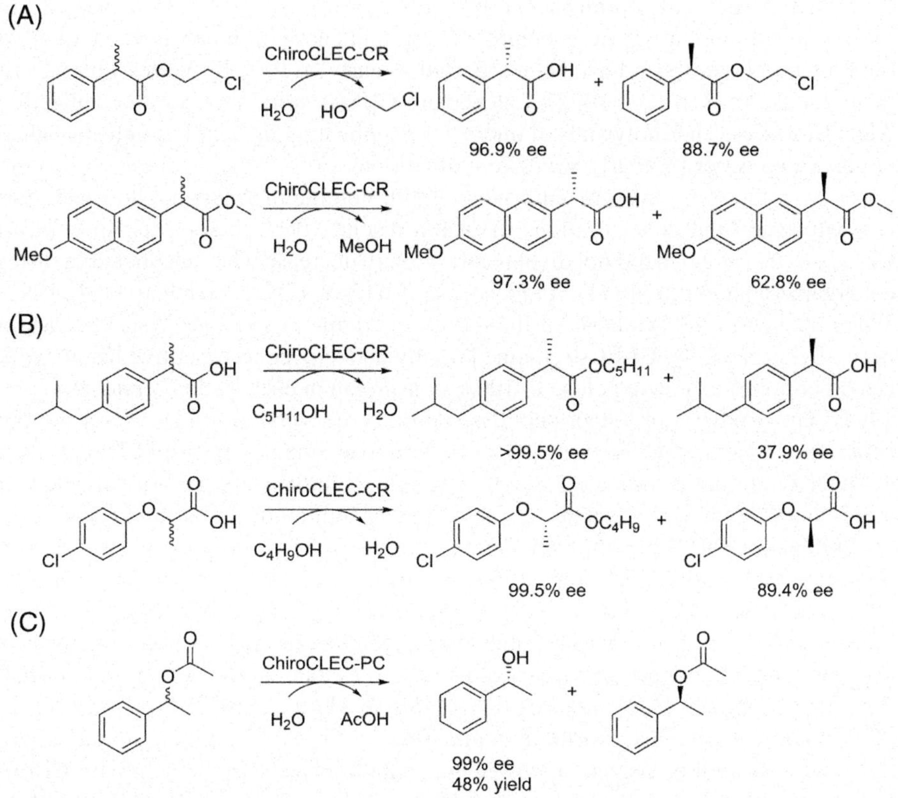


**FIGURE 4.7**

Examples for the application of lipase CLECs in chiral resolutions.

(A) Hydrolysis of racemic $\alpha$-carboxylic acid esters using CLECs of *Candida rugosa* lipase, (B) esterification of racemic carboxylic acids using CLECs of *Candida rugosa* lipase, and (C) hydrolysis of racemic 1-phenylethyl acetate using CLECs of *Pseudomonas cepacia* lipase.

substrate ester was used as a second organic phase with an organic (substrate)-to-aqueous phase ratio of 0.3 (corresponding to 30 kg of ester in the 100 L reaction). At a catalyst loading of 0.5 g/L (i.e., 50 g of CLECs in total), a conversion of 48.8% was achieved within 18 h with an enantiomeric excess of the product alcohol of > 99.9%. Afterwards, the CLECs were recovered by filtration and reused for another 8 batch runs. Although the initial specific activity in the reactor decreased from run 1 to run 9, mainly due to significant mass loss of the smaller CLECs during filtration, product enantiomeric excess stayed at > 99.9% throughout the repetitive batch runs. This example highlights the exquisite suitability of CLECs for application in industrial processes, even though separation of the rather small CLEC particles from the reaction medium requires more specialized equipment.

Next to lipases, also proteases, especially subtilisin Carlsberg and thermolysin, have been explored as CLEC preparations for application in biocatalysis. CLECs of the protease thermolysin have been reported as early as 1992 by Vertex Pharmaceuticals Inc (Clair & Navia, 1992). Later, these CLECs were also commercialized by Altus Biologics. Thermolysin is of interest for application in peptide synthesis, which requires low-water or nearly water-free conditions to shift the reaction equilibrium towards synthesis for maximized product yield. Soluble thermolysin, however, does not withstand high concentrations of water-miscible organic solvents. In contrast, CLECs of thermolysin did not display any activity decrease when incubated in different organic-aqueous mixtures (50:50, v/v) for 5 days at 40°C (Persichetti et al., 1995). When applied in the synthesis of the chiral aspartame precursor Z-Asp-PheOMe in neat ethyl acetate, the CLECs remained highly active over 18 repetitive batch cycles (20 h per cycle), achieving close to 100% conversion in each cycle (Persichetti et al., 1995). The synthesis of longer peptides (up to the conversion of the insulin B chain containing 30 amino acids) was also successful using the thermolysin CLECs, albeit a reduced initial rate compared to soluble thermolysin was observed for peptides of more than seven amino acids due to mass transfer limitations in the CLECs.

Similarly, CLECs of subtilisin Carlsberg have also been applied in peptide synthesis as well as the chiral resolution of natural and nonnatural amino acids via ester hydrolysis (Margolin & Navia, 2001; Wang et al., 1997). Moreover, their operational stability in repeated batch versus continuous flow in transesterification reactions employing neat organic solvents as well as their application in supercritical fluids has been assessed (Amorim Fernandes & Halling, 2002; Fernandes et al., 2005; Fontes et al., 2001; Noritomi et al., 2007). Surprisingly, a rapid initial activity decrease, followed by a second slower inactivation phase, for the subtilisin CLECs was observed in batch reactions and continuous flow, independent of the used organic solvent and the applied water activity (Amorim Fernandes & Halling, 2002; Fernandes et al., 2005). The actual cause for this rapid initial deactivation phase has not been discussed in the literature. However, it attests a lower operational stability for the CLECs of subtilisin Carlsberg compared to CLECs of other enzymes. It should be noted, however, that a significant pH-dependence of the catalytic activity of subtilisin Carlsberg has been reported. In contrast to lyophilized soluble enzyme, CLECs do not exhibit a pH memory effect in organic solvents, independent of their pretreatment, i.e., either washing with organic solvent or lyophilization, to remove excess water (Xu & Klibanov, 1996). This is especially relevant if pH control is required to adjust the protonation state of important active-site residues, such as the deprotonation of the histidine of the catalytic triad in subtilisin Carlsberg. Xu and Klibanov have demonstrated that in such cases, pairs of organic solvent-soluble acids and conjugate bases can be applied, which resulted in a marked activation of the subtilisin CLECs in organic media (Xu & Klibanov, 1996).

Interestingly, also CLECs of the glucoamylase from *Aspergillus niger* have been reported and applied in a PBR for the continuous hydrolysis of starch (Abraham et al., 2004). As observed for other CLECs, the glucoamylase CLECs exhibited a higher temperature stability compared to soluble enzyme, but also increased catalytic activity at higher substrate concentration. However, the $K_M$ of the CLECs for the substrate starch increased by 10-fold compared to the soluble enzyme. This is not surprising considering the diffusion limitations of the polymeric starch within the crystals. Using maltodextrin (a mixture of glucose monomers, dimers, and oligomers) instead, a high STY of 714.1 g $L^{-1}$ $h^{-1}$ was achieved in the PBR (Abraham et al., 2004).

Apart from hydrolases, also CLECs of oxidoreductases have been studied as immobilized biocatalysts. For example, Altus Biologics commercialized CLECs of horse liver alcohol dehydrogenase, an enzyme that has been widely applied in biocatalysis. CLEC generation was performed in the presence of the reduced nicotinamide adenine dinucleotide (NADH) cofactor to ensure maximal loading of the CLECs with this cofactor for later application in ketone reduction (St Clair N et al., 2000). Instead, when the CLECs were prepared in the absence of cofactor, full CLEC activity could not be reconstituted upon addition of external NADH. Interestingly, not only did the horse liver alcohol dehydrogenase display improved stability towards higher temperature or the presence of organic solvents (the latter being of particular importance for cofactor regeneration using, e.g., isopropanol as co-substrate) when immobilized as CLECs, but also for the cofactor NADH, a prolonged half-life was observed when bound to the CLECs in comparison to free cofactor in solution.

Moreover, CLECs of the laccase from *Trametes versicolor* have been applied in a PBR (containing 120 mg of CLECs) for the continuous oxidation of pyrogallol to purpurogallin (Fig. 4.8A), a biologically-active, natural compound from white oak (Roy & Abraham, 2006a, 2006b). Purpurogallin production was performed continuously over 650 h at 28°C without any obvious decrease in CLEC activity. The highest STY of 269 g $L^{-1}$ $h^{-1}$ was achieved at a residence time of 3.5 s using 5 mM substrate concentration. In a different example, CLECs of the same laccase were applied for the effective treatment of synthetic dye-containing waste effluents from different textile industries (Perveen et al., 2022).

Also, CLECs of the industrially relevant xylose isomerase from *Streptomyces rubiginosus* have been studied (Visuri, 1989). Initially called glucose isomerase, this enzyme is used in industry as a carrier-immobilized enzyme for the isomerization of glucose to fructose in the production of high-fructose corn syrup (Fig. 4.8B). In the large-scale production of this enzyme, crystallization is performed as a purification and concentration step (Visuri, 1985). Therefore, it was obvious to also investigate CLEC formation and to demonstrate the applicability of xylose isomerase CLECs in the isomerization of glucose (Visuri, 1989). Later, CLECs of xylose isomerase have been studied as well in a PBR setup for the synthesis of rare sugars (Jokela et al., 2002). While the enzyme displays high catalytic activity for glucose isomerization to yield fructose, its activity towards rare sugars is much lower, requiring higher enzyme loadings. That is why CLECs are beneficial over carrier-immobilized enzymes in this context. This way, the conversion of, e.g., L-lyxose or L- and D-mannose by xylose isomerase could be observed. Moreover, the isomerization of L-arabinose to L-ribulose and L-ribose (Fig. 4.8B) in the PBR was studied in more detail with the aim to maximize L-ribose formation. By recirculating a 500 mg $mL^{-1}$ L-arabinose solution through the PBR over 116 h at 50°C, an L-ribose concentration of 32 mg $mL^{-1}$ could

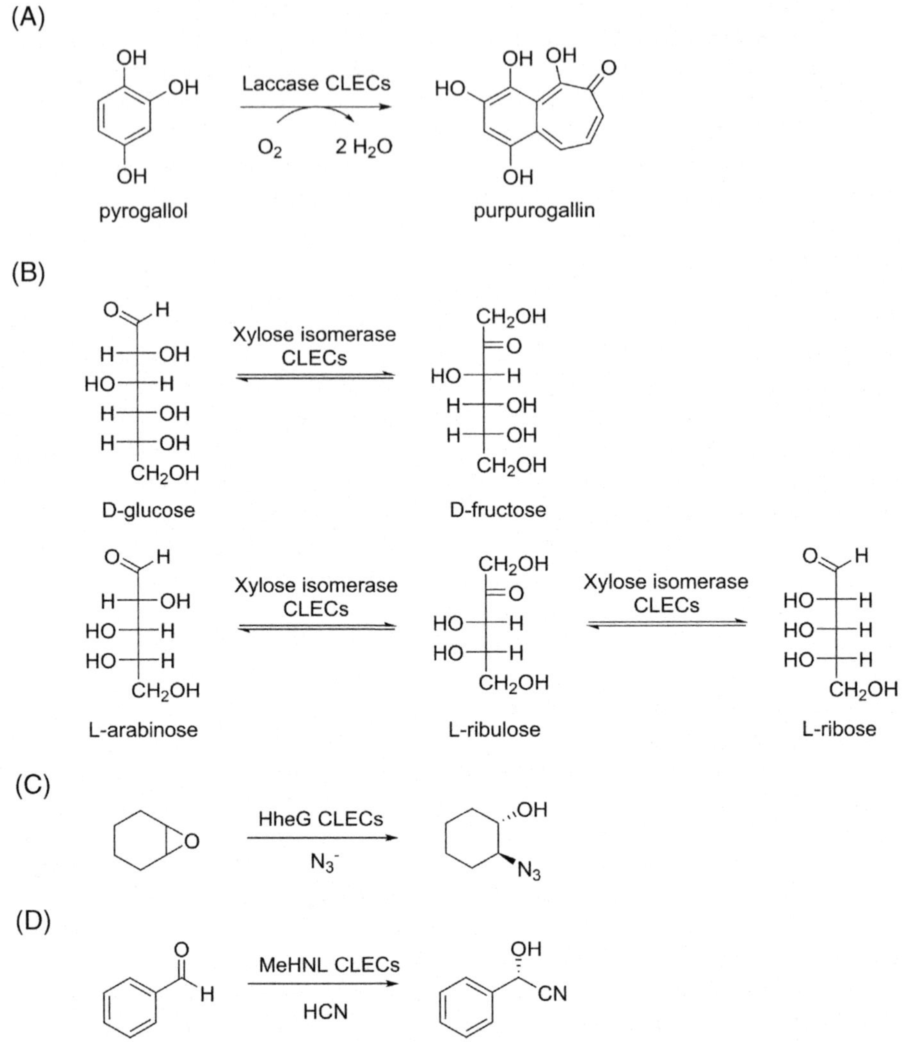


**FIGURE 4.8**

Further examples for the biocatalytic application of CLECs.

(A) Oxidation of pyrogallol using CLECs of the laccase from *Trametes versicolor*, (B) sugar isomerization using CLECs of xylose isomerase from *Streptomyces rubiginosus*, (C) enantioselective ring opening of cyclohexene oxide with azide using CLECs of halohydrin dehalogenases HheG, and (D) enantioselective $\alpha$-cyanohydrin formation using CLECs of the hydroxynitrile lyase from *Manihot esculenta*.

be achieved, corresponding to 6.4% of total sugars (Jokela et al., 2002). Interestingly, in plug-flow mode, a partial separation of the different sugar products in the effluent could be observed. Hence, the CLECs in the PBR not only served as a catalyst but also as a chiral matrix for simultaneous (partial) product separation (Jokela et al., 2002; Leisola et al., 2001).

In another example, the performance of the CLECs of the halohydrin dehalogenase HheG in batch, fed-batch, and continuous mode was compared for the ring opening of cyclohexene oxide with azide as the nucleophile (Fig. 4.8C) (Staar & Schallmey, 2023). As mentioned earlier, the enzyme was engineered previously to ensure efficient CLEC formation. While the CLECs displayed exquisite operational stability in repeated batch and continuous mode, considerably higher production rates (up to 23.9 kg product $g_{CLECs}^{-1}$ $L^{-1}$ $h^{-1}$) and product enantiomeric excesses (64% $ee_P$) were obtained using the CLEC-based PBR compared to batch and fed-batch mode (Staar & Schallmey, 2023; Staar et al., 2022a). Later on, CLECs of a more enantioselective variant of HheG were generated and also applied in the azidolysis of cyclohexene oxide, yielding the product (1*S*,2*S*)-2-azidocyclohexan-1-ol with 96% $ee_P$ (Staar et al., 2024).

Likewise, the operational stability of CLECs of the hydroxynitrile lyase from *Manihot esculenta* was studied for the reaction of benzaldehyde with hydrocyanic acid (HCN) in dibutylether, and compared to the same enzyme immobilized on Celite (Fig. 4.8D) (Costes et al., 2001). While the carrier-immobilized enzyme was rapidly inactivated over seven reaction cycles, the CLECs retained about 70% of their initial activity. However, upon CLEC formation, a large amount of activity compared to soluble enzyme was already lost. This was explained by an excessive cross-linking using GA in acetone as well as a suboptimal arrangement of the enzyme molecules within the crystal, resulting in obstruction of the enzyme active site.

### 4.4.2 Other applications

Similar to the biocatalytic application of CLECs, their use in biosensors for the detection of, e.g., pollutants and other compounds has been reported as well. Enzymes in biosensors need to be robust to withstand varying or even harsh conditions in the analyte solution and to exhibit sufficient shelf life. Therefore, enzymes for biosensor applications are usually immobilized. Using CLECs in this context offers the additional advantage of high local enzyme concentrations for improved biosensor performance. In literature, biosensor applications of CLECs include the use of glucose oxidase CLECs for quantification of glucose levels in blood (Luiz de Mattos, 2001), the use of urease CLECs for the measurement of urea concentrations in circulating body fluids as an early indication of renal disease (Navia & St. Clair, 1995), or the application of laccase and horseradish peroxidase CLECs for the detection of phenols and other organic pollutants (Roy & Abraham, 2004; Roy et al., 2005). Likewise, an amperometric biosensor based on CLECs of the organophosphate hydrolase from *Flavobacterium* sp. has been developed for the detection of organophosphate pesticides (Laothanachareon et al., 2008). The latter enzyme is also of interest for the detoxification of these pesticides and chemical nerve agents such as the thiophosphonate VX (Margolin & Navia, 2001). In a different approach, CLEC generation was performed in a microfluidic chip, named McCLEC, in combination with on-chip photonic detection, for potential application as a continuous biosensor (Conejero-Muriel et al., 2015, 2016). Impressively, the CLEC-containing microfluidic chips could be stored at room temperature for several months (up to 1 year) without obvious losses in enzymatic activity.

Another fascinating application of CLECs is their use in the oral administration of therapeutic enzymes. This is of special interest if the orally applied enzyme has to be protease-resistant and has to survive the low pH conditions of the stomach before activity exertion in the duodenum. The principle was originally demonstrated by Simi and Abraham in 2007 for subtilisin CLECs (Simi & Emilia Abraham, 2007). In 2014, a real case study was published by scientists of Eli Lilly on the example of Liprotamase (Hetrick et al., 2014). At that time, this was an oral drug candidate composed of a lipase, a protease, and an amylase, each of microbial origin, for the treatment of exocrine pancreatic insufficiency in patients with cystic fibrosis. The authors specifically focused on the lipase immobilized as CLECs as the structure and activity of this enzyme had to be preserved at low pH (as present in the stomach), while it had to be active at neutral to alkaline pH (as existing in the duodenum). Instead of using GA, they employed the lysine-specific crosslinker bis(sulfosuccinimidyl)suberate (Hetrick et al., 2014). By varying its concentration, the crosslinking degree was adjusted to obtain CLECs that are stable and insoluble at pH 4, while the CLECs dissolved rapidly at pH 6.5, as desired for high lipase activity.

Moreover, CLECs have gained significant attention as bioorganic zeolites to be used, e.g., as chromatographic material in chiral separations or as microporous scaffolds for the incorporation of transition metal complexes (Abe et al., 2016; Margolin & Navia, 2001). Vilenchik et al. (1998) demonstrated the possibility to use thermolysin CLECs for the size-dependent separation of PEG molecules of different molecular weights, the separation of (*S*)-ibuprofen and (*R*)-phenyllactic acid via differential adsorption to the CLEC material, and the chiral separation of racemic phenyllactic acid (Vilenchik et al., 1998). Later, the Leisola group further reported the use of xylose isomerase CLECs for the purification of xylitol from dilute or impure solutions and the chiral separation of D- and L-arabitol (Pastinen et al., 1998, 2000).

The combination of CLEC scaffolds with transition metal complexes for the generation of heterogeneous biohybrid catalysts was introduced more recently (Abe et al., 2018). In one example, hen egg white lysozyme CLECs were soaked with the ruthenium complex $[Ru(benzene)Cl_2]_2$ for application in enantioselective hydrogen-transfer reactions of acetophenone derivatives (Fig. 4.9A) (Tabe et al., 2014). The metal complex was shown to bind in the porous solvent channels of the CLECs through specific coordination via histidine side chains. Even though products with only moderate enantiomeric excesses were obtained in the hydrogen-transfer reactions, the observed enantioselectivity was solely induced by the chirality of the

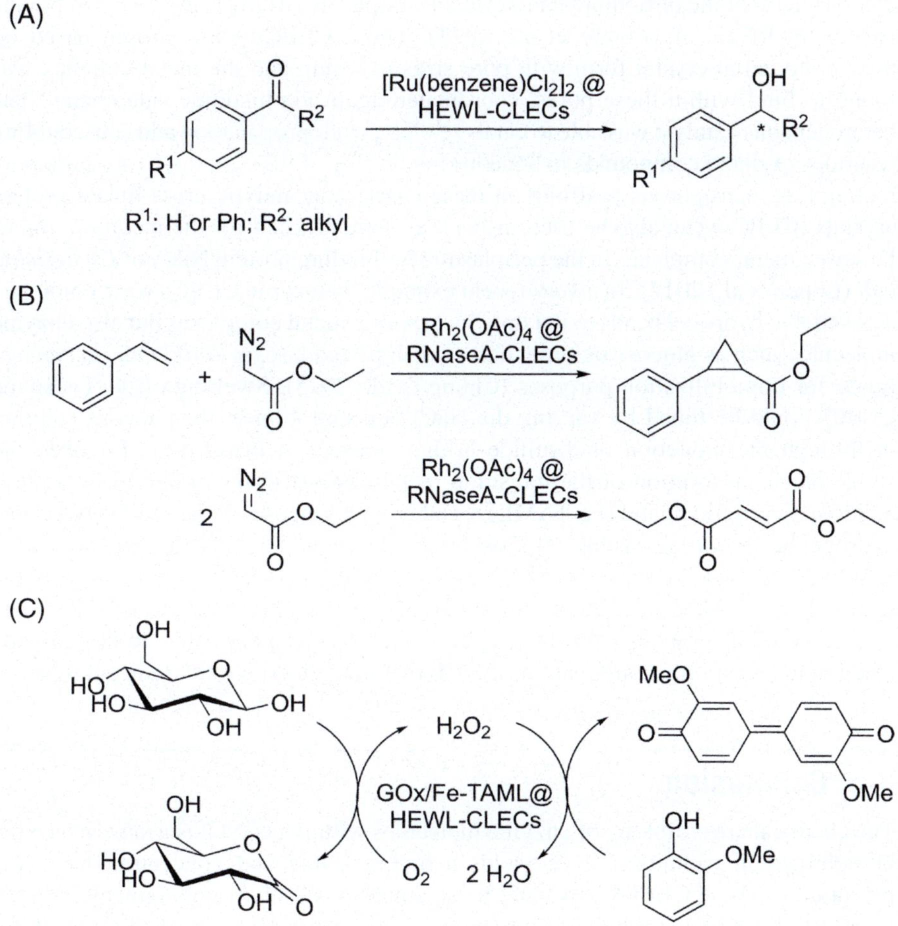


**FIGURE 4.9**

Examples for the application of CLEC-based heterogeneous biohybrid catalysts.

(A) Application of hen egg white lysozyme (HEWL)-CLECs soaked with the ruthenium catalyst $[Ru(benzene)Cl_2]_2$ in enantioselective hydrogen-transfer reactions of acetophenone derivatives, (B) olefin cyclopropanation and self-coupling reaction of ethyl diazoacetate catalyzed by the dirhodium paddlewheel complex $Rh_2(OAc)_4$ bound to CLECs of bovine pancreatic RNaseA, and (C) chemo-enzymatic cascade reaction for the oxidation of guaiacol catalyzed by a combination of glucose oxidase and the transition metal catalyst Fe-TAML, both bound to hydrogel beads derived from lysozyme CLECs.

CLECs. Moreover, different product enantiomers were preferentially gained depending on the crystal form of the applied lysozyme CLECs (please note that lysozyme is able to crystallize in different space groups, resulting in different arrangements of the enzyme molecules within the crystal). Later, the same group reported the incorporation of the dirhodium paddlewheel complex $Rh_2(OAc)_4$ in CLECs of bovine pancreatic RNase A (Loreto et al., 2023). These CLECs were chosen based on their preferential crystal form with pore sizes of 4 nm, and the metal complex was found to bind within these pores, coordinated again via histidine side chains. The heterogeneous catalyst was able to catalyze olefin cyclopropanation and self-coupling reactions of diazo compounds in aqueous media (Fig. 4.9B). As the CLECs are only required as microporous scaffolds in these cases, noncatalytic cross-linked protein crystals (CLPCs) can also be used instead, as demonstrated for the incorporation of Fe-based metal complexes in the periplasmic Ni-binding protein NikA of *Escherichia coli* (Lopez et al., 2017). In a very recent example, lysozyme CLECs were employed as swellable hydrogel beads to not only incorporate metal complexes but also enzyme molecules such as glucose oxidase, horseradish peroxidase, or *Burkholderia cepacia* lipase for immobilization purposes (Chung et al., 2025). Swelling of the lysozyme CLECs could be tuned by varying the NaCl concentration in the aqueous solution or through the reduction of disulfide bridges present in lysozyme. Moreover, the simultaneous adsorption of the transition metal catalyst iron complexed by a tetra-amido macrocyclic ligand (Fe-TAML) together with glucose oxidase in the lysozyme hydrogel beads was also achieved. This catalytic system was subsequently applied in a cascade reaction starting with glucose oxidation to provide the hydrogen peroxide required by Fe-TAML for guaiacol oxidation (Fig. 4.9C) (Chung et al., 2025). The authors could demonstrate that their hybrid catalyst was more active in this cascade reaction than a simple combination of Fe-TAML and glucose oxidase in solution.

## 4.5 Conclusion

The chemical cross-linking of enzyme molecules within a crystal is a superb enzyme immobilization technique, as it yields a highly stable and concentrated enzyme preparation. These CLECs withstand harsh conditions such as high temperature, pH extremes, as well as polar organic solvents. As such, they are ideally suited for application in biocatalysis, in biosensors, and beyond. Despite these great advantages of CLECs compared to other enzyme immobilization techniques, the scientific interest in the CLEC technology has diminished in recent years. This might mainly be caused by certain presumptions or prejudices, such as the requirement of highly purified proteins for crystallization, or that proteins are very difficult to crystallize in general. However, many examples exist in literature demonstrating that protein crystallization from crude extracts and on a large scale is possible. Commercial, ready-to-use crystallization screening kits facilitate the rapid identification of suitable crystallization conditions for an enzyme of interest. Simple enzyme engineering approaches, such as SER, can be applied to improve the crystallizability as well as the crystallization kinetics of an enzyme. Moreover, a broad range of chemical cross-linkers apart from GA is commercially available. Enzyme engineering can even be applied to install desired cross-linking sites on the enzyme surface, ensuring effective cross-linking with such alternative cross-linkers. Thus, nowadays, many options exist to facilitate

CLEC preparation for an enzyme of interest. Even though some possible applications of CLECs, which have been suggested in the literature (e.g., the use of CLECs as chiral separation matrices), have not prevailed, new fascinating applications have been added recently. This includes the use of cross-linked enzyme/protein crystals as scaffolds for the generation of heterogeneous biohybrid catalysts, and further potential applications of CLECs are likely still to come.

## References


Abe, S., Maity, B., & Ueno, T. (2016). Design of a confined environment using protein cages and crystals for the development of biohybrid materials. *Chemical Communications, 52*(39), 6496–6512. http://pubs.rsc.org/en/journals/journal/cc. doi:10.1039/c6cc01355d.

Abe, S., Maity, B., & Ueno, T. (2018). Functionalization of protein crystals with metal ions, complexes and nanoparticles. *Current Opinion in Chemical Biology, 43*, 68–76. http://www.elsevier.com/locate/cbi. doi:10.1016/j.cbpa.2017.11.015.

Abraham, T. E., Joseph, J. R., Bindhu, L. B. V., & Jayakumar, K. K. (2004). Crosslinked enzyme crystals of glucoamylase as a potent catalyst for biotransformations. *Carbohydrate Research, 339*(6), 1099–1104. doi:10.1016/j.carres.2004.01.011.

Amorim Fernandes, J. F., & Halling, P. J. (2002). Operational stability of high initial activity protease catalysts in organic solvents. *Biotechnology Progress, 18*(6), 1455–1457. doi:10.1021/bp020098g.

Ayala, M., Horjales, E., Pickard, M. A., & Vazquez-Duhalt, R. (2002). Cross-linked crystals of chloroperoxidase. *Biochemical and Biophysical Research Communications, 295*(4), 828–831. doi:10.1016/S0006-291X(02)00766-0.

Bar-Eli, A., & Katchalski, E. (1960). A water-insoluble trypsin derivative and its use as a trypsin column. *Nature, 188*(4753), 856–857. doi:10.1038/188856a0.

Bijelic, A., & Rompel, A. (2018). Polyoxometalates: More than a phasing tool in protein crystallography. *Chemical Texts, 4*(3). https://link.springer.com/journal/40828. doi:10.1007/s40828-018-0064-1.

Brady, D., Steenkamp, L., Skein, E., Chaplin, J. A., & Reddy, S. (2004). Optimisation of the enantioselective biocatalytic hydrolysis of naproxen ethyl ester using ChiroCLEC-CR. *Enzyme and Microbial Technology, 34*(3–4), 283–291. www.elsevier.com/locate/enzmictec. doi:10.1016/j.enzmictec.2003.11.002.

Chayen, N. E. (1998). Comparative studies of protein crystallization by vapour-diffusion and microbatch techniques. *Acta Crystallographica Section D: Biological Crystallography, 54*(1), 8–15. doi:10.1107/S0907444997005374.

Chung, J. S., Hartman, E. M., Mertick-Sykes, E. J., Pimentel, E. B., & Martell, J. D. (2025). Hyper-expandable cross-linked protein crystals as scaffolds for catalytic reactions. *ACS Applied Materials and Interfaces, 17*(1), 311–321. http://pubs.acs.org/journal/aamick. doi:10.1021/acsami.4c15051.

Clair, N. L., & Navia, M. A. (1992). Cross-linked enzyme crystals as robust biocatalysts. *Journal of the American Chemical Society, 114*(18), 7314–7316. doi:10.1021/ja00044a064.

Collins, A. M., Maslin, C., & Davies, R. J. (1998). Scale-up of a chiral resolution using cross-linked enzyme crystals. *Organic Process Research and Development, 2*(6), 400–406. http://pubs.acs.org/journal/oprdfk. doi:10.1021/op980032v.

Conejero-Muriel, M., Rodríguez-Ruiz, I., Verdugo-Escamilla, C., Llobera, A., & Gavira, J. A. (2016). Continuous sensing photonic lab-on-a-chip platform based on cross-linked enzyme crystals. *Analytical Chemistry, 88*(23), 11919–11923. http://pubs.acs.org/journal/ancham. doi:10.1021/acs.analchem.6b03793.

Conejero-Muriel, M., Rodríguez-Ruiz, I., Martínez-Rodríguez, S., Llobera, A., & Gavira, J. A. (2015). McCLEC, a robust and stable enzymatic based microreactor platform. *Lab on a Chip, 15*(20), 4083–4089. http://pubs.rsc.org/en/journals/journal/lc. doi:10.1039/c5lc00776c.

Cooper, D. R., Boczek, T., Grelewska, K., Pinkowska, M., Sikorska, M., Zawadzki, M., & Derewenda, Z. (2007). Protein crystallization by surface entropy reduction: Optimization of the SER strategy. *Acta Crystallographica. Section D, Biological crystallography., 63*(Pt 5), 636–645.

Costes, D., Wehtje, E., & Adlercreutz, P. (2001). Cross-linked crystals of hydroxynitrile lyase as catalyst for the synthesis of optically active cyanohydrins. *Journal of Molecular Catalysis – B Enzymatic, 11*(4–6), 607–612. doi:10.1016/S1381-1177(00)00057-6.

Durbin, S. D., & Feher, G. (1996). Protein crystallization. *Annual Review of Physical Chemistry, 47*(1), 171–204. doi:10.1146/annurev.physchem.47.1.171.

Fernandes, J. F. A., McAlpine, M., & Halling, P. J. (2005). Operational stability of subtilisin CLECs in organic solvents in repeated batch and in continuous operation. *Biochemical Engineering Journal, 24*(1), 11–15. www.elsevier.com/locate/bej. doi:10.1016/j.bej.2005.01.014.

Fernández-Penas, R., Verdugo-Escamilla, C., Martínez-Rodríguez, S., & Gavira, J. A. (2021). Production of cross-linked lipase crystals at a preparative scale. *Crystal Growth & Design, 21*(3), 1698–1707. doi:10.1021/acs.cgd.0c01608.

Filip, B., Kołodziej, M., Bochenek, R., Chutkowski, M., & Antos, D. (2024). Computational fluid dynamics for determining the interplay between stirring conditions and crystal size distribution in small laboratory devices. *Industrial and Engineering Chemistry Research, 63*(37), 16208–16219. http://pubs.acs.org/journal/iecred. doi:10.1021/acs.iecr.4c01816.

Fontes, N., Almeida, M. C., Garcia, S., Peres, C., Partridge, J., Halling, P. J., & Barreiros, S. (2001). Supercritical fluids are superior media for catalysis by cross-linked enzyme microcrystals of subtilisin Carlsberg. *Biotechnology Progress, 17*(2), 355–358. doi:10.1021/bp000148m.

Goldschmidt, L., Cooper, D. R., Derewenda, Z. S., & Eisenberg, D. (2007). Toward rational protein crystallization: A web server for the design of crystallizable protein variants. *Protein Science, 16*(8), 1569–1576. http://www.proteinscience.org/cgi/reprint/16/8/1569. doi:10.1110/ps.072914007.

Grob, P., Huber, M., Walla, B., Hermann, J., Janowski, R., Niessing, D., Hekmat, D., & Weuster-Botz, D. (2020). Crystal contact engineering enables efficient capture and purification of an oxidoreductase by technical crystallization. *Biotechnology Journal, 15*(11). http://onlinelibrary.wiley.com/journal/10.1002/(ISSN)1860-7314. doi:10.1002/biot.202000010.

Hebel, D., Huber, S., Stanislawski, B., & Hekmat, D. (2013a). Stirred batch crystallization of a therapeutic antibody fragment. *Journal of Biotechnology, 166*(4), 206–211. doi:10.1016/j.jbiotec.2013.05.010.

Hebel, D., Ürdingen, M., Hekmat, D., & Weuster-Botz, D. (2013b). Development and scale up of high-yield crystallization processes of lysozyme and lipase using additives. *Crystal Growth & Design, 13*(6), 2499–2506. doi:10.1021/cg400212p.

Hekmat, D., Hebel, D., Schmid, H., & Weuster-Botz, D. (2007). Crystallization of lysozyme: From vapor diffusion experiments to batch crystallization in agitated ml-scale vessels. *Process Biochemistry, 42*(12), 1649–1654. doi:10.1016/j.procbio.2007.10.001.

Hekmat, D. (2015). Large-scale crystallization of proteins for purification and formulation. *Bioprocess and Biosystems Engineering, 38*(7), 1209–1231. doi:10.1007/s00449-015-1374-y.

Hermann, J., Bischoff, D., Grob, P., Janowski, R., Hekmat, D., Niessing, D., Zacharias, M., & Weuster-Botz, D. (2021). Controlling protein crystallization by free energy guided design of interactions at crystal contacts. *Crystals, 11*(6), 588. doi:10.3390/cryst11060588.

Hermanson, G. T. (2013). *Bioconjugate techniques* (pp. 1–1146). Academic Press. doi:10.1016/C2009-0-64240-9.

Hetrick, E. M., Sperry, D. C., Nguyen, H. K., & Strege, M. A. (2014). Characterization of a novel cross-linked lipase: Impact of cross-linking on solubility and release from drug product. *Molecular Pharmaceutics, 11*(4), 1189–1200. http://pubs.acs.org/journal/mpohbp. doi:10.1021/mp4006529.

Jancarik, J., & Kim, S. H. (1991). Sparse matrix sampling: A screening method for crystallization of proteins. *Journal of Applied Crystallography, 24*(4), 409–411. doi:10.1107/s0021889891004430.

Jiménez-García, B., Elez, K., Koukos, P. I., Bonvin, A. M., Vangone, A., & Elofsson, A. (2019). PRODIGY-crystal: A web-tool for classification of biological interfaces in protein complexes. *Bioinformatics, 35*(22), 4821–4823. http://bioinformatics.oxfordjournals.org/. doi:10.1093/bioinformatics/btz437.

Jokela, J., Pastinen, O., & Leisola, M. (2002). Isomerization of pentose and hexose sugars by an enzyme reactor packed with cross-linked xylose isomerase crystals. *Enzyme and Microbial Technology, 31*(1–2), 67–76. doi:10.1016/S0141-0229(02)00074-1.

Kwan, T. O. C., Danson, A. E., Draper, P., Reardon, P., & Moraes, I. (2023). High-throughput protein crystallization via microdialysis. *Journal of Visualized Experiments, 2023*(193). www.jove.com. doi:10.3791/64744.

Lalonde, J. (1997). Practical catalysis with enzyme crystals. *Chemtech, 27*(2), 38–45.

Lalonde, J. J., Govardhan, C., Khalaf, N., Martinez, A. G., Visuri, K., & Margolin, A. L. (1995). Cross-linked crystals of *Candida rugosa* lipase: Highly efficient catalysts for the resolution of chiral esters. *Journal of the American Chemical Society, 117*(26), 6845–6852. doi:10.1021/ja00131a006.

Laothanachareon, T., Champreda, V., Sritongkham, P., Somasundrum, M., & Surareungchai, W. (2008). Cross-linked enzyme crystals of organophosphate hydrolase for electrochemical detection of organophosphorus compounds. *World Journal of Microbiology and Biotechnology, 24*(12), 3049–3055. doi:10.1007/s11274-008-9851-y.

Lee, T. S., Turner, M. K., & Lye, G. J. (2002). Mechanical stability of immobilized biocatatysts (CLECs) in dilute agitated suspensions. *Biotechnology Progress, 18*(1), 43–50. doi:10.1021/bp010131j.

Leisola, M., Jokela, J., Finell, J., & Pastinen, O. (2001). Simultaneous catalysis and product separation by cross-linked enzyme crystals. *Biotechnology and Bioengineering, 72*(4), 501–505. doi:10.1002/1097-0290(20010220)72:4⟨501::AID-BIT1013⟩3.0.CO;2-J.

Lopez, S., Rondot, L., Leprêtre, C., Marchi-Delapierre, C., Ménage, S., & Cavazza, C. (2017). Cross-linked artificial enzyme crystals as heterogeneous catalysts for oxidation reactions. *Journal of the American Chemical Society, 139*(49), 17994–18002. http://pubs.acs.org/journal/jacsat. doi:10.1021/jacs.7b09343.

Loreto, D., Maity, B., Morita, T., Nakamura, H., Merlino, A., & Ueno, T. (2023). Cross-linked crystals of dirhodium tetraacetate/RNase A adduct can be used as heterogeneous catalysts. *Inorganic Chemistry, 62*(19), 7515–7524. http://pubs.acs.org/journal/inocaj. doi:10.1021/acs.inorgchem.3c00852.

Lu, J., Wang, X. J., & Ching, C. B. (2002). Batch crystallization of soluble proteins: Effect of precipitant, temperature and additive. *Progress in Crystal Growth and Characterization of Materials, 45*(3), 201–217. doi:10.1016/S0960-8974(02)80003-1.

Luiz de Mattos, I. (2001). Evaluation of glucose biosensors based on Prussian Blue and lyophilised, crystalline and cross-linked glucose oxidases (CLEC®). *Talanta, 54*(5), 963–974. doi:10.1016/s0039-9140(01)00367-8.

Lusty, C. J. (1999). A gentle vapor-diffusion technique for cross-linking of protein crystals for cryocrystallography. *Journal of Applied Crystallography, 32*(1), 106–112. http://journals.iucr.org/j/issues/2006/03/00/isscont s.html. doi:10.1107/S002188989801053X.

Margolin, A. (1996). Novel crystalline catalysts. *Trends in Biotechnology, 14*(7), 223–230. doi:10.1016/0167-7799(96)10031-7.

Margolin A.L., Govardhan C.P., Visuri K.J., Uotila S.S. (2000). Patent Methods of preparing carbohydrate crosslinked glycoprotein crystals (US6500933B1). United States Patent and Trademark Office.

Margolin, A. L., & Navia, M. A. (2001). Protein crystals as novel catalytic materials. *Angewandte Chemie (International ed. in English), 40*(12), 2204–2222.

Mattson, G., Conklin, E., Desai, S., Nielander, G., Savage, M. D., & Morgensen, S. (1993). A practical approach to crosslinking. *Molecular Biology Reports, 17*(3), 167–183. doi:10.1007/bf00986726.

Migneault, I., Dartiguenave, C., Bertrand, M. J., & Waldron, K. C. (2004). Glutaraldehyde: Behavior in aqueous solution, reaction with proteins, and application to enzyme crosslinking. *BioTechniques, 37*(5), 790–802. https://www.future-science.com/journal/btn. doi:10.2144/04375rv01.

Nanev, C. N. (2007). Protein crystal nucleation: Recent notions. *Crystal Research and Technology, 42*(1), 4–12. doi:10.1002/crat.200610761.

Navarro, A., Wu, H. S., & Wang, S. S. (2009). Engineering problems in protein crystallization. *Separation and Purification Technology, 68*(2), 129–137. doi:10.1016/j.seppur.2009.05.002.

Navia M.A., St. Clair N.L. (1995). Patent Biosensors, extracorporeal devices and methods for detecting substances using crosslinked protein crystals (US6004768A). United States Patent and Trademark Office.

Neugebauer, P., & Khinast, J. G. (2015). Continuous crystallization of proteins in a tubular plug-flow crystallizer. *Crystal Growth & Design, 15*(3), 1089–1095. doi:10.1021/cg501359h.

Noritomi, H., Sasanuma, A., Kato, S., & Nagahama, K. (2007). Catalytic properties of cross-linked enzyme crystals in organic media. *Biochemical Engineering Journal, 33*(3), 228–231. doi:10.1016/j.bej.2006.10.024.

Nowotny, P., Hermann, J., Li, J., Krautenbacher, A., Klöpfer, K., Hekmat, D., & Weuster-Botz, D. (2019). Rational crystal contact engineering of *Lactobacillus brevis* alcohol dehydrogenase to promote technical protein crystallization. *Crystal Growth & Design, 19*(4), 2380–2387. doi:10.1021/acs.cgd.9b00067.

Pastinen, O., Jokela, J., Eerikäinen, T., Schwabe, T., & Leisola, M. (2000). Cross-linked glucose isomerase crystals as a liquid chromatographic separation material. *Enzyme and Microbial Technology, 26*(7), 550–558. doi:10.1016/S0141-0229(00)00137-X.

Pastinen, O., Visuri, K., & Leisola, M. (1998). Xylitol purification by cross-linked glucose isomerase crystals. *Biotechnology Techniques, 12*(7), 557–560. doi:10.1023/A:1008863700518.

Persichetti, R. A., Lalonde, J. J., Govardhan, C. P., Khalaf, N. K., & Margolin, A. L. (1996). *Candida rugosa* lipase: Enantioselectivity enhancements in organic solvents. *Tetrahedron Letters, 37*(36), 6507–6510. http://www.journals.elsevier.com/tetrahedron-letters/. doi:10.1016/0040-4039(96)01430-X.

Persichetti, R. A., Clair, N. L. S., Griffith, J. P., Navia, M. A., & Margolin, A. L. (1995). Cross-linked enzyme crystals (CLECs) of thermolysin in the synthesis of peptides. *Journal of the American Chemical Society, 117*(10), 2732–2737. doi:10.1021/ja00115a008.

Perveen, S., Noreen, S., Shahid, S., Mehboob, H., Aslam, S., Iqbal, H. M. N., & Bilal, M. (2022). Carrier-free cross-linked laccase crystals for biocatalytic degradation of textile industrial effluents. *Applied Biochemistry and Biotechnology, 194*(4), 1775–1789. doi:10.1007/s12010-021-03795-3.

Quiocho, F. A., & Richards, F. M. (1966). The enzymic behavior of carboxypeptidase A in the solid state. *Biochemistry, 5*(12), 4062–4076. doi:10.1021/bi00876a041.

Quiocho, F. A., Bishop, W. H., & Richards, F. M. (1967). Effects of changes in some solvent parameters on carboxypeptidase A in solution and in cross-linked crystals. *Proceedings of the National Academy of Sciences, 57*(3), 525–537. doi:10.1073/pnas.57.3.525.

Quiocho, F. A., & Richards, F. M. (1964). Intermolecular cross linking of a protein in the crystalline state: Carboxypeptidase A. *Proceedings of the National Academy of Sciences, 52*(3), 833–839. doi:10.1073/pnas.52.3.833.

RCSB PDB, (2026). Protein Data Bank. https://www.rcsb.org/.

Roy, J. J., Abraham, T. E., Abhijith, K. S., Kumar, P. V. S., & Thakur, M. S. (2005). Biosensor for the determination of phenols based on cross-linked enzyme crystals (CLEC) of laccase. *Biosensors and Bioelectronics, 21*(1), 206–211. doi:10.1016/j.bios.2004.08.024.

Roy, J. J., & Abraham, T. E. (2006a). Continuous biotransformation of pyrogallol to purpurogallin using cross-linked enzyme crystals of laccase as catalyst in a packed-bed reactor. *Journal of Chemical Technology and Biotechnology, 81*(11), 1836–1839. doi:10.1002/jctb.1612.

Roy, J. J., & Abraham, T. E. (2006b). Preparation and characterization of cross-linked enzyme crystals of laccase. *Journal of Molecular Catalysis B: Enzymatic, 38*(1), 31–36. www.elsevier.com/locate/molcatb. doi:10.1016/j.molcatb.2005.11.001.

Roy, J. J., & Abraham, T. E. (2004). Strategies in making cross-linked enzyme crystals. *Chemical Reviews, 104*(9), 3705–3721. doi:10.1021/cr0204707.

Rupp, B. (2009). *Biomolecular crystallography: Principles, practice, and application to structural biology*. CRC Press Inc.

Rupp, B. (2015). Origin and use of crystallization phase diagrams. *Acta Crystallographica. Section F, Structural Biology Communications, 71*(Pt 3), 247–260. doi:10.1107/S2053230ž1500374X.

Schmidt, S., Havekost, D., Kaiser, K., Kauling, J., & Henzler, H.-J. (2005). Crystallization for the downstream processing of proteins. *Engineering in Life Sciences, 5*(3), 273–276. doi:10.1002/elsc.200500116.

Schmit, J. D., & Dill, K. (2012). Growth rates of protein crystals. *Journal of the American Chemical Society, 134*(9), 3934–3937. doi:10.1021/ja207336r.

Simi, C. K., & Emilia Abraham, T. (2007). Encapsulation of crosslinked subtilisin microcrystals in hydrogel beads for controlled release applications. *European Journal of Pharmaceutical Sciences, 32*(1), 17–23. doi:10.1016/j.ejps.2007.05.003.

Smejkal, B., Helk, B., Rondeau, J. M., Anton, S., Wilke, A., Scheyerer, P., Fries, J., Hekmat, D., & Weuster-Botz, D. (2013). Protein crystallization in stirred systems-scale-up via the maximum local energy dissipation. *Biotechnology and Bioengineering, 110*(7), 1956–1963. doi:10.1002/bit.24845.

St Clair, N., Wang, Y. F., & Margolin, A. L. (2000). Cofactor-bound cross-linked enzyme crystals (CLEC) of alcohol dehydrogenase. *Angewandte Chemie (International ed. in English), 39*(2), 380–383.

Staar, M., Ahlborn, L., Estévez-Gay, M., Pallasch, K., Osuna, S., & Schallmey, A. (2024). A dynamic loop in halohydrin dehalogenase HheG regulates activity and enantioselectivity in epoxide ring opening. *ACS Catalysis, 14*(21), 15976–15987. http://pubs.acs.org/page/accacs/about.html. doi:10.1021/acscatal.4c04815.

Staar, M., Henke, S., Blankenfeldt, W., & Schallmey, A. (2022a). Biocatalytically active and stable cross-linked enzyme crystals of halohydrin dehalogenase HheG by protein engineering. *ChemCatChem, 14*(9), e202200145. http://onlinelibrary.wiley.com/journal/10.1002/(ISSN)1867-3899. doi:10.1002/cctc.202200145.

Staar, M., & Schallmey, A. (2023). Performance of cross-linked enzyme crystals of engineered halohydrin dehalogenase HheG in different chemical reactor systems. *Biotechnology and Bioengineering, 120*(11), 3210–3223. http://onlinelibrary.wiley.com/journal/10.1002/(ISSN)1097-0290. doi:10.1002/bit.28528.

Staar, M., Staar, S., & Schallmey, A. (2022b). Crystal contact engineering for enhanced cross-linking efficiency of HheG crystals. *Catalysts, 12*(12), 1553. doi:10.3390/catal12121553.

Tabe, H., Abe, S., Hikage, T., Kitagawa, S., & Ueno, T. (2014). Porous protein crystals as catalytic vessels for organometallic complexes. *Chemistry – An Asian Journal, 9*(5), 1373–1378. www.interscience.wiley.com. doi:10.1002/asia.201301347.

Takakura, T., Ito, T., Yagi, S., Notsu, Y., Itakura, T., Nakamura, T., Inagaki, K., Esaki, N., Hoffman, R. M., & Takimoto, A. (2006). High-level expression and bulk crystallization of recombinant L-methionine $\gamma$-lyase, an anticancer agent. *Applied Microbiology and Biotechnology, 70*(2), 183–192. doi:10.1007/s00253-005-0038-2.

Thomas, D. H., Rob, A., & Rice, D. W. (1989). A novel dialysis procedure for the crystallization of proteins. *Protein Engineering, Design and Selection,* 2(6), 489–491. doi:10.1093/protein/2.6.489.

Thomas, K. M, Kwon, S., & Lakerveld, R. (2021). Continuous protein crystallization in mixed-suspension mixed-product-removal crystallizers. *Crystal Growth & Design, 21*(2), 757–769. doi:10.1021/acs.cgd.0c00885.

Vilenchik, L. Z., Griffith, J. P., St. Clair, N., Navia, M. A., & Margolin, A. L. (1998). Protein crystals as novel microporous materials. *Journal of the American Chemical Society, 120*(18), 4290–4294. doi:10.1021/ja973449+.

Visuri K.J. (1985). Patent Stable glucose isomerase concentrate and a process for the preparation thereof (US4699882A). United States Patent and Trademark Office.

Visuri K.J. (1989). Patent cross-linked glucose isomerase (EP0341503A2). European Patent Office.

Walla, B., Bischoff, D., Janowski, R., von den Eichen, N., Niessing, D., & Weuster-Botz, D. (2021). Transfer of a rational crystal contact engineering strategy between diverse alcohol dehydrogenases. *Crystals, 11*(8), 975. doi:10.3390/cryst11080975.

Wang, Y. F., Yakovlevsky, K., Zhang, B., & Margolin, A. L. (1997). Cross-linked crystals of subtilisin: Versatile catalyst for organic synthesis. *Journal of Organic Chemistry, 62*(11), 3488–3495. doi:10.1021/jo9618334.

Wichmann, J., Mayer, J., Hintmann, M., Lukat, P., Blankenfeldt, W., & Biedendieck, R. (2023). Multistep engineering of a penicillin G acylase for systematic improvement of crystallization efficiency. *Crystal Growth & Design, 23*(5), 3230–3243. doi:10.1021/acs.cgd.2c01408.

Wong, S. S., & Wong, L. J. C. (1992). Chemical crosslinking and the stabilization of proteins and enzymes. *Enzyme and Microbial Technology, 14*(11), 866–874. doi:10.1016/0141-0229(92)90049-T.

Xu, K., & Klibanov, A. M. (1996). pH control of the catalytic activity of cross-linked enzyme crystals in organic solvents. *Journal of the American Chemical Society, 118*(41), 9815–9819. doi:10.1021/ja961596m.

Yan, E. K., Cao, H. L., Zhang, C. Y., Lu, Q. Q., Ye, Y. J., He, J., Huang, L. J., & Yin, D. C. (2015). Cross-linked protein crystals by glutaraldehyde and their applications. *RSC Advances, 5*(33), 26163–26174. http://pubs.rsc.org/en/journals/journalissues. doi:10.1039/c5ra01722j.

Zeppezauer, M., Eklund, H., & Zeppezauer, E. S. (1968). Micro diffusion cells for the growth of single protein crystals by means of equilibrium dialysis. *Archives of Biochemistry and Biophysics, 126*(2), 564–573. doi:10.1016/0003-9861(68)90443-8.

CHAPTER

# Cross-linked enzyme aggregates (CLEAs)

5

**Shilong Xing**[a] **and Yanjun Jiang**[b]

[a]*School of Chemical Engineering and Technology, Hebei University of Technology, Tianjin, China,*
[b]*National-Local Joint Engineering Laboratory for Energy Conservation in Chemical Process Integration and Resources Utilization, Hebei University of Technology, Tianjin, China*

## 5.1 Introduction

Supported by rapid advances in protein engineering, enzymatic catalysis has emerged as a viable alternative to established chemical catalysis in recent years. Leveraging its inherent sustainability, scalability, and cost-effectiveness, enzymatic catalysis finds broad application across diverse sectors, including pharmaceutical manufacturing, energy production, environmental remediation, and fine chemical synthesis (Bayer et al., 2025; Benítez-Mateos et al., 2022). However, the industrial application of free enzymes is hindered by inadequate stability, manifested as a pronounced propensity for activity loss under complex industrial conditions. Moreover, the water solubility of free enzymes makes it difficult to recover and reuse them from the reaction system. These disadvantages significantly increase the cost of enzymatic processes and hinder their large-scale application (Basso & Serban, 2019; Bié et al., 2022; Robescu & Bavaro, 2025).

To overcome these limitations, enzyme immobilization techniques were developed, as mentioned before, aiming to confine enzymes to a specific space, rendering them insoluble while retaining their catalytic activity, so that they can be easily recovered and reused (Guzik et al., 2014). In general, carrier-bound enzyme immobilization requires a solid carrier to load enzyme molecules, and enzymes are usually bound to the carrier by adsorption, entrapment, or covalent binding. Although carrier-bound immobilization has a long history and widespread use, it still has some inherent limitations. For example, the use of a carrier introduces additional noncatalytic material or mass, which leads to a lower unit volume activity and catalytic efficiency (Cao et al., 2003; Sheldon, 2011a). And the preparation of certain carriers is expensive and time-consuming.

In view of these inherent limitations of carrier-bound immobilization, researchers have developed a carrier-free immobilization technology. As described in Chapter 2, this method avoids the use of carriers and forms stable, insoluble catalytic entities through the cross-linking between enzyme molecules. Cross-linked enzyme crystals (CLECs) represent an early carrier-free immobilization method, originally proposed

**Carrier-Free Enzyme Immobilization: Fundamentals and Innovations in Biocatalysis.**
**DOI: https://doi.org/10.1016/B978-0-443-44139-4.00002-2**

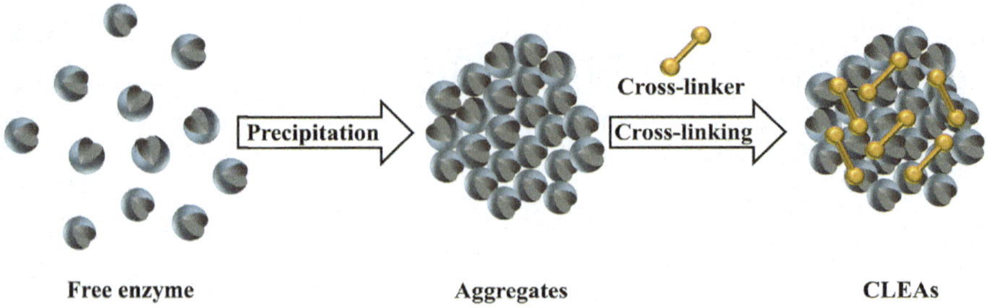


**FIGURE 5.1**

The preparation process of CLEAs. *CLEAs,* cross-linked enzyme aggregates.

in the 1960s to stabilize enzyme crystals for X-ray diffraction studies (Quiocho & Richards, 1964). By cross-linking enzyme crystals, their mechanical stability is increased while maintaining the activity and stability of the enzyme crystals (Velasco-Lozano et al., 2016). However, this technology requires purified enzymes to achieve the required enzyme crystals, thereby increasing the cost (Velasco-Lozano et al., 2022). Another submethod of carrier-free immobilization technology is cross-linked spray-dried enzymes (CSDEs), which are prepared by cross-linking spray-dried enzyme powders. Their dry form facilitates better storage and transportation. However, the loss of enzymatic activity induced by the spray-drying process poses a challenge, limiting the technology's widespread application (Bouguerra et al., 2024; Vivek et al., 2022). In search of a more cost-effective carrier-free immobilization strategy, an alternative technique—cross-linked enzyme aggregates (CLEAs)—was developed. The origins of this technology can be traced back to the 1960s, when Habeeb (1967) precipitated trypsin using ammonium sulfate and subsequently cross-linked it with glutaraldehyde (GA) to yield water-insoluble derivatives that retained activity. Avrameas and Ternynck (1969) further investigated the cross-linking of other proteins and utilized the derivatives prepared as immunoadsorbents. The technology was formally proposed and coined as "CLEAs" by Sheldon's group at the turn of this century (Cao et al., 2000). The preparation of CLEAs involves two simple steps (Fig. 5.1): the formation of physical aggregates by precipitation of enzymes and then cross-linking them such that a permanently insoluble and stable biocatalyst is formed. Because enzyme precipitation is a simple and well-established technique and allows the use of nonpurified enzymes, the entire preparation process of CLEAs is exceptionally straightforward and cost-effective. The CLEA technology was first commercialized by Professor Sheldon through the establishment of the company CLEA Technologies B.V. in 2002 (Sampaio et al., 2022). In addition, CLEAs have the advantages of high specific catalyst surface area and excellent operational stability (Chen et al., 2023a). Therefore, CLEAs technology has received wide attention since its birth, and has developed into one of the most important and widely studied carrier-free immobilization technologies.

Furthermore, the CLEAs technology demonstrates remarkable designability and flexibility. Through the co-precipitation and co-cross-linking of multiple enzymes, combined-CLEAs (combi-CLEAs) can be prepared for efficiently catalyzing multienzyme cascade reactions, enabling one-pot synthesis (Sheldon & Brady, 2018; Xu et al., 2018). This not only simplifies the process flow but may also enhance reaction efficiency by reducing mass transfer limitations. Similarly, by incorporating functional materials such as magnetic nanoparticles, magnetic CLEAs (M-CLEAs) can be produced, further facilitating the separation and recovery of the catalysts (de Sousa et al., 2020; Sheldon, 2019).

This chapter will provide a systematic elaboration of various aspects of CLEAs technology. First, an introduction will be given to the principles, advantages, and limitations of CLEAs, as well as the different parameters involved in their preparation and characterization. Also, a detailed description of combi-CLEAs for biocatalytic cascade reactions will be presented. Then, the applications of CLEAs across different fields will also be discussed. Finally, the current development and future challenges of CLEAs technology will also be discussed.

## 5.2 Principles, advantages, and limitations of cross-linked enzyme aggregates

The success of CLEAs technology lies in its simple yet clever design. It eliminates the need for carriers, instead utilizing enzyme molecules themselves as building blocks. Through the physical aggregation and cross-linking, a stable, highly efficient, and recyclable biocatalyst is produced. A profound understanding of the principle of CLEAs is essential for mastering, optimizing, and applying this technology.

### 5.2.1 The principle of cross-linked enzyme aggregates

The preparation of CLEAs involves two steps: precipitation and subsequent crosslinking.

The initial step in the preparation of CLEAs involves the formation of insoluble aggregates from enzymes dissolved in an aqueous solution. This is achieved by adding a precipitating agent to the aqueous system. Commonly used precipitating agents include salts, organic solvents, and nonionic polymers. The addition of a precipitating agent induces the aggregation of protein molecules by changing their hydration state or altering the electrostatic constant of the solution (Cui & Jia, 2015). The aggregates formed at this stage are maintained by noncovalent interactions and thus possess relatively unstable structures. Once the precipitating agent is removed or the solution conditions are altered, the aggregates may become soluble again.

Therefore, further cross-linking of aggregates is necessary to form stable, permanently insoluble immobilized enzymes (Kyomuhimbo & Brink, 2023).

Cross-linking of the aggregates is achieved by the use of cross-linkers. Different types of cross-linkers and their principles have been described in Chapter 3 and will not be repeated here. In summary, the cross-linking occurs between functional groups on the surface of enzyme molecules and oligomers of the cross-linker (Sheldon, 2011a). During this process, the preorganized superstructure and activity of the aggregates would be preserved (Nadar et al., 2016; Vivek et al., 2022), and the establishment of covalent interactions would prevent their dissociation and enhance robustness (Park et al., 2012; Sheldon et al., 2005).

### 5.2.2 Advantages and limitations of cross-linked enzyme aggregates

Based on the preparation process of CLEAs described above, simplicity and low cost are among the distinct advantages compared to other immobilization methods. Since precipitation is a commonly used protein purification method, the preparation of CLEAs can be regarded as combining purification and immobilization into a single operation (Reis et al., 2019). This significantly streamlines the manufacturing process and even enables the production of an immobilized enzyme directly from a crude enzyme extract. Furthermore, as a carrier-free immobilization method, CLEAs exhibit high volumetric activity and reaction yield, benefits derived from the absence of noncatalytic support materials. And, like other immobilization methods, CLEAs generally offer enhanced recyclability and significantly improved stability compared to their free enzymes, such as thermal stability, organic solvent tolerance, or pH stability (Ju et al., 2013; Kim & Kim, 2016; Sheldon et al., 2013; Tandjaoui et al., 2015). Another advantage of CLEAs technology lies in its designability and flexibility. For example, by co-precipitating multiple enzymes, it is easy to prepare combi-CLEAs for catalyzing cascade reactions; by incorporating magnetic nanoparticles, M-CLEAs that are easier to separate and recover can be prepared; and even by combining with other immobilization methods (such as adsorption and encapsulation), support materials for CLEAs can be provided, which can further improve their mechanical stability and provide larger particle size (Kim et al., 2007; Reshmi & Sugunan, 2013).

However, the CLEAs technology also has some limitations. For example, because of the low mechanical strength of CLEAs, they are difficult to withstand high shear stress and harsh conditions of stirring, which is not conducive to their application on an industrial scale (Del Arco et al., 2021; Ju et al., 2013). Moreover, in large-scale industrial applications, the size of CLEAs is a key parameter, as too small particle size increases the difficulty of filtration or centrifugation after the reactions, while too large particle size can lead to diffusion limitations, reducing its catalytic efficiency. However, controlling the size of CLEA particles is difficult (Velasco-Lozano et al., 2016).

**Table 5.1** The advantages and disadvantages of CLEAs.

| Feature | Advantages | Limitations |
|---|---|---|
| Enzyme loading and activity | Extremely high enzyme loading<br>High volume activity | Mass transfer limitations may exist for large substrate molecule |
| Preparation | Simple and economical | Parameters must be screened for each enzyme |
| Stability | Significantly enhanced higher stability against heat, protease and extreme pH | Low mechanical stability against shear stress and harsh conditions of stirring |
| Recovery | Easier to recover compared to free enzymes | Particle size and morphology can be irregular<br>difficult to fully recover |

CLEAs, *cross-linked enzyme aggregates.*

In addition to these already discussed, CLEAs also have other advantages and disadvantages, all of which are summarized in Table 5.1.

## 5.3 Characterization of cross-linked enzyme aggregates

Characterization of CLEAs is essential for enabling comparisons across studies and guiding rational process design and scale-up. Furthermore, the characterization of CLEAs is an important basis for optimizing the parameters of the preparation process. Therefore, the characterization parameters of CLEAs will be discussed first.

### 5.3.1 Structural property

The structural properties of CLEAs, particularly particle size and morphology, are critical to their functionality, impacting mass transfer, filterability, and overall catalytic efficiency. The typical particle size of CLEAs varies between 0.1 and 200 μm. In large-scale applications, smaller particles make filtration or centrifugation difficult after the reaction, while larger amorphous particles present mass transfer problems, which can reduce catalytic efficiency (Yamaguchi et al., 2011). Therefore, controlling the particle size requires a balance between these two factors. Methods for characterizing the structural properties of CLEAs typically include a laser diffraction particle size analyzer to determine the average particle diameter and size distribution of CLEAs, scanning and transmission electron microscopy to observe surface morphologies and aggregation, and Fourier Transform Infrared spectroscopy (FTIR) to analyze structural features and confirm cross-linking (Chen et al., 2023a; Morais et al., 2025). In combination, these methods provide a comprehensive structural assessment, guide the optimization of preparation parameters, and allow correlation of particle size and morphology with mass transfer behavior and filterability under operational conditions. Fig. 5.2 shows the morphologies of several CLEAs in the SEM image (Hong et al., 2021).

### 5.3.2 Catalytic performance

As catalysts, CLEAs are evaluated by their catalytic performance, an important characterization parameter directly linked to application feasibility and overall cost. Characterization of this performance is multifaceted and, as with free enzymes, typically encompasses activity, selectivity, and kinetic parameters. Immobilization as CLEAs can induce changes in enzyme conformation and in the micro-environment, which may alter multiple catalytic parameters (Aßmann et al., 2017); these effects are closely related to the preparation conditions. Accordingly, adjusting the preparation parameters of CLEAs can optimize catalytic activity and, in some cases, even change selectivity (Majumder et al., 2008; Wilson et al., 2006).

Activity recovery is a critical parameter in the process of enzyme immobilization, defined as the ratio of the activity observed after each processing step to the initial activity. It reflects the impact of the immobilization process on enzymatic function and serves as an important basis for evaluating the suitability of various process parameters. During the preparation of CLEAs, the assessment of activity recovery following each stage is of equal importance. The preparation of CLEAs comprises two steps: precipitation and cross-linking; numerous studies have utilized activity recovery as a primary indicator for optimizing the parameters involved in these steps. The enzyme precipitation step is pivotal to the final catalytic performance of the CLEAs. The efficacy of this step is intricately linked to numerous conditions, including the type and concentration of the precipitant, precipitation duration, temperature, and pH (Schoevaart et al., 2004; Sheldon, 2011a). The optimization of this step is usually grounded in the activity recovery measured following precipitation, such as the preparation of CLEAs for bromelain (Banerjee et al., 2020), $\beta$-galactosidase (Li et al., 2015), peroxidase (Tandjaoui et al., 2015), and acid urease (Zhang et al., 2016). In addition, precipitation efficiency is sometimes considered as well (Torabizadeh et al., 2014).

Activity recovery following cross-linking also serves as a key characterization parameter for CLEAs. It is generally defined as the ratio of the activity of the cross-linked CLEAs to that of the initial free enzyme, representing the overall enzymatic activity yield of the immobilization process. However, high activity recovery following the precipitation step does not necessarily guarantee high activity recovery

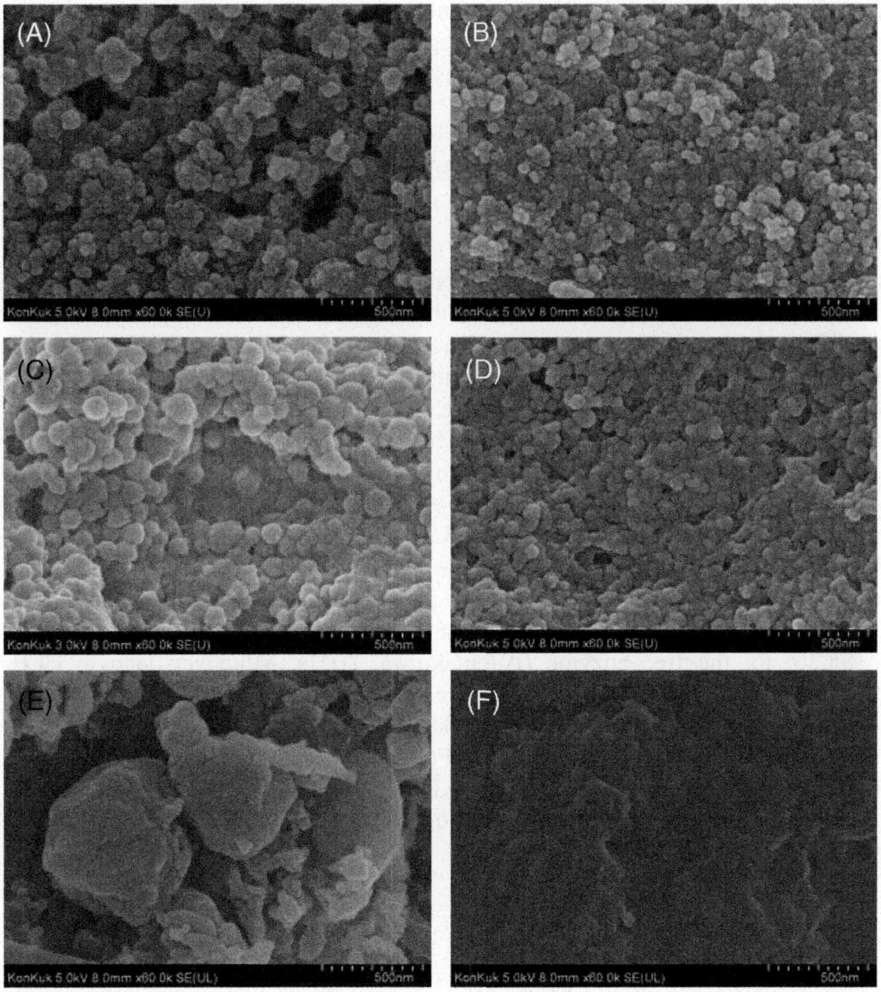


**FIGURE 5.2**

CLEAs morphology. SEM images of genipin-CLEAs (A), genipin-BSA-CLEAs (B), GA-CLEAs (C), GA-BSA-CLEAs (D), DP-CLEAs (E), and DP-BSA-CLEAs (F).

*Reproduced with permission from Hong, J., Jung, D., Park, S., Oh, Y., Oh, K. K., & Lee, S. H. (2021). Immobilization of laccase via cross-linked enzyme aggregates prepared using genipin as a natural cross-linker.* International Journal of Biological Macromolecules, 169*, 541–550. https://doi.org/10.1016/j.ijbiomac.2020.12.136.*

in the final cross-linked CLEAs. Consequently, some studies simultaneously monitor the post-cross-linking activity recovery during the screening of precipitation conditions to identify the optimal parameters. For instance, in the preparation of lipase CLEAs, Jiang et al. (2014a) concurrently determined the activity recovery both after precipitation and following cross-linking under various precipitant conditions. Similarly, Zhou et al. (2016) adopted this approach during the preparation of magnetic combined cross-linked enzyme aggregates combi-CLEAs of glucose oxidase (GOD) and horseradish peroxidase (HRP).

### 5.3.3 Stability

As an immobilization strategy, CLEAs are highly effective in enhancing enzyme stability. In contrast to physical adsorption or entrapment, the formation of covalent bonds between enzyme molecules and the cross-linker confers greater molecular rigidity and enhanced resistance to tertiary structure unfolding (Hojnik Podrepšek et al., 2019). Consequently, CLEAs typically exhibit superior performance under various denaturing conditions, including high temperatures, extreme pH, organic solvents, and the presence of proteases. Therefore, numerous studies have characterized the stability of CLEAs. For instance, nitrilase CLEAs remained stable at 4°C for 75 h, whereas the free enzyme lost 90% of its activity over the same period. Furthermore, the thermal stability of the CLEAs was enhanced several-fold at 25°C, 37°C, and 45°C compared to the soluble enzyme (Kumar et al., 2010). Similarly, laccase CLEAs demonstrated improved pH tolerance and thermal stability; after incubation at pH 5.5 for 8 h, the CLEAs retained over 95% of their activity, compared to 85.2% for the free enzyme (Ademakinwa, 2021). Moreover, laccase CLEAs prepared using natural serum as an additive (CLEAs-egg) exhibited not only enhanced thermal stability but also improved resistance to denaturants (urea and guanidine hydrochloride) and proteases (trypsin). Following incubation in 1 M GndHCl and 8 M urea for 1 h, the CLEAs-egg retained 68.5% and 91.1% of their initial activity, respectively, whereas the free enzyme retained only 53.4% and 81.3% (Jiang et al., 2014a).

### 5.3.4 Reusability

As immobilized enzymes, CLEAs exhibit reusability, which is another important property. Their insolubility facilitates separation and recovery from the reaction medium, thereby simplifying the downstream processing and is key to reducing process costs. In general, catalytic activity declines as the number of reuse cycles increases. This decrease can arise from several factors, including conformational changes caused by prolonged mechanical stirring, centrifugation, and filtration (Nawawi et al., 2020); notably, CLEAs show relatively low resistance to mechanical stress, which makes activity loss during these steps more likely. In addition, the leaching of the enzyme into the reaction medium is another cause of inactivation (Morshed et al., 2019). The rate of activity loss directly impacts costs, making it necessary to evaluate the reusability of CLEAs under repeated operations. Typically, this parameter is characterized under optimum reaction conditions by measuring the activity after each cycle.

The reusability of CLEAs is intrinsically linked to the parameters employed during the precipitation and cross-linking steps, as these factors dictate the structural characteristics of the CLEAs. Podrepšek et al. (2019) compared the reusability of cellulase CLEAs prepared with different precipitants. When ammonium sulfate was used, the CLEAs reached their half-life after 5 hydrolysis cycles. In contrast, the use of propanol significantly extended operational stability, with the half-life reached after 7 cycles. CLEAs prepared with methanol and ethanol exhibited even greater durability; after 7 cycles, the residual activities were 59% and 58%, respectively. Furthermore, cross-linking conditions also influence reusability. For instance, lipase CLEAs cross-linked with ethylene glycol-bis(succinic acid N-hydroxysuccinimide) (EG-NHS) demonstrated superior reusability compared to those prepared with GLA. After 5 reuse cycles, the EG-NHS-CLEAs retained 84.22% ± 3.96% of their initial activity, whereas the GLA-CLEAs retained 80.14% ± 2.57% (Rehman et al., 2016). In addition, the reusability of CLEAs is influenced by other factors, including the use of additives and various conditions within the precipitation and cross-linking steps. Table 5.2 summarizes the reusability of CLEAs prepared under various parameters.

## 5.4 Preparation of cross-linked enzyme aggregates

Although the preparation of CLEAs is relatively straightforward, it is necessary and tedious to optimize the various parameters involved in the preparation process to maximize their catalytic performance. Furthermore, as the optimum parameters can vary depending on the specific enzyme, tailored optimization is required for each individual enzyme (Roessl et al., 2010). At present, this optimization is largely empirical (Sheldon, 2007). This section will explore the optimization of preparation parameters and their influence on the performance of CLEAs, illustrated by specific research examples.

### 5.4.1 Precipitation of enzymes

Enzyme precipitation is the first step in the preparation of CLEAs, and this step is also considered an important factor affecting the activity recovery (Cui et al., 2013). The selection of precipitant is a primary consideration; in general, the precipitants used fall into three categories: inorganic salts, organic solvents, and polymers. The selection of precipitant depends on the enzyme and usually requires specific screening (Blanco-Llamero et al., 2021), for which activity recovery is the most commonly used screening basis. In addition, the precipitation step can also affect the structural property of CLEAs, so sometimes the particle size is also taken as a basis (Yu et al., 2006). Here, we provide some concrete examples of precipitant use, describing the selection for different enzymes and the observed effects, with the aim of offering a reference for the selection and use of precipitants in the preparation of CLEAs.

Inorganic salts are commonly employed as precipitants during the preparation of CLEAs, particularly ammonium sulfate. Numerous studies have selected ammonium

**Table 5.2** The reusability of CLEAs prepared under various parameters.

| Enzyme | Precipitants | Cross-linkers | Cross-linking time | Additives | Reusability | References |
|---|---|---|---|---|---|---|
| Cellulase | Ammonium sulfate | GA | | | Half-life reached after 5 cycles | Hojnik Podrepšek et al. (2019) |
| | Propanol | | | | Half-life reached after 7 cycles | |
| | Methanol | | | | 59% residual activity after 7 cycles | |
| | Ethanol | | | | 58% residual activity after 7 cycles | |
| | Polyethylene glycol | | 2 h | | 40% residual activity after 4 cycles | Perzon et al. (2017) |
| | Ammonium sulfate | | 2 h | | 10% residual activity after 1 cycles | |
| Lipase | Ammonium sulfate | EG-NHS | 150 min | | 84.22% ± 3.96% residual activity after 5 cycles | Rehman et al. (2016) |
| | | GA | | | 80.14 ± 2.57% residual activity after 5 cycles | |
| | Adsorbed onto octyl-functionalized mesocellular foams | Oxidized sodium alginate | 1.5 h | | Maintain high FAME yield after 5 repeated cycles (from 89% to 78%) | Jin et al. (2019) |
| | Ammonium sulfate | Dextran polyaldehyde | 2 h | | 52% residual activity after 8 cycles | Zhou et al. (2017) |
| | Ammonium sulfate | GA | 2 h | BSA | 75% residual activity after 8 cycles | Cui et al. (2016) |
| Acylase | Ammonium sulfate | GA | 2 h | Poly-lysine | 97% residual activity after 20 cycles | Ye et al. (2019) |
| Cellobiose 2-epimerase | Ammonium sulfate | GA | 1 h | Fructose or sucrose | About 60% residual activity after 10 cycles | Wang et al. (2018) |

sulfate as the optimal precipitant, for example, the preparation of CLEAs from lipase (Jiang et al., 2014a), α-amylase (Ullah et al., 2021), urease (Zeinali & Lenjannezhadian, 2018), glycerol dehydrogenase and reduced nicotinamide adenine dinucleotide (NADH) oxidase (Xu et al., 2020), bromelain (Banerjee et al., 2020), Acetyltransferase (You et al., 2019), and degrading enzymes (Bian et al., 2019). In addition, other inorganic salts, such as sodium sulfate, magnesium sulfate, and potassium phosphate, have also been screened as candidate precipitants (Sheldon, 2019; Yu et al., 2006), although they generally perform less effectively than ammonium sulfate.

Empirically, the use of organic solvents as precipitants is more likely to lead to enzyme inactivation. Nevertheless, the preparation of CLEAs using organic solvents has been widely reported. For example, ethanol has been employed for papain (Chen et al., 2017; Wang et al., 2010), cellulase (Hojnik Podrepšek et al., 2019), and lipase (Kartal et al., 2011); tert-butanol for penicillin G acylase (Kopp et al., 2014) and α-amylase (Torabizadeh et al., 2014); acetone for catalase (Jannat & Yang, 2020), and cutinase (Nikolaivits et al., 2017); isopropanol for lipase B (Diaz-Vidal et al., 2019).

Beyond inorganic salts and organic solvents, certain polymers have also been used as precipitants. Polyethylene glycol (PEG), in particular, has been demonstrated to stabilize the structure of certain enzymes (Prabhavathi Devi et al., 2009; Taboada-Puig et al., 2011; Xu et al., 2011). For the preparation of CLEAs for certain enzymes, using PEG as a precipitant may result in a higher activity recovery or enhanced stability. Representative examples include the immobilization of laccase (Bilal et al., 2021) and cellulase (Perzon et al., 2017).

In addition to selecting the precipitant type, adjusting the concentration of each precipitant can improve precipitation efficiency while maintaining high activity recovery, and thus is also treated as an optimization parameter. Table 5.3 summarizes the precipitants considered, their concentrations, precipitation efficiencies, and activity recovery. In addition, the precipitant identified as optimal during the precipitation step does not necessarily yield the best CLEAs after cross-linking (Sheldon, 2011a); therefore, screening precipitants on the basis of the performance of the CLEAs obtained after cross-linking provides a more rational design, albeit at the cost of increased experimental workload.

In the precipitation step, in addition to screening different precipitants and optimizing their concentrations, numerous factors can influence precipitation performance and the properties of the CLEAs ultimately obtained. These variables include the precipitation temperature and pH, the rate of precipitant addition, the precipitation time, and the stirring speed. All of these parameters should be considered during process design to produce CLEAs with superior performance.

### 5.4.2 Cross-linking of aggregates

After enzyme precipitation, the aggregates are stabilized by covalent cross-linking to yield a permanently insoluble, reusable biocatalyst. This step is critical for defining the catalytic efficiency, stability, and morphological characteristics of the CLEAs. Among available cross-linkers, glutaraldehyde (GA) is most widely used owing

**Table 5.3** A brief summary of the use of different precipitants.

| Precipitant type | Precipitants | Enzyme | Precipitant concentration (%) | Precipitation efficiency (%) | Activity recovery (%) | References |
|---|---|---|---|---|---|---|
| Ammonium sulfate | Ammonium sulfate | Lipase | 100 | | 97 | Jiang et al. (2014) |
| | Ammonium sulfate | $\alpha$-amylase | 60 | 63 | | Ullah et al. (2021) |
| | Ammonium sulfate | Urease | 80 | | | Zeinali and Lenjannezhadian (2018) |
| | Ammonium sulfate | Glycerol dehydrogenase | 100 | 98.75 | 127.14 | Xu et al. (2020) |
| | Ammonium sulfate | NADH oxidase | 60 | 92.66 | 136.84 | Xu et al. (2020) |
| | Ammonium sulfate | Bromelain | 60 | | 95 | Banerjee et al. (2020) |
| | Ammonium sulfate | Acetyltransferase | 60 | About 90 | 98.3 | You et al. (2019) |
| | Ammonium sulfate | Degrading enzymes | 70 | | 90 | Bian et al. (2019) |
| Organic solvents | Ethanol | Papain | 100 | | 97.4 | Wang et al. (2010) |
| | Ethanol | Papain | | | 94.58 | Chen et al. (2017) |
| | Ethanol | Cellulase | 90 | | 94 | Hojnik Podrepšek et al. (2019) |
| | Ethanol | Lipase | 90 | | 71 | Kartal et al. (2011) |
| | Tert-butanol | $\alpha$-amylase | 90 | 99.54 | | Torabizadeh et al. (2014) |
| | Acetone | Catalase | 25 | | 60 | Jannat and Yang (2020) |
| | Acetonitrile | Cutinase | 90 | | 75 | Nikolaivits et al. (2017) |
| | isopropanol | Lipase B | 90 | | 65.2 | Diaz-Vidal et al. (2019) |
| Polymers | PEG | Laccase | | | 72.85 | Bilal et al. (2021) |
| | PEG | Cellulase | 50 | | 110 | Perzon et al. (2017) |

to its low cost, high reactivity, and commercial availability. GA reacts primarily with the ε-amino groups of surface lysine residues via Schiff's base formation and Michael-type 1,4-addition, generating stable inter- and intramolecular bridges (Sheldon, 2011a). Glutaraldehyde has been extensively employed to prepare CLEAs of diverse enzymes, including α-l-rhamnosidase (Alvarenga et al., 2014), tyrosinase (Aytar & Bakir, 2008), lipase (Gupta et al., 2009), trypsin (Menfaatli & Zihnioglu, 2015), Glucoamylase (Li et al., 2016), and so on.

Although GA has a wide range of applications, its small molecular size and cytotoxicity limit its applicability in food and biomedical processes (Dong et al., 2010; Gao et al., 2015). Moreover, excessive cross-linking with GA can modify essential amino groups near the active site, leading to a significant loss of activity. To overcome these issues, several alternative cross-linkers have been explored. For instance, dextran polyaldehyde, a macromolecular dialdehyde, has been used to cross-link enzymes like lipase, penicillin acylase (Mateo et al., 2004), and nitrile hydratase (Zhou et al., 2017). This cross-linker provides larger pore structures, which reduce diffusion limitations and minimize active-site obstruction. CLEAs of lipase from *Burkholderia cepacia* prepared using dextran polyaldehyde (100 to 200 kDa) exhibited higher hydrolytic activity and stability than those prepared using GA (Caballero Valdés et al., 2011). Similarly, oxidized polysaccharides such as oxidized sodium alginate (Jin et al., 2019) or dialdehyde starch (Abd Rahman et al., 2021; Lee et al., 2025) have been successfully used as eco-friendly alternatives. For example, Jin et al. (2019) used oxidized sodium alginate as an alternative to GA, yielding CLEAs of *Candida antarctica* lipase B (CALB) with enhanced thermal stability and tolerance to organic solvents. In addition, genipin can react with primary amine groups under mild conditions and exhibits excellent biocompatibility and low toxicity, and has therefore been used as a natural cross-linker for proteins (Arteche Pujana et al., 2013; Klein et al., 2016). Genipin has been successfully used to prepare CLEAs of Trametes versicolor laccase; compared with counterparts prepared using GA or dextran polyaldehyde, CLEAs prepared by genipin exhibit smaller spherical particles and higher substrate affinity (Hong et al., 2021). Although macromolecular or otherwise milder cross-linkers have been associated with smaller activity losses for many enzymes, they are not universally applicable, and for some enzymes, GA remains the preferable choice. Consequently, the choice of cross-linker is enzyme-dependent. Nonetheless, these alternatives offer clear advantages in biocompatibility and environmental compatibility with low toxicity.

When the number of reactive ε-amino groups on an enzyme is limited, cross-linkers that react with phenolic and thiol residues can be used, such as *p*-benzoquinone (Wang et al., 2011), or reagents that exhibit potential reactivity toward free carboxyl groups, including EG-NHS, polyethyleneimine (PEI), and chitosan, can be considered. Rehman et al. (2016) prepared CLEAs of Pencillium notatum lipase (PNL) using EG-NHS and, relative to GA, observed a smaller effect on activity recovery together with a smaller size and a larger surface area. Carbodiimide activation of enzyme carboxyl groups followed by cross-linking with PEI or chitosan has likewise been used successfully to prepare CLEAs of lipases

(Velasco-Lozano et al., 2014) and laccases (Arsenault et al., 2011). In addition to covalent strategies, the mild and safe ionic cross-linker sodium tripolyphosphate (TPP) has recently been applied to CLEAs of *Aspergillus niger* lipase, obtaining maximum activity recovery (99.5% $\pm$ 0.634%) and cross-linking yield (88.4% $\pm$ 0.46%) under optimal process conditions, illustrating little effect on enzyme activity (Qian et al., 2020). Collectively, these newer cross-linkers broaden the options available for the CLEA cross-linking step.

In addition to the intrinsic properties of the cross-linker, the cross-linker concentration (or the cross-linker-to-enzyme molar ratio) and the cross-linking time are key determinants of the extent of cross-linking and directly influence the structural properties, stability, and catalytic performance of CLEAs (Cruz et al., 2012; Kim et al., 2013). Insufficient amount of cross-linker yields loosely bound aggregates prone to leaching, whereas excessive cross-linking produces overly rigid networks that restrict conformational flexibility and impede substrate diffusion (Hormigo et al., 2012). Accordingly, the extent of cross-linking should be controlled by tuning the cross-linker concentration (or the cross-linker-to-enzyme molar ratio) together with the cross-linking time, so as to strike an appropriate balance between stability and activity. In many cases, the optimization of the dosage of cross-linker and the cross-linking time both present a "bell-shaped curve." For example, in a study on cellulase CLEAs, both GA concentration and cross-linking time were optimized; the highest activity recovery was obtained at 2% (v/v) GA, with lower or higher concentrations leading to decreased activity, and the maximum activity recovery likewise occurred at a cross-linking time of 3.5 h (Li et al., 2022). Optimization of these parameters should also consider their impact on stability, since a greater degree of cross-linking may enhance stability even as it diminishes catalytic performance, and trade-offs should be made according to the requirements of the practical application. Table 5.4 summarizes the key parameters involved in the cross-linking process reported in the above studies.

In addition to the parameters discussed above, several additional factors influence the performance of CLEAs. The cross-linking reaction is typically pH dependent: the pH maintained during cross-linking affects both structure and catalytic behavior. For *Candida rugosa* lipase, alkaline conditions produced CLEAs with larger particle sizes, and because activity is highly sensitive to particle size, the highest activity was observed at 40–50 μm (Yu et al., 2006). Temperature likewise affects cross-linking outcomes. Higher temperatures accelerate the reaction and shorten the time required, so temperature and cross-linking time should be optimized in tandem, with consideration of the enzyme's intrinsic thermal stability. Control of mixing method and intensity during cross-linking can also prevent excessive aggregation without compromising structural integrity, thereby improving mass transfer and increasing activity; for example, Montoro-García et al. (2010) reported that replacing short vortexing with a reciprocating cell disruptor during the preparation of acetyl xylan esterase CLEAs yielded preparations with higher activity. Finally, the period of time left between the enzyme precipitation and the cross-linking step may influence the structural organization of the resulting CLEAs and can also be treated as a parameter for optimization (Pchelintsev et al., 2009).

**Table 5.4** A summary of the parameters involved in the cross-linking process.

| Reactive group | Cross-linkers | Enzyme | Cross-linking time | Cross-linker/ aldehyde group concentration | Activity recovery (%) | References |
|---|---|---|---|---|---|---|
| Amino groups | GA | *α*-L-Rhamnosidase | 1 h | 2% (v/v) | 81.5 | Alvarenga et al. (2014) |
| | GA | Tyrosinase | 3 h | 2% (v/v) | About 100 | Aytar and Bakir (2008) |
| | GA | Lipase | 12–14 h | About 6.25% (v/v) | 68.7 | Gupta et al. (2009) |
| | GA | Trypsin | 0.5 h | 1% (v/v) | | Menfaatli and Zihnioglu (2015) |
| | GA | Glucoamylase | 2 h | 0.5% (v/v) | 102 | Li et al. (2016) |
| | Dextran polyaldehyde | Penicillin G Acylase | 16 h | | 85–90 | Mateo et al. (2004) |
| | Dextran polyaldehyde | Lipase | 4 h | 1.7 g/L | | Caballero Valdés et al. (2011) |
| | Dextran polyaldehyde | nitrile hydratase | 2 h | 2.5 g/L | 49.64 | Zhou et al. (2017) |
| | Oxidized sodium alginate | Lipase | 1.5 h | 0.6 mM | | Jin et al. (2019) |
| | Dialdehyde starch | Levanase | 4 h | 0.8% (v/v) | 81.6 | Abd Rahman et al. (2021) |
| | Dialdehyde starch | *Is*PETase | 0.5 h | 0.5% (v/v) | 74.3 | Lee et al. (2025) |
| | Genipin | Laccase | 12 h | 1.8 mM | 18 | Hong et al. (2021) |
| | *p*-benzoquinone | Lipase | 2.5 h | 5 mM | | Wang et al. (2011) |
| Carboxyl groups | EG-NHS | Lipase | 2.5 h | 150 mM | 64.42 | Rehman et al. (2016) |
| | PEI | Lipases | 0.5 h | 3 g/L | 12 | Velasco-Lozano et al. (2014) |
| | Chitosan | Laccases | 24 h | 1.867 g/L | | Arsenault et al. (2011) |
| Ionic cross-linking | TPP | Lipases | 2 h | 8 mM | 99.5 | Qian et al. (2020) |

### 5.4.3 The use of additives

Apart from optimizing and controlling the parameters involved in precipitation and cross-linking, incorporating additives during these steps can enhance cross-linking efficiency or protect the enzyme's structure, thereby improving the performance of the CLEAs ultimately obtained.

Many commonly used cross-linkers react with amino groups on the enzyme surface; accordingly, enzymes with low amino group content often exhibit low cross-linking efficiency (Amaral-Fonseca et al., 2018). To overcome this limitation, a second protein can be incorporated as an amino group donor to promote the cross-linking reaction and improve activity recovery (Shah et al., 2006). Bovine serum albumin (BSA) is the most commonly used amino group feeder. For example, during the preparation of sucrose isomerase CLEAs, adding BSA at an appropriate concentration increased activity recovery from 19.5% to 31.2%, indicating a strong protective effect on the activity recovery of SIase. However, the amount of BSA added should not be excessive, because excessive BSA can reduce the number of cross-linked active enzymes and thereby dilute the activity of CLEAs (Chen et al., 2023b). Similar protective effects of BSA have been reported for many other enzymes, including xylanase (Abdul Wahab et al., 2019) and bovine pancreatic lipase (Cui et al., 2016). In addition, Hen egg white (HEW), which is rich in lysine residues, can also be utilized as an additive in the CLEA preparation process (Jiang et al., 2014b; Karimpil et al., 2011). Beyond the use of secondary proteins, the addition of free amino acids can also enhance cross-linking efficiency and consequently improve stability. For example, the addition of Arg, Lys, Ala, or Asp has been shown to enhance the stability of the resulting CLEAs (Mukherjee et al., 2016).

Beyond BSA, the addition of certain polymers can also provide more active amino groups. For example, trehalose synthase CLEAs were prepared using PEI and polyethylene glycol (PEG) as precipitants. The incorporation of PEI or PEG significantly enhanced both the pH tolerance and thermal stability of the resulting CLEAs (Zheng et al., 2014). And poly-lysine is another effective additive; its addition significantly increased the retained activity of penicillin G acylase in CLEAs (Ye et al., 2019). Apart from the addition of exogenous amino donors, enriching the surface amino content of enzymes via protein engineering serves as an alternative strategy. For instance, site-directed mutagenesis has been employed to increase the number of free amino groups in xylanase; the resulting CLEAs exhibited superior stability and reusability compared to those derived from the wild-type enzyme (Wahab et al., 2019).

In general, sugars can serve as protein protective agents by preserving the crucial water layer on the surface of protein and thereby protecting the protein secondary structure. Thus, sugars are often used as protective additives during the preparation of CLEAs. For example, Wang et al. (2018) developed a sugar-assisted strategy for cellobiose 2-epimerase CLEAs, in which activity recovery increased from 43.85% ± 2.56% of sugar-free CLEAs (Bla-CLEAs) to 64.82% ± 0.94% of fructose-assisted CLEAs (Fru-CLEAs) and 61.89% ± 3.64% of sucrose-assisted CLEAs (Suc-CLEAs). Trehalose decorated poly(amidoamine) (PAMAM) dendrimers have also been used to

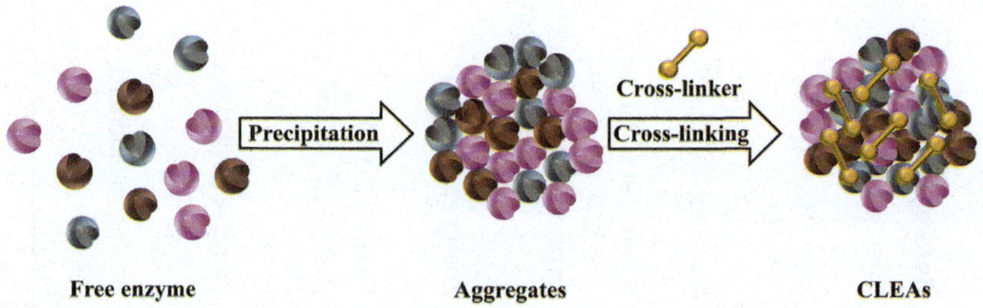


**FIGURE 5.3**

Preparation of combi-CLEAs.

stabilize combi-CLEAs of $\beta$-Galactosidase ($\beta$-Gal) and Glucose Isomerase, enhancing the high-temperature performance of $\beta$-Gal (Scott & Goddard, 2025). Another advantage of using sugars as additives is that they do not cause activity dilution. Table 5.5 summarizes the use of the additives described above.

## 5.5 Preparation of combi-cross-linked enzyme aggregates

By co-aggregating two or more enzymes and cross-linking them into a single CLEAs particle (Fig. 5.3), combined cross-linked enzyme aggregates (combi-CLEAs) that integrate multiple catalytic activities can be prepared ( Sheldon, 2019; Xu et al., 2018). Compared with CLEAs of a single enzyme, combi-CLEAs offer several advantages. For example, they can catalyze multistep consecutive reactions in one bioreactor without isolating intermediates, thereby reducing unit operations, solvent and reagent consumption, and by-product formation. They can also consume and convert intermediates in situ, shifting otherwise unfavorable equilibria toward products, which is particularly beneficial when certain intermediates are unstable (Gaggero & Albanese, 2021). Consequently, when cascade transformations are required or when a cofactor regeneration module must be incorporated, it is often preferable to employ combi-CLEAs. Although their preparation follows the same general principles as for single-enzyme CLEAs, the co-precipitation and cross-linking of multiple enzymes introduce additional considerations, including the ratio among enzymes and their compatibility with precipitation and cross-linking conditions (Periyasamy et al., 2016). This section outlines the additional parameters and influencing factors that merit attention in the preparation of combi-CLEAs.

### 5.5.1 Proportion of enzymes

During the preparation of combi-CLEAs, the proportion of every enzyme is an important parameter that exerts a pronounced influence on the overall efficiency of the cascade reaction, since each enzyme has its own catalytic rate. In most cases,

**Table 5.5** Summary of additives used in the preparation of CLEAs.

| Types | Additives | Enzyme | Effect | References |
|---|---|---|---|---|
| Albumin | BSA | sucrose isomerase | Increased activity recovery from 19.5% to 31.2% | Chen et al. (2023a) |
| | BSA | Xylanase | Thermal stability increased to 1.39 times that of the control, and reusability improved | Abdul Wahab et al. (2019) |
| | BSA | Lipase | The residual activity after eight reuse cycles increased from 20% to 75% | Cui et al. (2016) |
| | Hen egg white | Lipase; laccase | Higher stability compared to free enzymes | Karimpil et al. (2011); Jiang et al. (2014b) |
| | amino acid (Arg, Lys, Ala, or Asp) | Lipase or protease | Higher thermal stability and organic solvent tolerance than those without added amino acids | Mukherjee et al. (2016) |
| Polymers | PEI | Trehalose synthase | Significantly enhanced thermostability and pH tolerance | Zheng et al. (2014) |
| | PEG | | | |
| | Poly-lysine | Acylase | Improved activity recovery to ≈ 96% | Ye et al. (2019) |
| Sugars | Fructose | Cellobiose 2-epimerase | Activity recovery increased from 43.85% ± 2.56% to 64.82% ± 0.94% | Wang et al. (2018) |
| | Sucrose | Cellobiose 2-epimerase | Activity recovery increased from 43.85% ± 2.56% to 61.89% ± 3.64% | Wang et al. (2018) |
| | Trehalose | β-Galactosidase and Glucose Isomerase | β-Gal activity at high temperature increased by 58.6%, and stability improved 2.26-fold | Scott and Goddard (2025) |

CLEAs, *cross-linked enzyme aggregates.*

the optimal ratio cannot be reliably inferred from the activities of the free enzymes because multienzyme systems exhibit complex synergistic effects and the activities may shift upon conversion to CLEAs. Consequently, the optimal ratio is typically identified empirically through screening. For example, in combi-CLEAs of glucose oxidase (GOx) and HRP, weight ratios of GOx to HRP in the range of 1 to 600 were examined to determine the best proportion. When the GOx/HRP ratio was less than 150, HRP was present in excess and the overall reaction rate was limited by the GOx-catalyzed step. When the ratio was higher than 150, GOx became excessive and further increases in the ratio produced little change in the overall rate; when the ratio was higher than 400, the overall rate declined. The optimal ratio was therefore set at 150 (Nguyen & Yang, 2017). Beyond overall reaction efficiency, other criteria may also be considered. For the combi-CLEAs of levansucrase and inulosucrase for producing short to medium chain length levan-type fructooligosaccharides (LFOS), adjusting the enzyme ratio showed that combi-CLEAs prepared from 1:2 Lev-Inu showed the lowest recovered activity. However, the higher amount of Inu presumably inhibited the formation of high molecular weight levan polymers, resulting in the highest yield of LFOS. Therefore, this ratio is still chosen as the best ratio (Charoenwongpaiboon et al., 2023). In conclusion, optimization of enzyme proportions is necessary. Because the absolute enzyme concentrations can influence the optimal ratio, screening typically begins by fixing the concentration of one enzyme.

### 5.5.2 Compatibility of precipitation and cross-linking conditions

Another challenge in preparing combi-CLEAs is the compatibility of multiple enzymes with various parameters, because each enzyme has distinct physicochemical properties. Therefore, the parameters applied during precipitation and cross-linking must be carefully adjusted so that each enzyme functions effectively; in practice, this optimization essentially encompasses all of the CLEAs preparation parameters described above.

The selection of precipitant is the primary consideration in the precipitation step. A precipitant should be chosen that can efficiently precipitate all enzymes while avoiding inactivation of any component enzyme. Generally, this is evaluated by measuring, for each enzyme, activity recovery and precipitation efficiency with different precipitants (Talekar et al., 2013; Periyasamy et al., 2016). Gao et al. (2015; Zhou et al., 2016) prepared magnetic combi-CLEAs of glucose oxidase (GOD) and HRP. During the co-precipitation step, six different precipitants were evaluated for their effects on the activity recovery of each enzyme. With saturated ammonium sulfate as the precipitant, HRP showed high activity recovery (133.31%), whereas GOD recovered only 39.88%. In contrast, ethanol afforded high recoveries for both enzymes (127.51% for HRP and 106.83% for GOD) and was therefore identified as the more suitable precipitant. These results underscore the importance of optimizing the compatibility between the enzyme and the precipitant in combi-CLEA preparation.

Amadi et al. (2022) reported the preparation of combi-CLEAs from cellulase, xylanase, and pectinase using a three-phase partitioning (TPP) method. In this method, ammonium sulfate was added first, followed by t-butanol as the organic solvent. The salt expels the proteins from the solution, forming an interfacial precipitate between the lower aqueous phase and the upper organic phase. Using this approach, the residual activities of all three enzymes were higher than those obtained with precipitation by ammonium sulfate, n-propanol, or ethanol, demonstrating the effectiveness of this method for combi-CLEA preparation. Beyond the selection of precipitant, other parameters in the precipitation step must also be tuned for enzyme compatibility, including precipitant concentration, precipitation time, temperature, and pH.

Similarly, the use of cross-linkers must also take into account the compatibility of each enzyme. Glutaraldehyde remains the most commonly used cross-linker, but its concentration and the cross-linking time must be tightly controlled. In practice, if the enzymes require markedly different cross-linking conditions, compatibility can be improved by introducing protectants such as sugars or proteins (Li et al., 2018). Another option is to employ milder cross-linkers, for example, oxidized dextran (Li et al., 2018), which helps to mitigate activity losses. In some cases, the preparation conditions required by the enzymes are incompatible, which makes combi-CLEAs formation is difficult. In such situations, it may be advisable to modify the enzyme set or to prepare single-enzyme CLEAs separately and then use them as a mixture. For instance, Ramos et al. (2019) prepared CLEAs of porcine pancreas lipase (PPL) and *Thermomyces lanuginosus* lipase (TLL) separately and used them as a mixture. However, since the protocols to prepare CLEAs of each lipase were quite different, the two enzymes were not co-immobilized in a single CLEAs structure. Table 5.6 summarizes the preparation parameters of the combi-CLEAs discussed above.

## 5.6 Applications

The CLEAs technique is a simple and effective carrier-free immobilization technology. Due to advantages such as low preparation cost, high volumetric activity, and enhanced stability, CLEAs are increasingly widely used in the food, pharmaceutical, and chemical sectors (Yamaguchi et al., 2018). Accordingly, the representative applications of CLEAs, including typical examples in biomass conversion and environmental remediation, the food industry, and the production of pharmaceuticals and fine chemicals, will be briefly summarized in this section.

### 5.6.1 Applications in biomass conversion and environmental remediation

In order to reduce the use of fossil fuels and improve environmental sustainability, the bioconversion of renewable biomass into biofuels or other value-added products has been extensively investigated. In this context, CLEA technology has been widely applied to biomass conversion and the degradation of environmental wastes because of its simplicity and effectiveness in enhancing enzyme stability. For instance, the CLEAs of xylanase have been successfully employed to hydrolyze alkali

**Table 5.6** Preparation parameters of combi-CLEAs.

| Enzymes | Proportion of enzymes | Precipitants | Cross-linkers | Performance | References |
|---|---|---|---|---|---|
| Glucose oxidase (GOx)<br>Horseradish peroxidase (HRP) | GOx: HRP = 150:1 (weight ratio) | Acetonitrile | GA | The catalytic efficiency ($K_{cat}/K_m$) of the combi-CLEAs was 1.47-fold higher than that of the free enzyme mixture at the same ratio | Nguyen and Yang (2017) |
| Alpha amylase<br>Glucoamylase<br>Pullulanase | Alpha amylase: glucoamylase: pullulanase = 45 U: 45 U: 15 U | Ammonium sulfate | GA | 100%, 60% and 40% conversions were obtained with combi-CLEAs, separate CLEAs mixture and free enzyme mixture, respectively | Talekar et al. (2013) |
| Xylanase<br>cellulase<br>*β*-1,3-glucanase | | Ammonium sulfate and t-butanol (TPP method) | GA | Maximum hydrolysis reached 83.5% (48 h) for combi-CLEAs, compared to about 73% for free enzymes | Periyasamy et al. (2016) |
| Glucose oxidase (GOD)<br>Horseradish peroxidase (HRP) | GOD: HRP = 2.5: 1 (activity unit ratio) | Ethanol | GA | The maximum removal rate of DB38 by combi-CLEAs was 92.28%, higher than the 46.82% achieved by free enzymes | Zhou et al. (2016) |
| Cellulase<br>Xylanase<br>Pectinase | | Ammonium sulfate and t-butanol (TPP method) | GA | Complex substrates can be hydrolyzed simultaneously in a single reaction system | Amadi et al. (2022) |
| Enoate reductases (ERs)<br>Glucose dehydrogenase (GDH) | ERs: GDH = 11 U: 50 U | Ammonium sulfate | Oxidized dextran | Exhibit excellent thermal and pH stability, and superior reusability | Li et al. (2018) |

pretreated sugar cane bagasse to produce high-value functional oligosaccharides; compared with the free enzyme, the Xyl-CLEAs were more stable at 50°C and retained more than 50% of their initial activity after five consecutive reuse cycles. Moreover, immobilization altered its catalytic profile: whereas the free enzyme mainly produced xylose, the Xyl-CLEAs produced xylooligosaccharides, thereby enhancing its application potential (Hero et al., 2018). Lipase CLEAs have also been widely used to produce biodiesel from diverse oil feedstocks, including microalgae (Lai et al., 2012), vegetable mixes (López et al., 2014), and cooking waste (Guimarães et al., 2021).

Combi-CLEAs prepared from multiple enzymes have also shown promise for one-pot biomass conversion. For example, combi-CLEAs prepared from commercially available alpha amylase, glucoamylase, and pullulanase enabled one-pot hydrolysis of starch with 100% conversion, whereas the separate CLEAs mixture and free enzyme mixture reached only 60% and 40% conversion, respectively (Talekar et al., 2013).

In addition to biomass conversion, CLEAs have also been applied to environmental remediation. For example, CLEAs of lignin peroxidase (LiP) can efficiently degrade textile-reactive dyes and exhibit higher degradation rates, stability, and reusability than the free enzyme (Parveen et al., 2021). The CLEAs of Acrylamidase have also been used to remove acrylamide from industrial wastewater and retained 40% of their initial activity after three degradation cycles (Kulkarni et al., 2020). In addition, catalase has been immobilized as CLEAs to improve activity retention and stability for the degradation of $H_2O_2$ (Jannat & Yang, 2020; Nguyen & Yang, 2014; Tükel et al., 2013).

The accumulation of micropollutants (MPs) in water bodies poses serious risks to ecosystems and human health (Heeb et al., 2012; Kim et al., 2018). The biocatalytic membrane that integrates enzyme catalysis into membrane separation has emerged as an efficient solution for MPs removal. In this context, Liu et al. (2024) developed a strategy to immobilize CLEAs on a membrane using polydopamine-modified UiO-66-$NH_2$ as the intermediate layer (Fig. 5.4). UiO-66-$NH_2$, a kind of metal-organic frameworks (MOFs), enhances the mechanical stability of CLEAs. while polydopamine (PDA), with its porous structure and bioadhesion properties, adsorbs MPs and enriches them, thereby synergistically promoting CLEA-catalyzed degradation. The laccase catalytic membranes constructed by this strategy achieved a bisphenol A removal rate of 97.13%, and also had a good treatment effect on MPs like antibiotics and dye/salt mixture solution. Moreover, replacing laccase with phosphotriesterase (PTE) to fabricate PTE-CLEAs biocatalytic membrane enabled the removal of the organophosphorus pesticide methyl parathion in water at 94.04%, demonstrating the scalability of this strategy.

### 5.6.2 Applications in food industry

CLEAs also have applications in the food industry, particularly in winemaking and sweetener production. During the winemaking process, releasing aroma compounds

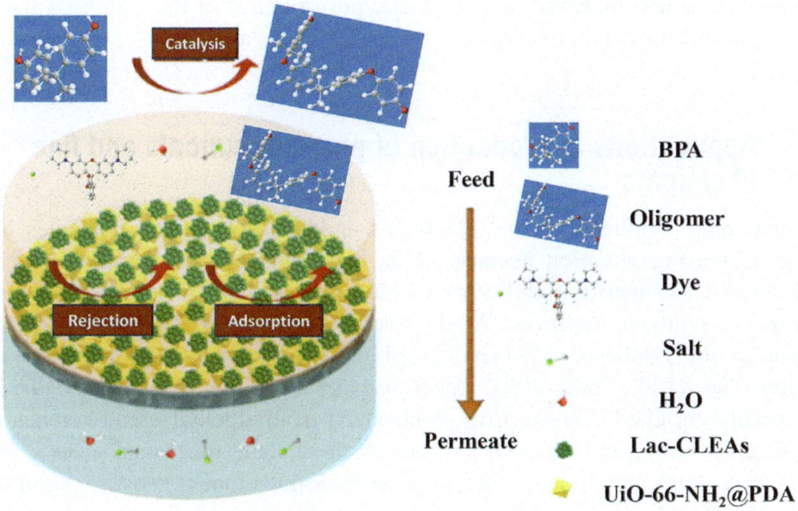


**FIGURE 5.4**

Biocatalytic membranes with crosslinked enzyme aggregates for micropollutant removal.

*Reproduced with permission from Liu, G., Wang, L., Zhao, A., Guo, L., Zhou, L., He, Y., Ma, L., Liu, Y., Gao, J., & Jiang, Y. (2024). Biocatalytic membranes with crosslinked enzyme aggregates for micropollutant removal.* Chemical Engineering Journal, 479, *147635. https://doi.org/10.1016/j.cej.2023.147635.*

from their glycosidic precursors is essential for developing the bouquet of white wines. Although glycosidases catalyze this reaction, commercial soluble enzyme preparations often suffer from rapid inactivation and low operational efficiency (Mateo et al., 2007). To overcome this, combi-CLEAs of $\beta$-D-glucosidase ($\beta$G) and $\alpha$-L-arabinofuranosidase (ARA) were prepared. In order to further control particle size and enhance recoverability for industrial applications, the aggregates were entrapped in polymeric chitosan beads. The resulting immobilized enzymes retained full activity after 91 days of incubation under winemaking conditions, making them suitable for long-term aroma enhancement in wines (Tavernini et al., 2020). Similarly, in the biocatalytic conversion of glucose to gluconic acid, cross-linked multienzymes (CLME) comprising glucose oxidase (GOD) and catalase (CAT) were prepared using genipin. Co-localization of enzymes promoted the decomposition of the inhibitory intermediate $H_2O_2$ generated by GOD, yielding a 10% higher glucose conversion after 15 h than a simple mixture of the free enzymes, demonstrating its kinetic advantage (Cui et al., 2017).

Immobilized enzymes have been widely used to produce functional oligosaccharides via starch hydrolysis. In this context, combi-CLEAs of cyclodextrin glucanotransferase (CGTase) and maltogenic amylase (Mag1) were successfully prepared for the production of maltooligosaccharides (MOS) from starch (Yip et al., 2023). Molecular docking analysis identified O-carboxymethyl chitosan (OCMCS) as the cross-linker, which improved the enzymatic performance of the combi-CLEAs, and the maximum MOS yield reached 777.46 mg/g (Yip et al., 2024).

### 5.6.3 Applications in production of pharmaceuticals and fine chemicals

The enzymatic synthesis of fine chemicals and pharmaceutical molecules has attracted increasing attention because of its excellent selectivity and mild reaction conditions. The superior stability of CLEAs makes them well-suited for catalyzing organic synthesis reactions. Many studies have prepared CLEAs of penicillin acylase for the synthesis of $\beta$-lactam antibiotics, effectively improving operational stability (Cao et al., 2000, 2001; Pchelintsev et al., 2009; Ye et al., 2019). Enantiomerically enriched cyanohydrins synthesized from aldehydes and ketones are of considerable value, and efficient derivatization of their functional groups enables access to a range of products, including $\alpha$-hydroxyketones, $\beta$-hydroxyamines, $\alpha$-aminonitriles, $\alpha$-hydroxyesters, and $\alpha$-fluoronitriles (Brunel & Holmes, 2004). Hydroxynitrile lyases in CLEA form effectively catalyze the hydrocyanation of 3-pyridinecarboxaldehyde in dichloromethane, which minimizes background racemic cyanide addition and enzyme-catalyzed product racemization. Enantiocomplementary R- and S-cyanohydrins were obtained in yields above 65%, with >93% (Roberge et al., 2007). Icariin is a rare and high-value isopentane flavonoid from *herba Epimedium*, reported to delay hepatic fibrosis and inhibit cancer cell proliferation, with relevance to hepatocellular carcinoma therapy (Fan et al., 2019; Gandhi et al., 2018). Liu et al. (2022) co-immobilized $\alpha$-L-rhamnosidase (Rha1) and $\beta$-glucosidase (Glu4) as CLEAs to synthesize icariin from epimedin C. Compared with the free enzymes, the CLEAs exhibited greater tolerance to sugars and produced 34.24 g/L icariin within 8 h at a substrate concentration of 100 g/L, indicating the great potential for industrialization.

In recent years, coupled systems that integrate photocatalysis or chemocatalysis with biocatalysis have drawn growing attention. This combination of complementary catalytic systems expands the reaction space accessible under milder and ultimately more sustainable conditions (Parisotto & Prandi, 2025). CLEAs technology has also found applications in the preparation of such photo-enzymatic or chemo-enzymatic integrated catalytic systems. For instance, a photo-enzymatic coupled catalytic system was constructed by immobilizing CLEAs of unspecific peroxygenases (UPOs) in Pd-loaded three-dimensional ordered macroporous titania (3DOM $TiO_2$) (Fig. 5.5). UPOs can use $H_2O_2$ as a co-substrate to catalyze various oxidation reactions for selective oxyfunctionalization of nonactivating C-H and C = C bonds. However, they are prone to inactivation at high $H_2O_2$ concentrations. The multifunctional 3DOM photocatalyst-assisted carrier can generate $H_2O_2$ in situ under photocatalytic conditions, which mitigates this issue. Moreover, immobilizing CLEAs within the three-dimensional ordered macroporous (3DOM) structure not only facilitates the contact of substrate and enzyme but also further stabilizes the CLEAs, increases enzyme loading,

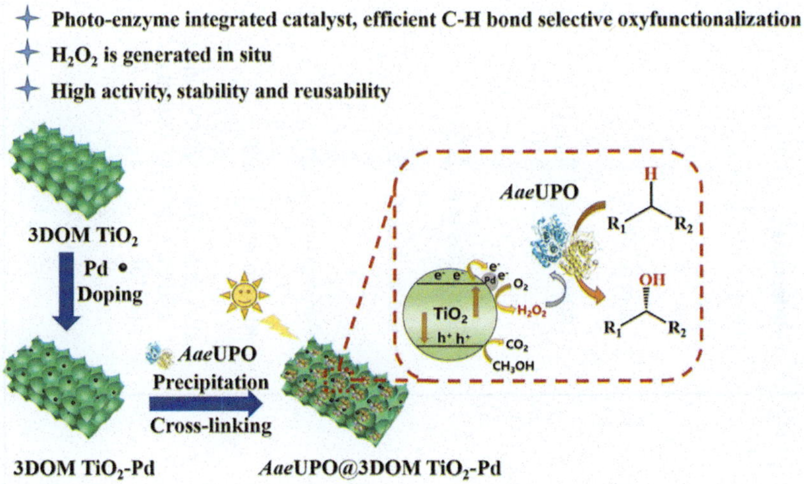


**FIGURE 5.5**

Schematic diagram of photo-enzyme catalyst AaeUPO@3DOM TiO2-Pd.

*Reproduced with permission from Deng, X., Zheng, X., Jia, F., Cao, C., Song, H., Jiang, Y., Liu, Y., Liu, G., Li, S., & Wang, L. (2023). Unspecific peroxygenases immobilized on Pd-loaded three-dimensional ordered macroporous (3DOM) titania photocatalyst for photo-enzyme integrated catalysis.* Applied Catalysis B: Environmental, 330, *122622. https://doi.org/10.1016/j.apcatb.2023.122622.*

and reduces leaching. Benefiting from these advantages, AaeUPO@3DOM $TiO_2$–Pd exhibited good stability and reusability, and demonstrates high catalytic efficiency and significantly improved conversion rate in the conversion of ethylbenzene to (R)-1-phenethyl alcohol (Deng et al., 2023). In a similar study, the CLEAs of AaeUPO were incorporated into a RhIII-complex within a hierarchical covalent organic framework (AaeUPO@Rh-macroCOF). The Rh-complex can generate $H_2O_2$ in situ using formate as a hydrogen donor, thereby constructing a chemo-enzymatic cascade catalytic system (Huang et al., 2026).

In conclusion, CLEAs technology is simple and mature, and has application potential in various fields. Table 5.7 summarizes the above applications.

## 5.7 Summary and perspectives

CLEAs represent one of the most widely used carrier-free immobilization techniques. They possess typical advantages of carrier-free immobilization, such as high enzyme loading and high volumetric activity. Additionally, CLEAs offer unique benefits, including simple preparation, low cost, no requirement for highly purified enzymes, and excellent stability. Moreover, CLEAs can be readily applied to prepare combi-CLEAs for co-immobilization of multiple enzymes, meeting the demands of various

**Table 5.7** A summary of the applications of CLEAs in various fields.

| Fields | Enzyme | Purpose | References |
|---|---|---|---|
| Biomass conversion and environmental remediation | Xylanase | Hydrolyze bagasse to produce high-value functional oligosaccharides | Hero et al. (2018) |
| | Lipase | Catalyze biodiesel production from microalgal oil | Lai et al. (2012) |
| | Lipase | Biodiesel production by esterification of FFAs and transesterification of nonedible vegetable oils | López et al. (2014) |
| | Lipase | Biolubricant production using waste cooking oil | Guimarães et al. (2021) |
| | lpha amylase, Glucoamylase, Pullulanase | Hydrolysis of starch | Talekar et al. (2013) |
| | Lignin peroxidase | Degradation of textile-reactive dyes | Parveen et al. (2021) |
| | Acrylamidase | Removal of acrylamide from industrial wastewater | Kulkarni et al. (2020) |
| | Catalase | Degradation of $H_2O_2$ | Jannat and Yang (2020), Nguyen and Yang (2014), Tükel et al. (2013) |
| | Laccase | Removal of Bisphenol A | Liu et al. (2024) |
| | phosphotriesterase | Removal of the organophosphorus pesticide methyl parathion | Liu et al. (2024) |

(*continued on next page*)

| | | | |
|---|---|---|---|
| Food industry | $\beta$-D-glucosidase, $\alpha$-L-arabinofuranosidase | Aroma enhancement in wines | Tavernini et al. (2020) |
| | Glucose oxidase, catalase | Conversion of glucose to gluconic acid | Cui et al. (2017) |
| | Cyclodextrin glucanotransferase | Produce functional oligosaccharides via starch hydrolysis | Yip et al. (2023), Yip et al. (2024) |
| Production of pharmaceuticals and fine chemicals | Penicillin acylase | Synthesis of $\beta$-lactam antibiotics | Cao et al. (2000), Pchelintsev et al. (2009), Ye et al. (2019), Cao et al. (2001) |
| | Hydroxynitrile lyases | Hydrocyanation of 3-pyridinecarboxaldehyde | Roberge et al. (2007) |
| | $\alpha$-L-rhamnosidase, $\beta$-glucosidase | Synthesize icariin from epimedin C | Liu et al. (2022) |
| | Unspecific peroxygenases | catalyze various oxidation reactions for selective oxyfunctionalization of nonactivating C-H and C = C bonds | Deng et al. (2023) |
| | Unspecific peroxygenases | selective C($sp^3$)-H oxyfunctionalization | Huang et al. (2026) |

cascade catalytic processes. Recently, the integration of CLEAs technology with other immobilization methods has led to the development of versatile composite catalytic systems, demonstrating broad application potential. Examples include the previously mentioned biocatalytic membrane systems incorporating CLEAs for wastewater treatment, as well as photo-enzymatic and chemo-enzymatic systems applied in diverse organic reaction processes. Thus, CLEAs provide an effective strategy for enhancing enzyme properties and have been widely employed in fields such as food processing, chemical synthesis, and environmental remediation.

Despite the widespread application and development of CLEAs technology, it still faces several challenges. For example: (1) Poor mechanical stability, which makes them susceptible to harsh stirring conditions and shear stress, posing challenges for large-scale applications. (2) Tendency to agglomerate during centrifugation or filtration steps, thereby increasing mass transfer limitations. (3) Uncontrollable cross-linking that may damage essential active sites of the enzyme, leading to inactivation.

To address these limitations, various strategies have been developed for improvement. In response to the poor mechanical stability, many studies have combined CLEAs with other immobilization methods to provide carrier support. In particular, the incorporation of magnetic nanoparticles enables the construction of magnetic cross-linked enzyme aggregates (M-CLEAs), which not only enhance the resistance to mechanical stress but also simplify downstream processing and improve recovery and reusability. A detailed introduction to M-CLEAs will be provided in the next chapter.

Regarding the issue of enzyme activity loss due to cross-linking, current solutions remain largely empirical. By adjusting various parameters involved in the cross-linking process, researchers aim to maximize the activity recovery rate. To further mitigate these challenges, a deeper characterization and understanding of structural changes in CLEAs are necessary. Several studies have already employed computational techniques such as molecular dynamics simulations to design and screen more effective cross- linkers. Additionally, the design of more efficient CLEAs reactors is essential for their large-scale application.

## References

Abd Rahman, N. H., Jaafar, N. R., Shamsul Annuar, N. A., Rahman, R. A., Abdul Murad, A. M., El-Enshasy, H. A., & Illias, R. M. (2021). Efficient substrate accessibility of cross-linked levanase aggregates using dialdehyde starch as a macromolecular cross-linker. *Carbohydrate Polymers, 267*, 118159. http://www.elsevier.com/wps/find/journaldescription.cws_home/405871/description#description. doi:10.1016/j.carbpol.2021.118159.

Abdul Wahab, M. K. H., El-Enshasy, H. A., Bakar, F. D. A., Murad, A. M. A., Jahim, J. M., & Illias, R. M. (2019). Improvement of cross-linking and stability on cross-linked enzyme aggregate (CLEA)-xylanase by protein surface engineering. *Process Biochemistry, 86*, 40–49. www.elsevier.com/inca/publications/store/4/2/2/8/5/7. doi:10.1016/j.procbio.2019.07.017.

Ademakinwa, A. N. (2021). A heat-resistant intracellular laccase immobilized via cross-linked enzyme aggregate preparation: Characterization, application in bisphenol A removal and phytotoxicity evaluation. *Journal of Hazardous Materials, 419*, 126480. doi:10.1016/j.jhazmat.2021.126480.

Alvarenga, A. E., Amoroso, M. J., Illanes, A., & Castro, G. R. (2014). Cross-linked $\alpha$-l-rhamnosidase aggregates with potential application in food industry. *European Food Research and Technology, 238*(5), 797–801. doi:10.1007/s00217-014-2157-4.

Amadi, O. C., Awodiran, I. P., Moneke, A. N., Nwagu, T. N., Egong, J. E., & Chukwu, G. C. (2022). Concurrent production of cellulase, xylanase, pectinase and immobilization by combined cross-linked enzyme aggregate strategy: Advancing tri-enzyme biocatalysis. *Bioresource Technology Reports, 18*, 101019. doi:10.1016/j.biteb.2022.101019.

Amaral-Fonseca, M., Kopp, W., Giordano, R. D. L. C., Fernández-Lafuente, R., & Tardioli, P. W. (2018). Preparation of magnetic cross-linked amyloglucosidase aggregates: Solving some activity problems. *Catalysts, 8*(11). https://www.mdpi.com/2073-4344/8/11/496/pdf. doi:10.3390/catal8110496.

Arsenault, A., Cabana, H., & Jones, J. P. (2011). Laccase-based CLEAs: Chitosan as a novel cross-linking agent. *Enzyme Research, 2011*(1), 376015. doi:10.4061/2011/376015.

Arteche Pujana, M., Pérez-Álvarez, L., Cesteros Iturbe, L. C., & Katime, I. (2013). Biodegradable chitosan nanogels crosslinked with genipin. *Carbohydrate Polymers, 94*(2), 836–842. doi:10.1016/j.carbpol.2013.01.082.

Aßmann, M., Stöbener, A., Mügge, C., Gaßmeyer, S. K., Hilterhaus, L., Kourist, R., Liese, A., & Kara, S. (2017). Reaction engineering of biocatalytic (S)-naproxen synthesis integrating in-line process monitoring by Raman spectroscopy. *Reaction Chemistry and Engineering, 2*(4), 531–540. www.rsc.org/journals-books-databases/about-journals/reaction-chemistry-engineering/. doi:10.1039/c7re00043j.

Avrameas, S., & Ternynck, T. (1969). The cross-linking of proteins with glutaraldehyde and its use for the preparation of immunoadsorbents. *Immunochemistry, 6*(1), 53–66. doi:10.1016/0019-2791(69)90178-5.

Aytar, B. S., & Bakir, U. (2008). Preparation of cross-linked tyrosinase aggregates. *Process Biochemistry, 43*(2), 125–131. doi:10.1016/j.procbio.2007.11.001.

Banerjee, S., Arora, A., Vijayaraghavan, R., & Patti, A. F. (2020). Extraction and crosslinking of bromelain aggregates for improved stability and reusability from pineapple processing waste. *International Journal of Biological Macromolecules, 158*, 318–326. www.elsevier.com/locate/ijbiomac. doi:10.1016/j.ijbiomac.2020.04.220.

Basso, A., & Serban, S. (2019). Industrial applications of immobilized enzymes—A review. *Molecular Catalysis, 479*, 110607. doi:10.1016/j.mcat.2019.110607.

Bayer, T., Wu, S., Snajdrova, R., Baldenius, K., & Bornscheuer, U. T. (2025). An update: Enzymatic synthesis for industrial applications. *Angewandte Chemie International Edition, 64*(27), e202505976. http://onlinelibrary.wiley.com/journal/10.1002/(ISSN)1521-3773. doi:10.1002/anie.202505976.

Benítez-Mateos, A. I., Roura Padrosa, D., & Paradisi, F. (2022). Multistep enzyme cascades as a route towards green and sustainable pharmaceutical syntheses. *Nature Chemistry, 14*(5), 489–499. http://www.nature.com/nchem/index.html. doi:10.1038/s41557-022-00931-2.

Bian, H., Cao, M., Wen, H., Tan, Z., Jia, S., & Cui, J. (2019). Biodegradation of polyvinyl alcohol using cross-linked enzyme aggregates of degrading enzymes from *Bacillus niacini*. *International Journal of Biological Macromolecules, 124*, 10–16. www.elsevier.com/locate/ijbiomac. doi:10.1016/j.ijbiomac.2018.11.204.

Bié, J., Sepodes, B., Fernandes, P. C. B., & Ribeiro, M. H. L. (2022). Enzyme immobilization and co-immobilization: Main framework, advances and some applications. *Processes, 10*(3), 494. https://www.mdpi.com/2227-9717/10/3/494/pdf. doi:10.3390/pr10030494.

Bilal, M., Noreen, S., Asgher, M., & Parveen, S. (2021). Development and characterization of cross-linked laccase aggregates (Lac-CLEAs) from *Trametes versicolor* IBL-04 as ecofriendly biocatalyst for degradation of dye-based environmental pollutants. *Environmental Technology & Innovation, 21*, 101364. doi:10.1016/j.eti.2021.101364.

Blanco-Llamero, C., García-García, P., & Señoráns, F. J. (2021). Cross-linked enzyme aggregates and their application in enzymatic pretreatment of microalgae: Comparison between CLEAs and Combi-CLEAs. *Frontiers in Bioengineering and Biotechnology, 9*, 794672. http://journal.frontiersin.org/journal/bioengineering-and-biotechnology#archive. doi:10.3389/fbioe.2021.794672.

Bouguerra, O. M., Wahab, R. A., Huyop, F., Al-Fakih, A. M., Mahmood, W. M. A. W., Mahat, N. A., & Sabullah, M. K. (2024). An overview of crosslinked enzyme aggregates: Concept of development and trends of applications. *Applied Biochemistry and Biotechnology, 196*, 5711–5739. doi:10.1007/s12010-023-04809-y.

Brunel, J. M., & Holmes, I. P. (2004). Chemically catalyzed asymmetric cyanohydrin syntheses. *Angewandte Chemie International Edition, 43*(21), 2752–2778. doi:10.1002/anie.200300604.

Caballero Valdés, E., Wilson Soto, L., & Aroca Arcaya, G. (2011). Influence of the pH of glutaraldehyde and the use of dextran aldehyde on the preparation of cross-linked enzyme aggregates (CLEAs) of lipase from *Burkholderia cepacia*. *Electronic Journal of Biotechnology, 14*(3). http://10.2225/vol14-issue3-fulltext-1.

Cao, L., van Langen, L. M., van Rantwijk, F., & Sheldon, R. A. (2001). Cross-linked aggregates of penicillin acylase: Robust catalysts for the synthesis of $\beta$-lactam antibiotics. *Journal of Molecular Catalysis B: Enzymatic, 11*(4-6), 665–670. doi:10.1016/S1381-1177(00)00078-3.

Cao, L., Van Rantwijk, F., & Sheldon, R. A. (2000). Cross-linked enzyme aggregates: A simple and effective method for the immobilization of penicillin acylase. *Organic Letters, 2*(10), 1361–1364. http://pubs.acs.org/journal/orlef7. doi:10.1021/ol005593x.

Cao, L., Langen, L., & Sheldon, R. A. (2003). Immobilised enzymes: Carrier-bound or carrier-free? *Current Opinion in Biotechnology, 14*(4), 387–394. doi:10.1016/s0958-1669(03)00096-x.

Charoenwongpaiboon, T., Wangpaiboon, K., Field, R. A., Prousoontorn, M., & Pichyangkura, R. (2023). Cross-linked enzyme aggregates (combi-CLEAs) derived from levansucrase and variant inulosucrase are highly efficient catalysts for the synthesis of levan-type fructooligosaccharides. *Molecular Catalysis, 535*, 112827. doi:10.1016/j.mcat.2022.112827.

Chen, N., Chang, B., Shi, N., Yan, W., Lu, F., & Liu, F. (2023a). Cross-linked enzyme aggregates immobilization: Preparation, characterization, and applications. *Critical Reviews in Biotechnology, 43*(3), 369–383. http://www.tandfonline.com/loi/ibty20. doi:10.1080/07388551.2022.2038073.

Chen, N., Chang, B., Shi, N., Lu, F., & Liu, F. (2023b). Robust and recyclable cross-linked enzyme aggregates of sucrose isomerase for isomaltulose production. *Food Chemistry, 399*, 134000. doi:10.1016/j.foodchem.2022.134000.

Chen, Z., Wang, Y., Liu, W., Wang, J., & Chen, H. (2017). A novel cross-linked enzyme aggregates (CLEAs) of papain and neutrase-production, partial characterization and application. *International Journal of Biological Macromolecules, 95*, 650–657. www.elsevier.com/locate/ijbiomac. doi:10.1016/j.ijbiomac.2016.11.083.

Cruz, J., Barbosa, O., Rodrigues, R. C., Fernandez-Lafuente, R., Torres, R., & Ortiz, C. (2012). Optimized preparation of CALB-CLEAs by response surface methodology: The necessity to employ a feeder to have an effective crosslinking. *Journal of Molecular Catalysis B: Enzymatic, 80*, 7–14. doi:10.1016/j.molcatb.2012.04.013.

Cui, C., Chen, H., Chen, B., & Tan, T. (2017). Genipin cross-linked glucose oxidase and catalase multi-enzyme for gluconic acid synthesis. *Applied Biochemistry and Biotechnology, 181*(2), 526–535. http://www.springer.com/humana+press/journal/12010. doi:10.1007/s12010-016-2228-z.

Cui, J. D., & Jia, S. R. (2015). Optimization protocols and improved strategies of cross-linked enzyme aggregates technology: Current development and future challenges. *Critical Reviews in Biotechnology, 35*(1), 15–28. http://www.tandfonline.com/loi/ibty20. doi:10.3109/07388551.2013.795516.

Cui, J. D., Li, L. L., & Bian, H. J. (2013). Immobilization of cross-linked phenylalanine ammonia lyase aggregates in microporous silica gel. *PLoS One, 8*(11), e80581. http://www.plosone.org/article/fetchObject.action?uri=info%3Adoi%2F10.1371%2Fjournal.pone.0080581&representation=PDF. doi:10.1371/journal.pone.0080581China.

Cui, J. D., Liu, R. L., & Li, L. B. (2016). A facile technique to prepare cross-linked enzyme aggregates of bovine pancreatic lipase using bovine serum albumin as an additive. *Korean Journal of Chemical Engineering, 33*(2), 610–615. http://www.springerlink.com/content/120599/. doi:10.1007/s11814-015-0190-z.

Del Arco, J., Alcántara, A. R., Fernández-Lafuente, R., & Fernández-Lucas, J. (2021). Magnetic micro-macro biocatalysts applied to industrial bioprocesses. *Bioresource Technology, 322*, 124547. doi:10.1016/j.biortech.2020.124547.

Deng, X., Zheng, X., Jia, F., Cao, C., Song, H., Jiang, Y., Liu, Y., Liu, G., Li, S., & Wang, L. (2023). Unspecific peroxygenases immobilized on Pd-loaded three-dimensional ordered macroporous (3DOM) titania photocatalyst for photo-enzyme integrated catalysis. *Applied Catalysis B: Environmental, 330*, 122622. doi:10.1016/j.apcatb.2023.122622.

de Sousa, M., Silva Gurgel, B., Pessela, B. C., & Gonçalves, L. R. B. (2020). Preparation of CLEAs and magnetic CLEAs of a recombinant L-arabinose isomerase for D-tagatose synthesis. *Enzyme and Microbial Technology, 138*, 109566. www.elsevier.com/locate/enzmictec. doi:10.1016/j.enzmictec.2020.109566.

Diaz-Vidal, T., Armenta-Perez, V. P., Rosales-Rivera, L. C., Mateos-Díaz, J. C., & Rodríguez, J. A. (2019). Cross-linked enzyme aggregates of recombinant *Candida antarctica* lipase B for the efficient synthesis of olvanil, a nonpungent capsaicin analogue. *Biotechnology Progress, 35*(4), e2807. http://onlinelibrary.wiley.com/journal/10.1021/(ISSN)1520-6033. doi:10.1002/btpr.2807.

Dong, T., Zhao, L., Huang, Y., & Tan, X. (2010). Preparation of cross-linked aggregates of aminoacylase from *Aspergillus melleus* by using bovine serum albumin as an inert additive. *Bioresource Technology, 101*(16), 6569–6571. http://www.sciencedirect.com. doi:10.1016/j.biortech.2010.03.061.

Fan, Y., Li, S., Ding, X., Yue, J., Jiang, J., Zhao, H., Hao, R., Qiu, W., Liu, K., Li, Y., Wang, S., Zheng, L., Ye, B., Meng, K., & Xu, B. (2019). First-in-class immune-modulating small molecule icaritin in advanced hepatocellular carcinoma: Preliminary results of safety, durable survival and immune biomarkers. *BMC Cancer, 19*(1), 279. http://www.biomedcentral.com/bmccancer/. doi:10.1186/s12885-019-5471-1.

Gaggero, N., & Albanese, D. C. M. (2021). Cascade biocatalytic processes through combined cross-linked enzyme aggregates. *Current Organic Chemistry, 25*(21), 2666–2675. doi:10.2174/1385272825666210809125858.

Gandhi, G. R., Neta, M. T. S. L., Sathiyabama, R. G., Quintans, J. D. S. S., de Oliveira e Silva, A. M., Araújo, A. A. D. S., Narain, N., Júnior, L. J. Q., & Gurgel, R. Q. (2018). Flavonoids as Th1/Th2 cytokines immunomodulators: A systematic review of studies on animal models. *Phytomedicine, 44*, 74–84. www.urbanfischer.de/journals/phytomed. doi:10.1016/j.phymed.2018.03.057.

Gao, J., Wang, Q., Jiang, Y., Gao, J., Liu, Z., Zhou, L., & Zhang, Y. (2015). Formation of nitrile hydratase cross-linked enzyme aggregates in mesoporous onion-like silica: Preparation and catalytic properties. *Industrial and Engineering Chemistry Research, 54*(1), 83–90. http://pubs.acs.org/journal/iecred. doi:10.1021/ie503018m.

Guimarães, J. R., Miranda, L. P., Fernandez-Lafuente, R., & Tardioli, P. W. (2021). Immobilization of Eversa® transform via CLEA technology converts it in a suitable biocatalyst for biolubricant production using waste cooking oil. *Molecules, 26*(1), 193. https://www.mdpi.com/1420-3049/26/1/193. doi:10.3390/molecules26010193.

Gupta, P., Dutt, K., Misra, S., Raghuwanshi, S., & Saxena, R. K. (2009). Characterization of cross-linked immobilized lipase from thermophilic mould Thermomyces lanuginosa using glutaraldehyde. *Bioresource Technology, 100*(18), 4074–4076. doi:10.1016/j.biortech.2009.03.076.

Guzik, U., Hupert-Kocurek, K., & Wojcieszyńska, D. (2014). Immobilization as a strategy for improving enzyme properties—Application to oxidoreductases. *Molecules, 19*(7), 8995–9018. doi:10.3390/molecules19078995.

Habeeb, A. F. S. A. (1967). Preparation of enzymically active, water-insoluble derivatives of trypsin. *Archives of Biochemistry and Biophysics, 119*, 264–268. doi:10.1016/0003-9861(67)90453-5.

Heeb, F., Singer, H., Pernet-Coudrier, B., Qi, W., Liu, H., Longrée, P., Müller, B., & Berg, M. (2012). Organic micropollutants in rivers downstream of the megacity Beijing: Sources and mass fluxes in a large-scale wastewater irrigation system. *Environmental Science and Technology, 46*(16), 8680–8688. doi:10.1021/es301912q.

Hero, J. S., Romero, C. M., Pisa, J. H., Perotti, N. I., Olivaro, C., & Martinez, M. A. (2018). Designing cross-linked xylanase aggregates for bioconversion of agroindustrial waste biomass towards potential production of nutraceuticals. *International Journal of Biological Macromolecules, 111*, 229–236. doi:10.1016/j.ijbiomac.2017.12.166.

Hojnik Podrepšek, G., Knez, Ž., & Leitgeb, M. (2019). Activation of cellulase cross-linked enzyme aggregates (CLEAs) in $scCO_2$. *The Journal of Supercritical Fluids, 154*, 104629. doi:10.1016/j.supflu.2019.104629.

Hong, J., Jung, D., Park, S., Oh, Y., Oh, K. K., & Lee, S. H. (2021). Immobilization of laccase via cross-linked enzyme aggregates prepared using genipin as a natural cross-linker. *International Journal of Biological Macromolecules, 169*, 541–550. www.elsevier.com/locate/ijbiomac. doi:10.1016/j.ijbiomac.2020.12.136.

Hormigo, D., García-Hidalgo, J., Acebal, C., De la Mata, I., & Arroyo, M. (2012). Preparation and characterization of cross-linked enzyme aggregates (CLEAs) of recombinant poly-3-hydroxybutyrate depolymerase from Streptomyces exfoliatus. *Bioresource Technology, 115*, 177–182. doi:10.1016/j.biortech.2011.09.035.

Huang, F., Zhang, Q., Li, H., Liu, Y., Liu, G., Ma, L. I., Zhou, L., He, Y., Yue, X., & Jiang, Y. (2026). Formate-driven chemoenzymatic oxyfunctionalization of inert C($sp^3$)-H bonds on unspecific peroxygenase-CLEAs immobilized RhIII-complex functionalized hierarchical covalent organic frameworks. *Molecular Catalysis, 588*, 115525. doi:10.1016/j.mcat.2025.115525.

Jannat, M., & Yang, K. L. (2020). A millifluidic device with embedded cross-linked enzyme aggregates for degradation of $H_2O_2$. *ACS Applied Materials and Interfaces, 12*(5), 6768–6775. http://pubs.acs.org/journal/aamick. doi:10.1021/acsami.9b21480.

Jiang, Y., Shi, L., Huang, Y., Gao, J., Zhang, X., & Zhou, L. (2014a). Preparation of robust biocatalyst based on cross-linked enzyme aggregates entrapped in three-dimensionally ordered macroporous silica. *ACS Applied Materials and Interfaces, 6*(4), 2622–2628. doi:10.1021/am405104b.

Jiang, Y., Wang, Q. I., He, Y., Zhou, L., & Gao, J. (2014b). Co-aggregation of laccase and nature egg white: A simple method to prepare stable and recyclable biocatalyst. *Applied Biochemistry and Biotechnology, 172*, 2496–2506. doi:10.1007/s12010-013-0697-x.

Jin, W., Xu, Y., & Yu, X. W. (2019). Formation lipase cross-linked enzyme aggregates on octyl-modified mesocellular foams with oxidized sodium alginate. *Colloids and Surfaces B: Biointerfaces, 184*, 110501. www.elsevier.com/locate/colsurfb. doi:10.1016/j.colsurfb.2019.110501.

Ju, H., Jang, E., Ryu, B. H., & Kim, T. D. (2013). Characterization and preparation of highly stable aggregates of a novel type of hydrolase (BL28) from *Bacillus licheniformis*. *Bioresource Technology, 128*, 81–86. www.elsevier.com/locate/biortech. doi:10.1016/j.biortech.2012.10.016.

Karimpil, J. J., Melo, J. S., & D'Souza, S. F. (2011). Hen egg white as a feeder protein for lipase immobilization. *Journal of Molecular Catalysis B: Enzymatic, 71*(3-4), 113–118. doi:10.1016/j.molcatb.2011.04.006.

Kartal, F., Janssen, M. H. A., Hollmann, F., Sheldon, R. A., & Kilinc, A. (2011). Improved esterification activity of *Candida rugosa* lipase in organic solvent by immobilization as cross-linked enzyme aggregates (CLEAs). *Journal of Molecular Catalysis B: Enzymatic, 71*(3-4), 85–89. doi:10.1016/j.molcatb.2011.04.002.

Kim, M. H., Park, S., Kim, Y. H., Won, K., & Lee, S. H. (2013). Immobilization of formate dehydrogenase from *Candida boidinii* through cross-linked enzyme aggregates. *Journal of Molecular Catalysis B: Enzymatic, 97*, 209–214. www.elsevier.com/locate/molcatb. doi:10.1016/j.molcatb.2013.08.020.

Kim, M., Kim, J., Lee, J., Jia, H., Na, H. B., Youn, J. K., Kwak, J. H., Dohnalkova, A., Grate, J. W., Wang, P., Hyeon, T., Park, H. G., & Chang, H. N. (2007). Crosslinked enzyme aggregates in hierarchically-ordered mesoporous silica: A simple and effective method for enzyme stabilization. *Biotechnology and Bioengineering, 96*(2), 210–218. doi:10.1002/bit.21107.

Kim, S., Chu, K. H., Al-Hamadani, Y. A. J., Park, C. M., Jang, M., Kim, D. H., Yu, M., Heo, J., & Yoon, Y. (2018). Removal of contaminants of emerging concern by membranes in water and wastewater: A review. *Chemical Engineering Journal, 335*, 896–914. www.elsevier.com/inca/publications/store/6/0/1/2/7/3/index.htt. doi:10.1016/j.cej.2017.11.044.

Kim, Y. J., & Kim, Y. W. (2016). Optimizing the preparation conditions and characterization of cross-linked enzyme aggregates of a monoamine oxidase. *Food Science and Biotechnology, 25*(5), 1421–1425. http://www.springerlink.com/content/n3031612t262/. doi:10.1007/s10068-016-0221-5.

Klein, M. P., Hackenhaar, C. R., Lorenzoni, A. S. G., Rodrigues, R. C., Costa, T. M. H., Ninow, J. L., & Hertz, P. F. (2016). Chitosan crosslinked with genipin as support matrix for application in food process: Support characterization and β-d-galactosidase immobilization. *Carbohydrate Polymers, 137*, 184–190. http://www.elsevier.com/wps/find/journaldescription.cws_home/405871/description#description. doi:10.1016/j.carbpol.2015.10.069.

Kopp, W., Da Costa, T. P., Pereira, S. C., Jafelicci, M., Giordano, R. C., Marques, R. F. C., Araújo-Moreira, F. M., & Giordano, R. L. C. (2014). Easily handling penicillin G acylase magnetic cross-linked enzymes aggregates: Catalytic and morphological studies. *Process Biochemistry, 49*(1), 38–46. www.elsevier.com/inca/publications/store/4/2/2/8/5/7. doi:10.1016/j.procbio.2013.09.024.

Kulkarni, N. H., Muley, A. B., Bedade, D. K., & Singhal, R. S. (2020). Cross-linked enzyme aggregates of arylamidase from Cupriavidus oxalaticus ICTDB921: Process optimization, characterization, and application for mitigation of acrylamide in industrial wastewater. *Bioprocess and Biosystems Engineering, 43*(3), 457–471. https://rd.springer.com/journal/449. doi:10.1007/s00449-019-02240-4.

Kumar, S., Mohan, U., Kamble, A. L., Pawar, S., & Banerjee, U. C. (2010). Cross-linked enzyme aggregates of recombinant *Pseudomonas putida* nitrilase for enantioselective nitrile hydrolysis. *Bioresource Technology, 101*(17), 6856–6858. doi:10.1016/j.biortech.2010.03.084.

Kyomuhimbo, H. D., & Brink, H. G. (2023). Applications and immobilization strategies of the copper-centred laccase enzyme; a review. *Heliyon, 9*(2). http://www.journals.elsevier.com/heliyon/. doi:10.1016/j.heliyon.2023.e13156.

Lai, J. Q., Hu, Z. L., Sheldon, R. A., & Yang, Z. (2012). Catalytic performance of cross-linked enzyme aggregates of *Penicillium expansum* lipase and their use as catalyst for biodiesel production. *Process Biochemistry, 47*(12), 2058–2063. doi:10.1016/j.procbio.2012.07.024.

Lee, Y. L., Jaafar, N. R., Huyop, F., Bakar, F. D. A., Rahman, R. A., & Illias, R. M. D. (2025). Functionalization of amylopectin as a strategy to improve polyethylene terephthalate hydrolase-cross-linked enzyme aggregate (IsPETase-CLEA) in plastic degradation. *International Journal of Biological Macromolecules, 306*, 141492. https://www.sciencedirect.com/science/journal/01418130. doi:10.1016/j.ijbiomac.2025.141492.

Li, H., Xiao, W., Xie, P., & Zheng, L. (2018). Co-immobilization of enoate reductase with a cofactor-recycling partner enzyme. *Enzyme and Microbial Technology, 109*, 66–73. www.elsevier.com/locate/enzmictec. doi:10.1016/j.enzmictec.2017.09.013.

Li, L., Li, G., Cao, L.-C., Ren, G.-H., Kong, W., Wang, S.-D., Guo, G.-S., & Liu, Y.-H. (2015). Characterization of the cross-linked enzyme aggregates of a novel β-galactosidase, a potential catalyst for the synthesis of galacto-oligosaccharides. *Journal of Agricultural and Food Chemistry, 63*(3), 894–901.

Li, T., Gong, X., Yang, G., Li, Q., Huang, J., Zhou, N., & Jia, X. (2022). Cross-linked enzyme aggregates (CLEAs) of cellulase with improved catalytic activity, adaptability and reusability. *Bioprocess and Biosystems Engineering, 45*(5), 865–875. https://rd.springer.com/journal/449. doi:10.1007/s00449-022-02704-0.

Li, X. D., Wu, J., Jia, D. C., Wan, Y. H., Yang, N., & Qiao, M. (2016). Preparation of cross-linked glucoamylase aggregates immobilization by using dextrin and xanthan gum as protecting agents. *Catalysts, 6*(6), 77. http://www.mdpi.com/2073-4344/6/6/77/pdf. doi:10.3390/catal6060077.

Liu, F., Wei, B., Cheng, L., Zhao, Y., Liu, X., Yuan, Q., & Liang, H. (2022). Co-immobilizing two glycosidases based on cross-linked enzyme aggregates to enhance enzymatic properties for achieving high titer icaritin biosynthesis. *Journal of Agricultural and Food Chemistry, 70*(37), 11631–11642. http://pubs.acs.org/journal/jafcau. doi:10.1021/acs.jafc.2c04253.

Liu, G., Wang, L., Zhao, A., Guo, L., Zhou, L., He, Y., Ma, L., Liu, Y., Gao, J., & Jiang, Y. (2024). Biocatalytic membranes with crosslinked enzyme aggregates for micropollutant removal. *Chemical Engineering Journal, 479*, 147635. doi:10.1016/j.cej.2023.147635.

López, C., Cruz-Izquierdo, Á., Picó, E. A., García-Bárcena, T., Villarroel, N., Llama, M. J., & Serra, J. L. (2014). Magnetic biocatalysts and their uses to obtain biodiesel and biosurfactants. *Frontiers in Chemistry, 2*, 72. http://journal.frontiersin.org/article/10.3389/fchem.2014.00072/full. doi:10.3389/fchem.2014.00072.

Majumder, A. B., Mondal, K., Singh, T. P., & Gupta, M. N. (2008). Designing cross-linked lipase aggregates for optimum performance as biocatalysts. *Biocatalysis and Biotransformation, 26*(3), 235–242. doi:10.1080/10242420701685601.

Mateo, C., Palomo, J. M., Fernandez-Lorente, G., Guisan, J. M., & Fernandez-Lafuente, R. (2007). Improvement of enzyme activity, stability and selectivity via immobilization techniques. *Enzyme and Microbial Technology, 40*(6), 1451–1463. doi:10.1016/j.enzmictec.2007.01.018.

Mateo, C., Palomo, J. M., Van Langen, L. M., Van Rantwijk, F., & Sheldon, R. A. (2004). A new, mild cross-linking methodology to prepare cross-linked enzyme aggregates. *Biotechnology and Bioengineering, 86*(3), 273–276. doi:10.1002/bit.20033.

Menfaatli, E., & Zihnioglu, F. (2015). Carrier free immobilization and characterization of trypsin. *Artificial Cells, Nanomedicine and Biotechnology, 43*(2), 140–144. http://www.tandfonline.com/loi/ianb20#.VmugQbfovcs. doi:10.3109/21691401.2013.853178.

Montoro-García, S., Gil-Ortiz, F., Navarro-Fernández, J., Rubio, V., García-Carmona, F., & Sánchez-Ferrer, A. (2010). Improved cross-linked enzyme aggregates for the production of desacetyl $\beta$-lactam antibiotics intermediates. *Bioresource Technology, 101*(1), 331–336. doi:10.1016/j.biortech.2009.08.016.

Morais, J. R. F., Costa, I. O., Padilha, C. E. A., Rios, N. S., & Santos, E. S. D. (2025). Improving reusability of biocatalysts by exploiting cross-linked enzyme aggregates (CLEAs) with commercial cellulolytic cocktails for hydrolysis of green coconut waste. *Sustainability (Switzerland), 17*(9), 4221. http://www.mdpi.com/journal/sustainability/. doi:10.3390/su17094221.

Morshed, M. N., Behary, N., Bouazizi, N., Guan, J., Chen, G., & Nierstrasz, V. (2019). Surface modification of polyester fabric using plasma-dendrimer for robust immobilization of glucose oxidase enzyme. *Scientific Reports, 9*(1), 52087. www.nature.com/srep/index.html. doi:10.1038/s41598-019-52087-8.

Mukherjee, J., Majumder, A. B., & Gupta, M. N. (2016). Adding an appropriate amino acid during crosslinking results in more stable crosslinked enzyme aggregates. *Analytical Biochemistry, 507*, 27–32. doi:10.1016/j.ab.2016.05.012.

Nadar, S. S., Muley, A. B., Ladole, M. R., & Joshi, P. U. (2016). Macromolecular cross-linked enzyme aggregates (M-CLEAs) of $\alpha$-amylase. *International Journal of Biological Macromolecules, 84*, 69–78. www.elsevier.com/locate/ijbiomac. doi:10.1016/j.ijbiomac.2015.11.082.

Nawawi, N. N., Hashim, Z., Manas, N. H. A., Azelee, N. I. W., & Illias, R. M. (2020). A porous-cross linked enzyme aggregates of maltogenic amylase from *Bacillus lehensis* G1:

Robust biocatalyst with improved stability and substrate diffusion. *International Journal of Biological Macromolecules, 148*, 1222–1231. www.elsevier.com/locate/ijbiomac. doi:10.1016/j.ijbiomac.2019.10.101.

Nguyen, L. T., & Yang, K. L. (2014). Uniform cross-linked cellulase aggregates prepared in millifluidic reactors. *Journal of Colloid and Interface Science, 428*, 146–151. http://www.elsevier.com/inca/publications/store/6/2/2/8/6/1/index.htt. doi:10.1016/j.jcis.2014.04.033.

Nguyen, L. T., & Yang, K. L. (2017). Combined cross-linked enzyme aggregates of horseradish peroxidase and glucose oxidase for catalyzing cascade chemical reactions. *Enzyme and Microbial Technology, 100*, 52–59. www.elsevier.com/locate/enzmictec. doi:10.1016/j.enzmictec.2017.02.007.

Nikolaivits, E., Makris, G., & Topakas, E. (2017). Immobilization of a cutinase from *Fusarium oxysporum* and application in pineapple flavor synthesis. *Journal of Agricultural and Food Chemistry, 65*(17), 3505–3511. http://pubs.acs.org/journal/jafcau. doi:10.1021/acs.jafc.7b00659.

Parisotto, S., & Prandi, C. (2025). Integrating visible-light photocatalysis and enzymes for the stereoselective synthesis of alcohols and amines. *Advanced Synthesis and Catalysis, 367*(15), e9603. http://onlinelibrary.wiley.com/journal/10.1002/(ISSN)1615-4169. doi:10.1002/adsc.9603.

Park, J. M., Kim, M., Lee, H. J., Jang, A., Min, J., & Kim, Y. H. (2012). Enhancing the production of rhodobacter sphaeroides-derived physiologically active substances using carbonic anhydrase-immobilized electrospun nanofibers. *Biomacromolecules, 13*(11), 3780–3786. doi:10.1021/bm3012264.

Parveen, S., Asgher, M., & Bilal, M. (2021). Lignin peroxidase-based cross-linked enzyme aggregates (LiP-CLEAs) as robust biocatalytic materials for mitigation of textile dyes-contaminated aqueous solution. *Environmental Technology & Innovation, 21*, 101226. doi:10.1016/j.eti.2020.101226.

Pchelintsev, N. A., Youshko, M. I., & Švedas, V. K. (2009). Quantitative characteristic of the catalytic properties and microstructure of cross-linked enzyme aggregates of penicillin acylase. *Journal of Molecular Catalysis B: Enzymatic, 56*(4), 202–207. doi:10.1016/j.molcatb.2008.05.006.

Periyasamy, K., Santhalembi, L., Mortha, G., Aurousseau, M., & Subramanian, S. (2016). Carrier-free co-immobilization of xylanase, cellulase and $\beta$-1,3-glucanase as combined cross-linked enzyme aggregates (combi-CLEAs) for one-pot saccharification of sugarcane bagasse. *RSC Advances, 6*(39), 32849–32857. http://pubs.rsc.org/en/journals/journalissues. doi:10.1039/c6ra00929h.

Perzon, A., Dicko, C., Çobanoğlu, Ö., Yükselen, O., Eryilmaz, J., & Dey, E. S. (2017). Cellulase cross-linked enzyme aggregates (CLEA) activities can be modulated and enhanced by precipitant selection. *Journal of Chemical Technology and Biotechnology, 92*(7), 1645–1649. http://onlinelibrary.wiley.com/journal/10.1002/(ISSN)1097-4660. doi:10.1002/jctb.5160.

Podrepšek, G. H., Knez, Ž., & Leitgeb, M. (2019). Activation of cellulase cross-linked enzyme aggregates (CLEAs) in $scCO_2$. *The Journal of Supercritical Fluids, 154*, 104629. doi:10.1016/j.supflu.2019.104629.

Prabhavathi Devi, B. L. A., Guo, Z., & Xu, X. (2009). Characterization of cross-linked lipase aggregates. *Journal of the American Oil Chemists' Society, 86*(7), 637–642. doi:10.1007/s11746-009-1401-8.

Qian, J. Q., Zhao, C. Y., Ding, J., Chen, Y., & Guo, H. (2020). Preparation of nano-enzyme aggregates by crosslinking lipase with sodium tripolyphosphate. *Process Biochemistry, 97*, 19–26. doi:10.1016/j.procbio.2020.06.026.

Quiocho, F. A., & Richards, F. M. (1964). Intermolecular cross linking of a protein in the crystalline state: Carboxypeptidase-A. *PNAS, 52*(3), 833–839. doi:10.1073/pnas.52.3.833.

Ramos, M. D., Miranda, L. P., Fernandez-Lafuente, R., Kopp, W., & Tardioli, P. W. (2019). Improving the yields and reaction rate in the ethanolysis of soybean oil by using mixtures of lipase CLEAs. *Molecules, 24*(23), 4392. https://www.mdpi.com/1420-3049/24/23/4392/pdf. doi:10.3390/molecules24234392.

Rehman, S., Bhatti, H. N., Bilal, M., & Asgher, M. (2016). Cross-linked enzyme aggregates (CLEAs) of Pencilluim notatum lipase enzyme with improved activity, stability and reusability characteristics. *International Journal of Biological Macromolecules, 91*, 1161–1169. www.elsevier.com/locate/ijbiomac. doi:10.1016/j.ijbiomac.2016.06.081.

Reis, C., Sousa, E., Serpa, J., Oliveira, R., Oliveira, R., & Santos, J. (2019). Design of immobilized enzyme biocatalysts: Drawbacks and opportunities. *Química Nova, 42*(7). doi:10.21577/0100-4042.20170381.

Reshmi, R., & Sugunan, S. (2013). Improved biochemical characteristics of crosslinked β-glucosidase on nanoporous silica foams. *Journal of Molecular Catalysis B: Enzymatic, 85-86*, 111–118. doi:10.1016/j.molcatb.2012.08.007.

Roberge, C., Fleitz, F., Pollard, D., & Devine, P. (2007). Asymmetric synthesis of cyanohydrin derived from pyridine aldehyde with cross-linked aggregates of hydroxynitrile lyases. *Tetrahedron Letters, 48*(8), 1473–1477. doi:10.1016/j.tetlet.2006.12.053.

Robescu, M. S., & Bavaro, T. (2025). A comprehensive guide to enzyme immobilization: All you need to know. *Molecules, 30*(4), 939. http://www.mdpi.com/journal/molecules. doi:10.3390/molecules30040939.

Roessl, U., Nahálka, J., & Nidetzky, B. (2010). Carrier-free immobilized enzymes for biocatalysis. *Biotechnology Letters, 32*(3), 341–350. doi:10.1007/s10529-009-0173-4.

Sampaio, C. S., Angelotti, J. A. F., Fernandez-Lafuente, R., & Hirata, D. B. (2022). Lipase immobilization via cross-linked enzyme aggregates: Problems and prospects – A review. *International Journal of Biological Macromolecules, 215*, 434–449. doi:10.1016/j.ijbiomac.2022.06.139.

Schoevaart, R., Wolbers, M. W., Golubovic, M., Ottens, M., Kieboom, A. P. G., van Rantwijk, F., van der Wielen, L. A. M., & Sheldon, R. A. (2004). Preparation, optimization, and structures of cross-linked enzyme aggregates (CLEAs). *Biotechnol Bioeng, 87*(6), 754–762. doi:10.1002/bit.20184.

Scott, J. G., & Goddard, J. M. (2025). Trehalose decorated nanostructures stabilize combined cross-linked enzyme aggregates (Combi-CLEAs) of β-galactosidase and glucose isomerase. *International Journal of Biological Macromolecules, 302*, 140390. https://www.sciencedirect.com/science/journal/01418130. doi:10.1016/j.ijbiomac.2025.140390.

Shah, S., Sharma, A., & Gupta, M. N. (2006). Preparation of cross-linked enzyme aggregates by using bovine serum albumin as a proteic feeder. *Analytical Biochemistry, 351*(2), 207–213. http://www.elsevier.com/inca/publications/store/6/2/2/7/8/1/index.htt. doi:10.1016/j.ab.2006.01.028.

Sheldon, R. A. (2011). Characteristic features and biotechnological applications of cross-linked enzyme aggregates (CLEAs). *Applied Microbiology and Biotechnology, 92*(3), 467–477. doi:10.1007/s00253-011-3554-2.

Sheldon, R. A., & Brady, D. (2018). The limits to biocatalysis: Pushing the envelope. *Chemical Communications, 54*(48), 6088–6104. http://pubs.rsc.org/en/journals/journal/cc. doi:10.1039/c8cc02463d.

Sheldon, R. A., Schoevaart, R., & Van Langen, L. M. (2005). Cross-linked enzyme aggregates (CLEAs): A novel and versatile method for enzyme immobilization (a review). *Biocatalysis and Biotransformation, 23*(3-4), 141–147. doi:10.1080/10242420500183378.

Sheldon, R. A., van Pelt, S., Kanbak-Aksu, S., Rasmussen, J. A., & Janssen, M. H. A. (2013). Cross-linked enzyme aggregates (CLEAs) in organic synthesis. *Aldrichimica Acta, 46*(3), 81–93. http://www.sigmaaldrich.com/content/dam/sigma-aldrich/docs/Aldrich/Brochure/1/acta-46-3.pdf. Netherlands.

Sheldon, R. A. (2007). Enzyme immobilization: The quest for optimum performance. *Advanced Synthesis Catalysis, 349*(8-9), 1289–1307. doi:10.1002/adsc.200700082.

Sheldon, R. A. (2019). CLEAs, combi-CLEAs and "smart" magnetic CLEAs: Biocatalysis in a bio-based economy. *Catalysts, 9*(3), 261. doi:10.3390/catal9030261.

Taboada-Puig, R., Junghanns, C., Demarche, P., Moreira, M. T., Feijoo, G., Lema, J. M., & Agathos, S. N. (2011). Combined cross-linked enzyme aggregates from versatile peroxidase and glucose oxidase: Production, partial characterization and application for the elimination of endocrine disruptors. *Bioresource Technology, 102*(11), 6593–6599. doi:10.1016/j.biortech.2011.03.018.

Talekar, S., Pandharbale, A., Ladole, M., Nadar, S., Mulla, M., Japhalekar, K., Pattankude, K., & Arage, D. (2013). Carrier free co-immobilization of alpha amylase, glucoamylase and pullulanase as combined cross-linked enzyme aggregates (combi-cleas): A tri-enzyme biocatalyst with one pot starch hydrolytic activity. *Bioresource Technology, 147*, 269–275. www.elsevier.com/locate/biortech. doi:10.1016/j.biortech.2013.08.035.

Tandjaoui, N., Tassist, A., Abouseoud, M., Couvert, A., & Amrane, A. (2015). Preparation and characterization of cross-linked enzyme aggregates (CLEAs) of *Brassica rapa* peroxidase. *Biocatalysis and Agricultural Biotechnology, 4*(2), 208–213. http://www.journals.elsevier.com/biocatalysis-and-agricultural-biotechnology/. doi:10.1016/j.bcab.2014.12.002.

Tavernini, L., Ottone, C., Illanes, A., & Wilson, L. (2020). Entrapment of enzyme aggregates in chitosan beads for aroma release in white wines. *International Journal of Biological Macromolecules, 154*, 1082–1090. www.elsevier.com/locate/ijbiomac. doi:10.1016/j.ijbiomac.2020.03.031.

Torabizadeh, H., Tavakoli, M., & Safari, M. (2014). Immobilization of thermostable $\alpha$-amylase from *Bacillus licheniformis* by cross-linked enzyme aggregates method using calcium and sodium ions as additives. *Journal of Molecular Catalysis B: Enzymatic, 108*, 13–20. www.elsevier.com/locate/molcatb. doi:10.1016/j.molcatb.2014.06.005.

Tükel, S. S., Hürrem, F., Yildirim, D., & Alptekin, O. (2013). Preparation of crosslinked enzyme aggregates (CLEA) of catalase and its characterization. *Journal of Molecular Catalysis B: Enzymatic, 97*, 252–257. doi:10.1016/j.molcatb.2013.09.007.

Ullah, H., Pervez, S., Ahmed, S., Haleem, K. S., Qayyum, S., Niaz, Z., Nawaz, M. A., Nawaz, F., Subhan, F., & Tauseef, I. (2021). Preparation, characterization and stability studies of cross-linked $\alpha$-amylase aggregates (CLAAs) for continuous liquefaction of starch. *International Journal of Biological Macromolecules, 173*, 267–276. www.elsevier.com/locate/ijbiomac. doi:10.1016/j.ijbiomac.2021.01.057.

Velasco-Lozano, S., López-Gallego, F., Vázquez-Duhalt, R., Mateos-Díaz, J. C., Guisán, J. M., & Favela-Torres, E. (2014). Carrier-free immobilization of lipase from candida rugosa with polyethyleneimines by carboxyl-activated cross-linking. *Biomacromolecules, 15*(5), 1896–1903. http://pubs.acs.org/journal/bomaf6. doi:10.1021/bm500333v.

Velasco-Lozano, S., Rocha-Martin, J., & Santos, J. C. S. D. (2022). Editorial: Designing carrier-free immobilized enzymes for biocatalysis. *Frontiers in Bioengineering and Biotechnology, 10*, 924743. http://journal.frontiersin.org/journal/bioengineering-and-biotechnology#archive. doi:10.3389/fbioe.2022.924743.

Velasco-Lozano, S., López-Gallego, F., Mateos-Díaz, J. C., & Favela-Torres, E. (2016). Cross-linked enzyme aggregates (CLEA) in enzyme improvement—A review. *Biocatalysis, 1*(1). doi:10.1515/boca-2015-0012.

Vivek, C., Diksha, K., Kumar, D. V., Singh, K. S., Devendra, S., Kumar, D. V., & Himanshu, P. (2022). An insight in developing carrier-free immobilized enzymes. *Frontiers in Bioengineering and Biotechnology, 10*, 794411. doi:10.3389/fbioe.2022.794411.

Wahab, M. K. H. A., El-Enshasy, H. A. A., Bakar, F. D. A., Murad, A. M. A., Jahim, J. M., & Illias, R. M. (2019). Improvement of cross-linking and stability on cross-linked enzyme aggregate (CLEA)-xylanase by protein surface engineering. *Process Biochemistry, 86*, 40–49. doi:10.1016/j.procbio.2019.07.017.

Wang, A., Zhang, F., Chen, F., Wang, M., Li, H., Zeng, Z., Xie, T., & Chen, Z. (2011). A facile technique to prepare cross-linked enzyme aggregates using p-benzoquinone as cross-linking agent. *Korean Journal of Chemical Engineering, 28*(4), 1090–1095. doi:10.1007/s11814-010-0476-0.

Wang, M., Wang, H., Feng, Y., Xu, Q., Admassu, H., Yang, R., & Hua, X. (2018). Preparation and characterization of sugar-assisted cross-linked enzyme aggregates (CLEAs) of recombinant cellobiose 2-epimerase from *Caldicellulosiruptor saccharolyticus* (CsCE). *Journal of Agricultural and Food Chemistry, 66*(29), 7712–7721. http://pubs.acs.org/journal/jafcau. doi:10.1021/acs.jafc.8b02333.

Wang, M. F., Qi, W., Su, R. X., & He, Z. M. (2010). Preparation and properties of cross-linked enzyme aggregates of papain. *Gaodeng Xuexiao Huaxue Xuebao/Chemical Journal of Chinese Universities, 31*(9), 1774–1779.

Wilson, L., Fernández-Lorente, G., Fernández-Lafuente, R., Illanes, A., Guisán, J. M., & Palomo, J. M. (2006). CLEAs of lipases and poly-ionic polymers: A simple way of preparing stable biocatalysts with improved properties. *Enzyme and Microbial Technology, 39*(4), 750–755. doi:10.1016/j.enzmictec.2005.12.011.

Xu, D. Y., Yang, Y., & Yang, Z. (2011). Activity and stability of cross-linked tyrosinase aggregates in aqueous and nonaqueous media. *Journal of Biotechnology, 152*(1-2), 30–36. doi:10.1016/j.jbiotec.2011.01.014.

Xu, M. Q., Li, F. L., Yu, W. Q., Li, R. F., & Zhang, Y. W. (2020). Combined cross-linked enzyme aggregates of glycerol dehydrogenase and NADH oxidase for high efficiency in situ NAD+ regeneration. *International Journal of Biological Macromolecules, 144*, 1013–1021. www.elsevier.com/locate/ijbiomac. doi:10.1016/j.ijbiomac.2019.09.178.

Xu, M. Q., Wang, S. S., Li, L. N., Gao, J., & Zhang, Y. W. (2018). Combined cross-linked enzyme aggregates as biocatalysts. *Catalysts, 8*(10), 460. https://www.mdpi.com/2073-4344/8/10/460/pdf. doi:10.3390/catal8100460.

Yamaguchi, H., Miyazaki, M., Asanomi, Y., & Maeda, H. (2011). Poly-lysine supported cross-linked enzyme aggregates with efficient enzymatic activity and high operational stability. *Catalysis Science and Technology, 1*(7), 1256–1261. doi:10.1039/c1cy00084e.

Yamaguchi, H., Kiyota, Y., & Miyazaki, M. (2018). Techniques for preparation of cross-linked enzyme aggregates and their applications in bioconversions. *Catalysts, 8*(5), 174. doi:10.3390/catal8050174.

Ye, J., Li, A., Chu, T., Pan, X., & He, B. (2019). Poly-lysine supported cross-linked enzyme aggregates of penicillin G acylase and its application in synthesis of $\beta$-lactam antibiotics.

*International Journal of Biological Macromolecules, 140*, 423–428. www.elsevier.com/locate/ijbiomac. doi:10.1016/j.ijbiomac.2019.08.021.

Yip, Y. S., Jaafar, N. R., Rahman, R. A., Puspaningsih, N. N. T., Jailani, N., & Illias, R. M. (2024). Improvement of combined cross-linked enzyme aggregates of cyclodextrin glucanotransferase and maltogenic amylase by functionalization of cross-linker for maltooligosaccharides synthesis. *International Journal of Biological Macromolecules, 273*, 133241. https://www.sciencedirect.com/science/journal/01418130. doi:10.1016/j.ijbiomac.2024.133241.

Yip, Y. S., Manas, N. H. A., Jaafar, N. R., Rahman, R. A., Puspaningsih, N. N. T., & Illias, R. M. (2023). Combined cross-linked enzyme aggregates of cyclodextrin glucanotransferase and maltogenic amylase from *Bacillus lehensis* G1 for maltooligosaccharides synthesis. *International Journal of Biological Macromolecules, 242*, 124675. https://www.sciencedirect.com/journal/international-journal-of-biological-macromolecules. doi:10.1016/j.ijbiomac.2023.124675.

You, L. F., Wei, T., Zheng, Q. W., Jiang, B. H., Lin, J. F., & Guo, L. Q. (2019). Optimization of baccatin III production by cross-linked enzyme aggregate of taxoid 10$\beta$-O-acetyltransferase. *Molecular Biotechnology, 61*(7), 498–505. http://www.springer.com/humana+press/journal/12033. doi:10.1007/s12033-019-00179-1.

Yu, H. W., Chen, H., Wang, X., Yang, Y. Y., & Ching, C. B. (2006). Cross-linked enzyme aggregates (CLEAs) with controlled particles: Application to *Candida rugosa* lipase. *Journal of Molecular Catalysis B: Enzymatic, 43*(1-4), 124–127. doi:10.1016/j.molcatb.2006.07.001.

Zeinali, M., & Lenjannezhadian, H. (2018). Degradation of urea by entrapped cross-linked urease aggregates: A combinatorial approach to urease stabilization for environmental and industrial applications. *International Journal of Environmental Science and Technology, 15*(1), 49–56. doi:10.1007/s13762-017-1337-8.

Zhang, Q., Zha, X., Zhou, N., & Tian, Y. (2016). Preparation of crosslinked enzyme aggregates (CLEAs) of acid urease with urethanase activity and their application. *Journal of Basic Microbiology, 56*(4), 422–431. doi:10.1002/jobm.201500498.

Zheng, J., Chen, Y., Yang, L., Li, M., & Zhang, J. (2014). Preparation of cross-linked enzyme aggregates of trehalose synthase via co-aggregation with polyethyleneimine. *Applied Biochemistry and Biotechnology, 174*(6), 2067–2078. http://www.springer.com/humana+press/journal/12010. doi:10.1007/s12010-014-1104-y.

Zhou, L., Mou, H., Gao, J., Ma, L., He, Y., & Jiang, Y. (2017). Preparation of cross-linked enzyme aggregates of nitrile hydratase ES-NHT-118 from E. coli by macromolecular cross-linking agent. *Chinese Journal of Chemical Engineering, 25*(4), 487–492. doi:10.1016/j.cjche.2016.08.027.

Zhou, L., Tang, W., Jiang, Y., Ma, L., He, Y., & Gao, J. (2016). Magnetic combined cross-linked enzyme aggregates of horseradish peroxidase and glucose oxidase: An efficient biocatalyst for dye decolourization. *RSC Advances, 6*(93), 90061–90068. http://pubs.rsc.org/en/journals/journalissues. doi:10.1039/c6ra12009a.

CHAPTER

# Magnetic cross-linked enzyme aggregates (mCLEAs) 6

**Santosh Sethi and Virendra Rathod**
*Department of Chemical Engineering, Institute of Chemical Technology, Mumbai, Maharashtra, India*

## 6.1 Introduction

Enzyme immobilization is a well-established technique used to improve the stability, reusability, and cost effectiveness of biocatalysts in industrial processes. Traditionally, enzymes are immobilized by attaching them to solid support such as polymers, silica, or inorganic materials. These carrier-bound systems facilitate enzyme recovery and repeated use while often enhancing operational stability. However, despite these advantages, conventional immobilization methods exhibit several limitations (Cui & Jia, 2013). The presence of a solid carrier introduces mass-transfer resistance, requiring substrate to diffuse through the support matrix before reaching the enzyme active sites. Hence, the use of solid supports lowers the exposure of active enzymes and reduces the catalytic performance. Along with the lower catalytic performance, it also adds material cost, stability of the materials, and enzyme leaching, etc.

To address the limitations associated with carrier-bound enzyme immobilization, carrier-free strategies have gained attention. One of the most widely studied approaches is the preparation of cross-linked enzyme aggregates (CLEAs) (Cui et al., 2014). CLEAs are generally produced through a two-step approach: enzyme precipitation using agents such as salts, water-miscible organic solvents, or nonionic polymers, followed by chemical cross-linking of the resulting aggregates with glutaraldehyde, genipin, or dialdehyde polysaccharides to form an insoluble enzyme network through reactions with surface amino groups (Talekar et al., 2013). As a result, CLEAs typically show high enzyme loading and improved stability under thermal, chemical, and solvent conditions, making them suitable for several biocatalytic applications in food, pharmaceutical, and other specialty chemicals. (Asgher et al., 2014). However, their large-scale industrial application is often limited by challenges related to catalyst recovery and internal mass transfer (Wang et al., 2011).

Magnetic cross-linked enzyme aggregates (mCLEAs), a new generation of biocatalytic systems, have been developed that combine magnetic nanoparticles within the CLEA structure, thereby retaining the high catalytic efficiency of conventional CLEAs while enabling easy handling and rapid recovery. Upon application of an external magnetic field, mCLEAs can be easily separated from the reaction mixture

**Carrier-Free Enzyme Immobilization: Fundamentals and Innovations in Biocatalysis.**
**DOI: https://doi.org/10.1016/B978-0-443-44139-4.00007-1**

without the need for filtration or centrifugation, reducing processing time and energy while improving reusability (Mehde et al., 2018a).

Additionally, the use of magnetic nanoparticles in CLEAs enhances the mechanical stability of mCLEAs, allowing them to better withstand agitation and repeated operational cycles. The magnetic responsiveness also opens opportunities for advanced reactor designs, such as continuous-flow and microreactor systems, where the enzyme can be precisely positioned and efficiently reused (Kim et al., 2010; Melo et al., 2023). The development of mCLEAs represents a significant advancement in enzyme immobilization technology, which illustrates the evolution from carrier-bound enzymes to carrier-free CLEAs and, most recently, to magnetically responsive mCLEAs. This progression highlights the field's ongoing efforts to design biocatalysts that are more efficient, durable, and process-friendly bioprocesses (Del Arco et al., 2021).

The growing interest in mCLEAs also aligns closely with the principles of green and sustainable chemistry. Their compatibility with aqueous systems and mild operating conditions further enhances their environmental appeal, positioning them as key enablers of next-generation biocatalyst technologies. This chapter aims to discuss the fundamental principles of CLEA formation and magnetic integration, different synthesis approaches, and cross-linking strategies. The key factors influencing the catalytic performances, stability, and reusability of mCLEAs will also be highlighted. Additionally, it also considers current industrial applications, current challenges, and future research directions.

## 6.2 Synthesis strategies for magnetic cross-linked enzyme aggregates

The development of mCLEAs marks an important advancement in carrier-free enzyme immobilization, as it effectively integrates the high catalytic performance of mCLEAs with the potential advantage of magnetic separability. This integrated system allows for high enzyme loading while enabling rapid, simple, and efficient recovery of the biocatalyst using an external magnetic field.

### 6.2.1 Formation mechanisms

#### *6.2.1.1 Co-aggregation of enzymes and functionalization of magnetic nanoparticles*

Co-aggregation is the most widely and commonly adopted method for the synthesis of mCLEAs. This method involves the simultaneous aggregation of the enzyme molecule and magnetic nano-particles (MNPs) from the solution in the presence of a precipitating agent. During this process, enzyme molecules get trapped within the forming aggregate matrix. This causes the formation of enzyme aggregates surrounding the nanoparticles through intermolecular interaction, resulting in a stable enzyme -nanoparticle-complex (Bilal et al., 2018; Liu et al., 2018). MNPs, which are usually iron-oxide-based substances such as $Fe_3O_4$ or $Fe_2O_3$, interact with enzyme molecules through hydrophobic, electrostatic forces, and hydrogen bonding. These integrations are greatly enhanced, and the homogeneity of aggregation is improved when MNPs are surface functionalized with amino, hydroxyl, or carboxyl groups. Serving as nucleation sites, functionalized MNPs encourage consistent enzyme clustering and inhibit excessive nanoparticle agglomeration. The resultant enzyme-nanoparticle aggregates have a three-dimensional, porous structure with a magnetic core dispersed throughout the protein network. In order to achieve mechanical stability and uniform magnetic responsiveness, this spatial integration is essential. The necessity for controlled aggregations is highlighted by the fact that poor co-aggregation can result in heterogeneous particle size, weak magnetic separations, or enzyme leaching during repeated use (Bilal et al., 2019; Zakharchenko et al., 2018).

### 6.2.2 Cross-linking chemistry

Following co-aggregation, chemical cross-linking stabilizes the enzyme-MNP assembly, transforming weakly associated clusters into insoluble and mechanically stable mCLEAs. Crosslinking occurs primarily between functional groups on enzyme surfaces, most commonly amino groups of lysine residues and complementary reactive groups on other enzyme molecules or functionalized MNPs (Hojnik Podrepšek et al., 2022; Talekar et al., 2012).

### 6.2.3 Types of crosslinkers

1. Glutaraldehyde (GA)—It is the most frequently used linker due to its high reactivity, low cost, and effectiveness in forming Schiff base linkages. In mCLEAs, glutaraldehyde serves a dual role: cross-linking enzyme molecules to one another and covalently anchoring enzymes to amino-functionalized MNPs. However, excessive glutaraldehyde can induce enzyme over-rigidity, block active sites, and reduce catalytic activity (Noori et al., 2023; Tudorache et al., 2013).
2. Dialdehyde polysaccharides—It is a polymeric aldehyde derived from polysaccharides, including plants, animals, and microorganisms. In the preparation of mCLEAs, these polymers act as multifunctional cross-linkers due to the presence of highly reactive and readily participate in Schiff base cross-linking, a well-established and widely employed strategy involving reactions with amino-coating biomaterials (Zhang et al., 2025). It also provides adequate space between enzyme molecules, thereby minimizing steric hindrance and improving the retention of catalytic activity. Various dialdehyde polysaccharides are used as cross-linkers for immobilization, which include dextran polysaccharide, dextran polyaldehyde, chitosan, and pectin (Abd Rahman et al., 2021; Nadar & Rathod, 2016).
3. Genipin—It is a naturally derived cross-linker commonly used in the preparation of mCLEAs. It is obtained through chemical oxidation followed by reduction and hydrolysis, or via $\beta$-D-glucosidase hydrolysis of geniposide extracted from

the fruits of *Genipa americana* and *Gardenia jasminoides*. During the mCLEA synthesis, genipin reacts efficiently with primary amino groups present on enzyme surfaces under mild reaction conditions, resulting in the formation of a stable cross-linked enzyme network anchored onto magnetic particles (Hong et al., 2021). Compared to conventional cross-linkers such as glutaraldehyde, genipin exhibits significantly lower cytotoxicity, approximately 5000–10,000 times lower than glutaraldehyde, while maintaining good cross-linking efficiency. These properties make genipin suitable for the immobilization of sensitive enzymes and environmentally stable applications (Danait-Nabar & Singhal, 2024).

## 6.3 Synthesis methods

To prepare mCLEAs, a variety of synthesis techniques have been developed. These techniques are typically divided into 2 categories: in-situ or one-step and two-step immobilization techniques. The nature of the enzymes, their stability under processing conditions, and the suitable application all play a major role in the synthesis route selection.

### 6.3.1 One-step synthesis

The most popular and widely used technique for preparing mCLEA is the one-step synthesis method. This technique uses a single reaction mixture that contains the enzyme, MNPs, precipitant, and cross-linker, as depicted in Fig. 6.1. Magnetically recoverable enzyme aggregates are produced by the sequential process of enzyme precipitation, co-aggregation with MNPs, and cross-linking. This method is appealing for scale-up because it offers several benefits, such as ease of use, shorter processing time, and reduced material handling. However, because quick cross-linking can result in uneven aggregates or decreased enzyme activity, this technique necessitates control over reaction kinetics (Escalante Morales et al., 2023; Su et al., 2018).

### 6.3.2 Two-step synthesis approach

Conventionally, CLEAs are first synthesized using enzyme precipitation and crosslinking in the absence of MNPs. On the other hand, in the two-step synthesis method, MNPs are then added to the preformed CLEAs by physical adsorption, covalent attachment, or secondary cross-linking (Fig. 6.2). This approach enables independent optimization of CLEA formation, which is especially useful for enzymes that are susceptible to precipitation conditions or nanoparticle surfaces. Post-immobilization tuning of magnetic properties is also made possible by two-step synthesis. Nevertheless, MNPs are frequently confined to the surface of CLEAs, which can lead to nanoparticle detachment during repeated catalytic cycles and reduced magnetic responsiveness (Abdulhamid et al., 2021; Rai et al., 2021; Zhao et al., 2025b).

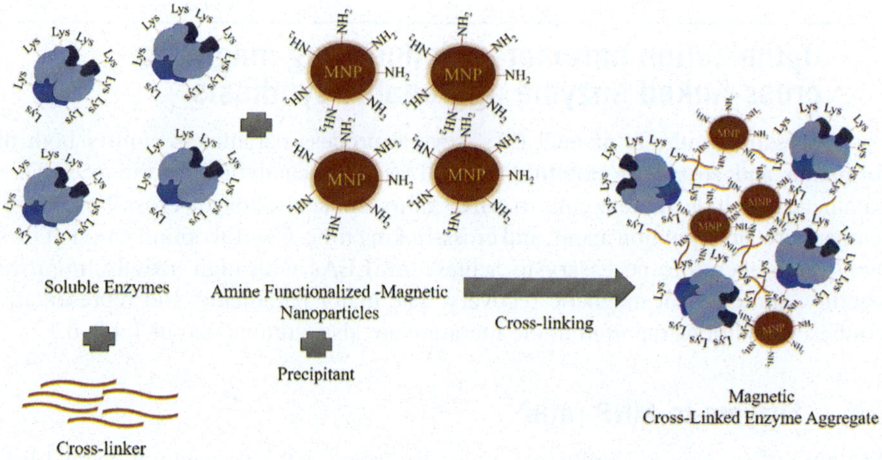


**FIGURE 6.1 One-step synthesis of mCLEAs.**

Schematic representation of the one-step synthesis of magnetic cross-linked enzyme aggregates (mCLEAs), illustrating simultaneous enzyme precipitation, magnetic nanoparticle incorporation, and cross-linking. CLEAs; cross-linked enzyme aggregates, mCLEAs; Magnetic cross-linked enzyme aggregates.

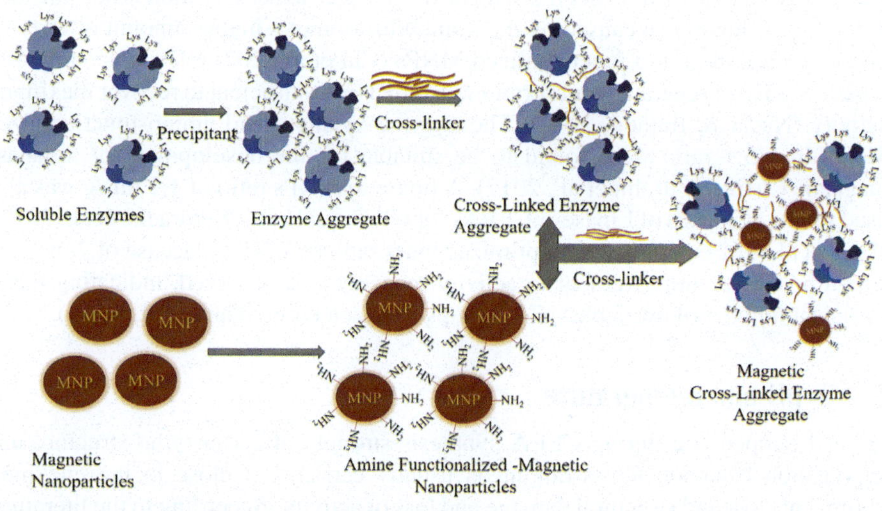


**FIGURE 6.2 Two-step synthesis of mCLEAs.**

Schematic representation of the two-step synthesis of magnetic cross-linked enzyme aggregates (mCLEAs), involving sequential enzyme aggregation/cross-linking and magnetic nanoparticle incorporation. *mCLEAs*; Magnetic cross-linked enzyme aggregates.

## 6.4 Optimization parameters influencing magnetic cross-linked enzyme aggregates synthesis

For successful synthesis of mCLEAs, several process parameters control both the formation and structural stability, as well as their catalytic performance. These parameters include the enzyme to MNP ratio, concentration of cross-linkers, pH, temperature, precipitation agent, and cross-linking time. Careful optimization of these variables is therefore necessary to achieve mCLEAs with high activity, improved stability, and efficient magnetic recovery. The major parameters and representative synthesis conditions reported in the literature are also summarized in Table 6.1.

### 6.4.1 Enzyme-to-MNP ratio

The ratio of enzyme to MNP is an important factor in the preparation of mCLEAs. If the amount of MNPs is too low, magnetic separation becomes inefficient, and recovery of the biocatalyst is difficult. Excessive MNPs reduce the enzyme content and can limit substrate diffusion within the aggregates. A suitable enzyme-to-MNP ratio always ensures easy magnetic recovery while maintaining good catalytic activity and reusability of the mCLEAs. As reported by Chi et al. (2021), the optimized enzyme to amino functionalized MNPs ratio was 0.02:1, and a significant reduction in the activity recovery was observed at a ratio of 0.03:1. This observation is possibly due to the stearic hindrance caused by the immobilization of a higher amount of enzyme on the surface of amino functionalized MNPs (Chi et al., 2021). For glucoamylase-based mCLEAs preparation, a 1:1 ratio was found to be sufficient to recover maximum activity (Nadar & Rathod, 2016). The enzyme xylanase and amino-functionalized MNPs in a 5:1 ratio were found to be suitable for the development of xylanase magnetic CLEAs (Purohit et al., 2017). A lipase to MNPs ratio of 1.2 mL/g was also used for the synthesis of Lipase-mCLEAs for GlyC synthesis (Tudorache et al., 2013). Wang et al. (2021) reported the optimized mass ratio of MNP to laccase of 7, beyond which no further improvement in activity recovery was observed, indicating that a sufficient number of available -$NH_2$ groups has been used (Wang et al., 2021).

### 6.4.2 pH and temperature

pH and temperature during CLEA synthesis strongly affect enzyme structure and aggregation behavior. Cross-linking is usually carried out close to the enzyme's optimal pH to avoid structural damage and loss of activity. According to the literature, moderate temperatures are preferred, as they support effective cross-linking while preserving enzyme stability. Unfavorable pH or temperature conditions may cause enzyme denaturation, leading to reduced catalytic activity. Reported studies indicated that mCLEAs synthesis was generally performed under mild pH and temperature conditions to maintain enzyme stability during the cross-linking. For example, mCLEAs synthesis at pH 6.0 and temperature 25°C using ammonium sulfate has been reported,

**Table 6.1** Summary of reported synthesis parameters of mCLEAs.

| Enzyme | Enzyme/ MNP ratio | pH | Temperature (°C) | Precipitating agent | Crosslinker concentration | Crosslinking time (h) | Reference |
|---|---|---|---|---|---|---|---|
| $\beta$-galactosidase | 0.1:1 | 6.0 | 25 | Ammonium sulphate | 120 mM | 6 | Rai et al. (2021) |
| Xylanase | 5:1 | | 25 | Ammonium sulphate | 250 mM | 8.36 | Purohit et al. (2017) |
| Transpeptidase | 0.02:1 | 9.0 | 30 | Iso-propanol | 20 mM | 1.5 | Chi et al. (2021) |
| Lipase | 5:1 | 7.0 | 30 | Ammonium sulphate | 40 mM | 3 | Badoei-dalfard et al. (2019) |
| Lipase | 1.2:1 | | 20 | Dimethyl carbonate (DMC) | 3.6% | 2 | Tudorache et al. (2013) |
| Laccase | 1:7 | 7.0 | 25 | Ammonium sulphate | 0.16% | 1 | Wang et al. (2021) |
| Glucoamylase | 1:1 | 5.0 | 28 ± 2 | Ammonium sulphate | 1.5% | 1 | Nadar and Rathod (2016) |

mCLEAs; *Magnetic cross-linked enzyme aggregates.*

while another study demonstrated effective preparation at pH 9.0 and 30°C (Chi et al., 2021; Purohit et al., 2017; Rai et al., 2021). Additionally, several studies reported mCLEA formation at pH 7.0 with a temperature range of 25°C to 30°C (Badoei-dalfard et al., 2019; Tudorache et al., 2013; Wang et al., 2021). Overall, these findings indicate that near-neutral to slightly alkaline pH and moderate temperatures were generally favored for the formation of stable mCLEAs while preserving enzyme activity.

### 6.4.3 Effect of precipitation agents

The precipitation agents are used to induce enzyme aggregation by lowering protein solubility without significantly affecting enzyme structure. Commonly used agents include inorganic salts, organic solvents, and water-soluble polymers. Ammonium sulfate and polyethylene glycol (PEG) are widely preferred because they promote gentle aggregation and help form stable, porous mCLEAs with minimal loss. Organic solvents such as ethanol, acetone, and isopropyl alcohol can cause rapid aggregation; excessive exposure may partially denature the enzyme. Therefore, careful selection and optimization of the precipitating agent are essential to obtain mCLEAs with good structural stability and catalytic efficiency. Several studies reported ammonium sulfate as an effective precipitant for lipase-based mCLEAs due to higher activity retention and immobilization efficiency (Cruz-Izquierdo et al., 2014). Tudorache et al. (2013) reported that 45% (v/v) dimethyl carbonate (DMC) was the optimal precipitating agent after screening through several agents such as ammonium sulfate, 2-propanol, PEG, and ethanol. The lipase precipitated with DMC also exhibited high catalytic efficiency for glycerol carbonate (GlyC) synthesis (Tudorache et al., 2013). In other reports, acetone (approx. 60%) and alcohol such as ethanol or 2-propanol were identified as optimal precipitants for enzymes, including cellulase, beta-galactosidase, and laccase, depending on enzyme type and synthesis conditions (Hojnik Podrepšek et al., 2022; Safari et al., 2026).

### 6.4.4 Cross-linker concentration

Cross-linker concentration determines the rigidity and porosity of the enzyme aggregate network. Low cross-linker level may lead to structural instability and enzyme leaching, while excessive cross-linking can reduce enzyme flexibility and block substrate access. Fine-tuning of cross-linker concentration is therefore critical for maximizing activity retention and operational stability. GA is the most widely used cross-linking agent. Several studies have reported that moderate GA concentration improved the stability and activity retention of immobilized enzymes. For example, increased GA concentration up to 1% enhanced activity recovery, whereas higher concentrations caused a decline due to excessive cross-linking, reduced enzyme flexibility, and limited substrate diffusion to active sites (Zhao et al., 2025b). Similar trends have been observed in various enzymes. Cellulase CLEAs prepared with ethanol and 0.625% GA showed about 94% relative activity, while higher GA levels resulted in partial enzyme inactivation (Ifko et al., 2023). Immobilization studies on beta-galactosidase, lipase, and transglutaminase also reported optimal GA as a cross-linking agent in the range of 1%–2% (Hojnik Podrepšek et al., 2022). A study on immobilization of xylanase enzyme on mCLEAs with the cross-linker concentration of 200 mM was considered the optimal cross-linker concentration. The decrease in activity after 200 mM was due to a high cross-linking density (Safari et al., 2026). Previous studies have also shown that GA concentrations exceeding 5 mM are more suitable for preparing mCLEAs. Higher concentration of GA induced the conformational changes that altered the shape of enzyme active sites, and below 5 mM, lower activity was observed, probably due to insufficient covalent cross-linking between enzyme and MNPs (Chi et al., 2021). Therefore, careful optimization of cross-linker concentration is essential to achieve a balance between structural stability and catalytic efficiency.

### 6.4.5 Cross-linking time

During the crosslinking stage, multifunctional cross-linkers react with functional groups on enzyme molecules to stabilize the aggregated structure formed during the precipitation. The duration of this reaction causes the extent of intermolecular bonding and, consequently, the rigidity and porosity of the resulting enzyme aggregates. Insufficient cross-linking time often leads to weakly stabilized enzyme aggregates that are prone to structural disintegration, enzyme leaching, and poor reusability during the catalytic cycles. Excessive cross-linking can also result in over-rigid structures, reduced pore accessibility, and partial masking of enzyme active sites, thereby diminishing catalytic activity. Tudorache et al. (2013) prepared mCLEAs varying the cross-linking time (1–17 h) and observed a marginal increase in cross-linking beyond 2 h, which was considered optimal (Tudorache et al., 2013). Another study on *Candida antarctica* lipase B (CALB) mCLEAs showed that extending the cross-linking time from 5 to 24 h increased activity retention from about 65% to nearly 100%, indicating the need for sufficient time to establish stable covalent bonds between the enzyme and support (Cruz-Izquierdo et al., 2014). Similar trends were reported in other immobilization systems, where enzyme activity significantly increased after longer reaction times (12–24 h), suggesting that adequate cross-linking duration is essential for achieving stable and active enzyme aggregates (Safari et al., 2026).

## 6.5 Characterization technique for magnetic cross-linked enzyme aggregates

Characterization of mCLEAs is essential to understand the relationship between structure, properties, and catalytic performance as well as confirming their suitability for industrial and biotechnological applications. Unlike traditional immobilized enzymes,

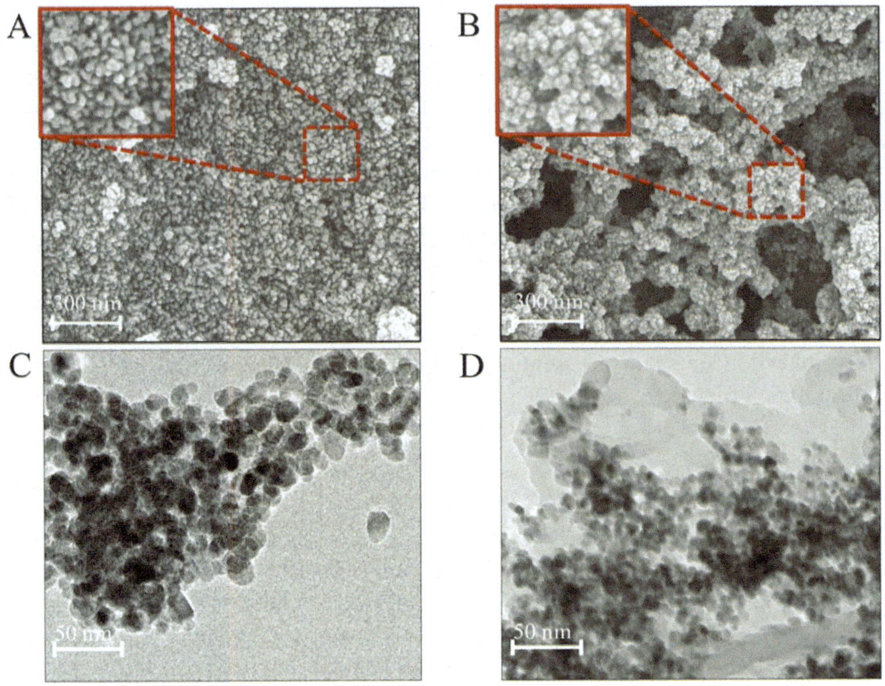


**FIGURE 6.3 SEM and TEM analysis of $Fe_3O_4$ nanoparticles and mCLEAs.**

SEM images of $Fe_3O_4$ nanoparticles (A) and mCLEAs (B), with insets showing magnified views of selected regions. TEM images of carboxyl-functionalized $Fe_3O_4$ (C), and mCLEAs (D). *mCLEAs*; Magnetic cross-linked enzyme aggregates.

*Reproduced with permission from Wei, B., Liu, F., Liu, X., Cheng, L., Yuan, Q., Gao, H., & Liang, H. (2022). Enhancing stability and by-product tolerance of β-glucuronidase based on magnetic cross-linked enzyme aggregates.* Colloids and Surfaces B: Biointerfaces, 210, *112241. https://doi.org/10.1016/j.colsurfb.2021.112241.*

mCLEAs represent hybrid biocatalytic systems composed of enzyme aggregates integrated with magnetic nanoparticles. As a result, their evaluation requires a multidisciplinary approach that combines morphological, structural, magnetic, and catalytic analysis. The Following section summarizes a few characterization techniques used for mCLEAs.

## 6.5.1 Morphological characterization by Scanning electron microscopy and Transmission electron microscopy

As shown in Fig. 6.3A and B, this method is widely used to examine the aggregate behavior, particle size distribution, and surface morphology of mCLEAs. Similar to CLEAs, the Scanning electron microscopy (SEM) images usually show regular, porous, and cluster-like structures. When compared to nonmagnetic CLEAs, the presence of MNPs frequently results in increased surface roughness and improved structural integrity. SEM analysis offered qualitative evidence of successful enzyme aggregation and cross-linking, as well as uniformity of particle formation under different synthesis conditions (Bhattacharya & Pletschke, 2014; Zhou et al., 2016). Changes in aggregation size can be directly linked to variations in synthesis parameters such as agitation speed or cross-linker concentration. In general, SEM is widely used to confirm enzymatic functionalization on the support by comparing the photomicrographs of pristine support with those of the enzyme support composite after immobilization.

Transmission electron microscopy (TEM) is used to visualize the internal structure of mCLEAs and the presence of magnetic particles and their distribution within the enzyme matrix. In TEM analysis, an electron beam passes through a very thin sample to visualize its internal structure. Areas containing elements with higher electron density scatter more electrons and appear darker in the images, while less dense regions allow more electrons to pass through and therefore appear brighter (Lin et al., 2021). Effective MNPs integration is also confirmed by TEM images, which usually display dark, electron-dense magnetic cores embedded within a proteinaceous network. Additionally, it is useful for distinguishing between MNPs trapped internally and surface-attached, as observed in Fig. 6.3C and D. Using TEM, the crystalline structure of the particle can also be determined. As a result, their evaluation requires through electron diffraction patterns (Lucena et al., 2019)

## 6.5.2 Structural and physicochemical characterization

### *6.5.2.1 Fourier transform infrared spectroscopy*

Fourier transform infrared spectroscopy (FTIR) spectroscopy is a primary tool for confirming chemical interactions and cross-linking in mCLEAs. It is also important to identify if the functional group is on the surface of MNPs before and after functionalization with enzymes, as shown in Fig. 6.4A. Protein secondary structure can be inferred from characteristic absorption bands that correspond to amide (~1650 $cm^{-1}$) and amide II (1540 $cm^{-1}$) regions (Chi et al., 2021). Enzyme aggregation and cross-linking were also indicated by shifts or intensity changes in these bands following immobilization. Additionally, Fe-O vibrations in the 500–600 $cm^{-1}$ range confirmed the presence of iron oxide nanoparticles, and variations in $C = N$ stretching vibrations frequently verify the creation of Schiff base linkages during the (Panwar et al., 2017).

### *6.5.2.2 X-ray diffraction*

X-ray diffraction (XRD) is used to evaluate the crystalline structure, phase composition, and average crystalline size of MNPs present in mCLEAs (Lucena et al., 2020). This technique is particularly useful for determining whether the magnetic core retains its structural integrity during immobilization and cross-linking, as observed in Fig. 6.3B. The appearance of characteristic diffraction peaks corresponding to

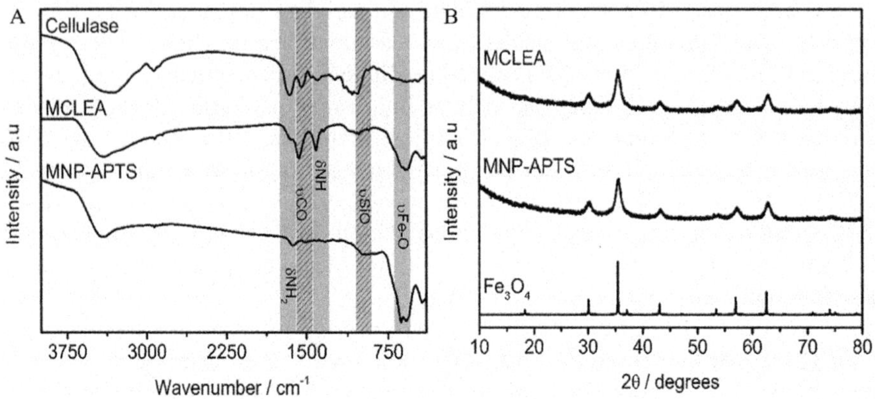


**FIGURE 6.4 FTIR and XRD characterization of MNP-APTS, mCLEAs.**

FTIR spectra of MNP-APTS, mCLEAs, and free cellulase enzyme (A) and Simulated XRD patterns of MNP, MNP-APTS, and mCLEAs (B). *mCLEAs*; Magnetic cross-linked enzyme aggregates.

*Reproduced with permission from Lucena, G. N., dos Santos, C. C., Pinto, G. C., Piazza, R. D., Guedes, W. N., Júnior, M. J., de Paula, A. V., Marques, R. F. C. (2020). Synthesis and characterization of magnetic cross-linked enzyme aggregate and its evaluation of the alternating magnetic field (AMF) effects in the catalytic activity.* Journal of Magnetism and Magnetic Materials, 516, *167326. https://doi.org/10.1016/j.jmmm.2020.167326.*

$Fe_3O_4$ or $Fe_2O_3$ confirms the presence and stability of magnetic phases. In most cases, XRD patterns recorded before and after immobilization show similar peak positions, suggesting that the synthesis process does not significantly alter the crystal structure of the nanoparticles within mCLEAs (Chi et al., 2021; Sandler et al., 2019).

### 6.5.2.3 X-ray photoelectron spectroscopy

Another powerful characterization technique, X-ray photoelectron spectroscopy (XPS), is widely used for surface analysis for the characterization of enzymatic biocomposites. This method is based on the photoelectric effect and enables elemental analysis of the outer surface of the materials to a depth of about 10 nm. When the samples are exposed to a lower-energy X-ray beam, photoelectrons are emitted and collected according to their kinetic energies, which are characteristic of their atomic origin (Bagus et al., 2013). In the case of MNPs, Amino functionalized MNPs, and mCLEAs, surface elemental analysis, as shown in Fig. 6.5 , revealed the appearance of nitrogen in amino MNPs, confirming successful (3-aminopropyl) triethoxysilane (APTES) functionalization, while the increased nitrogen content in MCLEAs indicated the effective incorporation of the enzyme along with amino functionalized MNPs (Hu et al., 2021). XPS enables the identification of all elements and also provides insights into chemical bonding and valence states (Lucena et al., 2019).

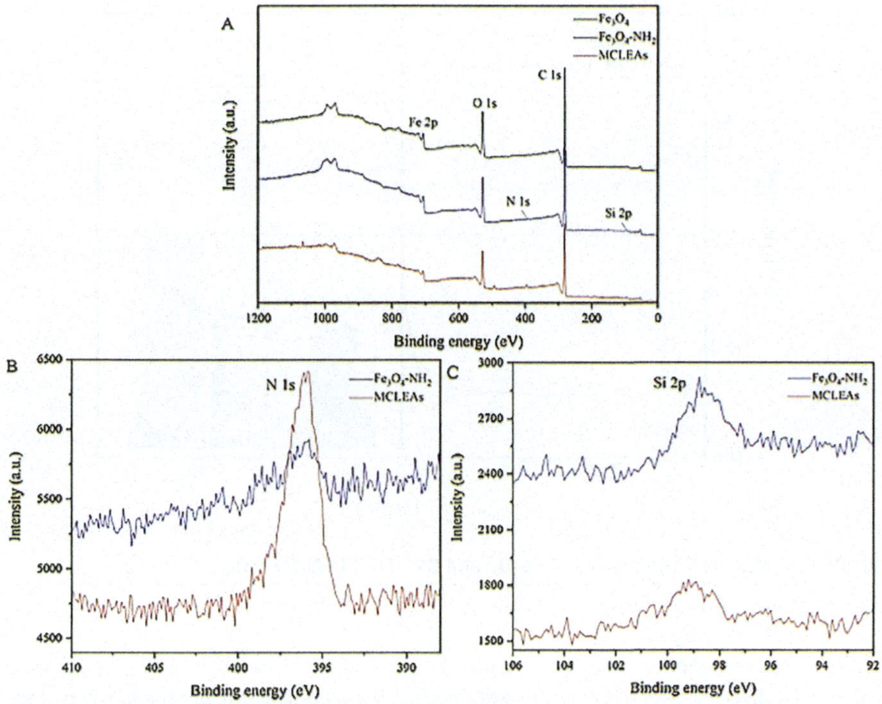


**FIGURE 6.5 XPS characterization of MNPs and mCLEAs.**

XPS survey spectrum of MNPs, MNPs -$NH_2$ and mCLEAs (A), XPS for N 1s electrons (B), and XPS for Si 2p electrons of MNPs -NH2 and mCLEAs (C). *mCLEAs*; Magnetic cross-linked enzyme aggregates.

*Reproduced with permission from: Hu, Q., Chen, G., Han, J., Wang, L., Cui, X., Wang, P., Chang, C., & Fu, Q. (2021). Determination of sarcosine based on magnetic cross-linked enzyme aggregates for diagnosis of prostate cancer.* Biochemical Engineering Journal, 172, *108039. https://doi.org/10.1016/j.bej.2021.108039.*

### 6.5.2.4 Vibrating sample magnetometry

Magnetic measurements play an important role in understanding and validating the performance of mCLEAs, particularly in terms of their ease of recovery and reuse (Fig. 6.6 ). Vibrating sample magnetometry (VSM) is one of the most common methods for measuring key magnetic parameters such as saturation magnetization (Ms), remanence (Mr), and coercivity (Hc). For practical uses, a high Ms is good because it makes magnetic separation more efficient. Low Mr and Hc values show super-magnetic behavior that allows the particles to redisperse easily even after the external magnetic field is removed, thus preventing the aggregation. After the enzyme immobilization, the Ms usually goes down. This is mostly because nonmagnetic materials like proteins or cross-linkers were added (Gama Cavalcante et al., 2024; Zhao et al., 2025a, 2025b). For example, several studies have reported that although

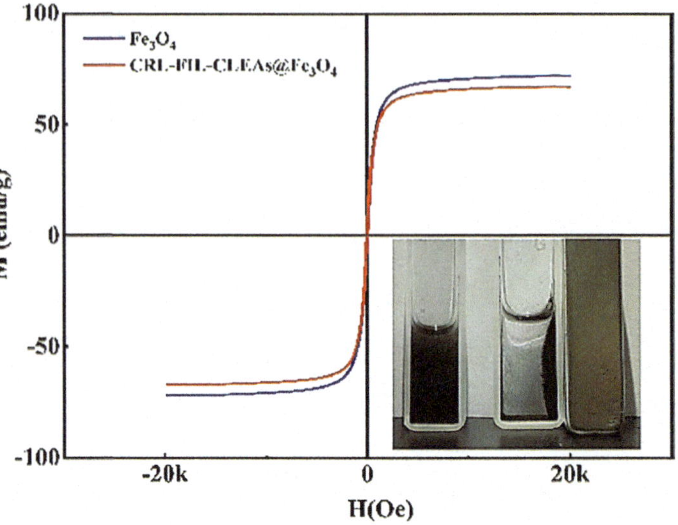


**FIGURE 6.6 Magnetization curve of $Fe_3O_4$ and CRL-FIL-CLEAs@$Fe_3O_4$.**

Magnetization curves of $Fe_3O_4$ and CRL-FIL-CLEAs@$Fe_3O_4$ measured using vibrating sample magnetometry (VSM).

*Reproduced with permission from Zou, B., Liu, F., Yan, Y., Feng, T., & Dowuona, J. N. N. (2022). Magnetic cross-linked lipase aggregates coupled with ultrasonic pretreatment for efficient synthesis of phytosterol oleate.* Molecular Catalysis, 531, *112649. https://doi.org/10.1016/j.mcat.2022.112649.*

the magnetization decreases after forming mCLEAs, the material still retains sufficient magnetic strength to be rapidly separated from the reaction medium using an external magnetic field (Farhan et al., 2021; Zhao et al., 2025b; Zou et al., 2022). Hysteresis studies further support this behavior, showing negligible Mr and Hc at room temperature, which confirms that the super-magnetic nature of the nanoparticle is largely preserved even after immobilization through mCLEAs (Cruz-Izquierdo et al., 2014; da Rocha et al., 2024; Polatoğlu & Yardım, 2022; Zou et al., 2022). Overall, VSM serves not only as a tool to assess magnetic particles but also as an indirect way to confirm successful immobilization and structural modification. The ability of mCLEAs to maintain strong magnetic responsiveness along with good dispersion stability highlights their suitability for repeated use, making them highly attractive for various industrial applications (Gama Cavalcante et al., 2024; Mehde et al., 2018b; Polatoğlu & Yardım, 2022).

### 6.5.3 Evaluation of enzyme performance

The catalytic performance of mCLEAs is commonly assessed by evaluating parameters such as catalytic activity, kinetic behavior, stability under different conditions, reusability, and recyclability. These characteristics are typically compared with those

**Table 6.2** Comparative performance characteristics of free enzyme, CLEAs, and mCLEAs.

| Parameters | Free enzyme | CLEAs | mCLEAs |
|---|---|---|---|
| Catalytic activity | Highest | Slightly reduced due to diffusion effects | Comparable to CLEAs |
| $K_m$ | Lower | Slightly higher | Slightly higher due to diffusion limitations |
| $V_{max}$ | Maximum | May decrease due to restricted enzyme flexibility | Comparable to CLEAs |
| pH stability | Narrow pH tolerance | Broder pH stability comparable to free enzyme | Wider pH tolerance |
| Thermal stability | Low and denatures at elevated temperatures | Improved compared to free enzyme | High thermal tolerance |
| Recovery | Difficult | Difficult require filtra-tion/centrifugation | Easy separation using external magnetic field |
| Reusability | Limited or not usable | Moderate | Excellent reusability with multiple catalytic cycles |

CLEAs; *cross-linked enzyme aggregates,* mCLEAs; *Magnetic cross-linked enzyme aggregates.*

of the corresponding free enzymes and conventional CLEAs to understand how immobilization and the incorporation of magnetic nanoparticles influence the enzyme performance in mCLEAs. A concise comparison is also presented in Table 6.2.

### *6.5.3.1 Enzyme activity assays*

The catalytic activity of an mCLEA is typically evaluated as an initial step to assessing its performance. This is typically carried out using standard substrate-specific assays, where the activity of mCLEAs is compared with that of the corresponding free enzyme and conventional nonmagnetic CLEAs. Such a comparison helps to understand how factors such as enzyme aggregation, the incorporation of magnetic nanoparticles, and applied cross-linking conditions influence overall enzymatic performance. Studies have reported that mCLEAs often exhibit higher catalytic activity than conventional CLEAs. For instance, Ifko et al. (2023) observed improved activity of mCLEAs towards carboxymethyl cellulose (CMC) compared with CLEAs. Similarly, Lucena et al. (2020) reported enhanced enzymatic performance of cellulase mCLEAs relative to cellulase CLEAs when evaluated using the same substrate. Activity retention was typically reported as a percentage of the free enzyme's initial activity. Although immobilization often leads to some loss of activity due to diffusion limitation and restricted enzyme mobility, the well-optimized mCLEAs are still able to retain a large fraction of their native catalytic efficiency (Mehde et al., 2018a; Nadar & Rathod, 2016). It has been observed that the presence of magnetic nanoparticles also helps improve dispersion and structural stability, which can partially compensate for activity loss typically observed in immobilized systems (Papatola et al., 2024).

#### 6.5.3.2 Kinetic parameters ($K_m$ and $V_{max}$)

Analysis of kinetic parameters such as $K_m$ and $V_{max}$ is commonly used to understand how enzyme behavior changes after immobilization. Michalis–Menten kinetic studies often show an increase in apparent $K_m$ value for mCLEAs, which is generally attributed to the diffusion barrier or reduced substrate accessibility to the active sites (Papatola et al., 2024). Variation in $V_{max}$ is also observed and may result from limited enzyme flexibility or partial blockage of catalytic regions within the aggregate. A comparative study between free cellulase and its immobilized forms, CLEAs and mCLEAs, showed ($0.012 \pm 0.0018$ mM, $0.055 \pm 0.0102$ mM, and $0.037 \pm 0.0012$ mM, respectively) that the immobilized enzymes exhibited a higher apparent $K_m$, suggesting lower affinity towards the cellulose substrate. In addition, the maximum reaction rate ($V_{max}$) of CLEAs and mCLEAs ($1.12 \pm 0.0012$ and $1.17 \pm 0.0023$ μmol/min, respectively) was found to be lower than that of the free enzyme. This decrease was mainly attributed to steric limitations that arise when immobilized enzymes interact with large polymeric substrates (Ifko et al., 2023). In another study, glucoamylase-based CLEAs were prepared and compared with the free form. The $V_{max}$ of mCLEAs ($4.4554 \pm 0.1659$ mol $min^{-1}$) was slightly lower than that of the free enzyme ($4.5991 \pm 0.2980$ mol $min^{-1}$), indicating that increased rigidity and compactness of the enzyme around MNPs hindered substrate diffusion to the active site after cross-linking; however, mCLEAs exhibited a 1.51-fold higher $V_{max}$ than the free enzyme (Nadar & Rathod, 2016). In the case of mCLEAs-lipase, the $V_{max}$ value of both the forms was the same, and the $K_m$ value was slightly lower than that of the free form, indicating a greater affinity. That is possibly due to the enzyme's conformational change after immobilization, which caused the protein to orient its active sites toward the substrate (Badoei-dalfard et al., 2019).

#### 6.5.3.3 Thermal and pH stability

Evaluating thermal and pH stability is essential for understanding how well mCLEAs perform under practical operating conditions. Immobilization typically improves enzyme stability by restricting conformational mobility and protecting against denaturation. The stability is typically assessed by exposing the biocatalyst to different temperatures and pH values and determining the remaining catalytic activity. According to Abdulhamid et al. (2021), the developed biocatalytic systems improved the thermal stability of the enzyme across a temperature range of 45°C–70°C, with lipase-mCLEAs demonstrating the highest stability. A comparable trend was also observed under varying pH conditions, where both lipase-mCLEAs and mCLEA maintained good stability within a pH range of 5–8 (Abdulhamid et al., 2021). Similar results were also observed in another study (Ifko et al., 2023). As a result, mCLEAs often exhibit wider pH tolerance and greater resistance to thermal inactivation than free enzymes.

This enhanced stability is one of the most significant advantages of mCLEAs for industrial biocatalytic applications (Hu et al., 2021; Safari et al., 2026).

#### 6.5.3.4 Magnetic separation and recovery efficiency

Magnetic separation and recovery tests are central to the characterization of mCLEAs and distinguish them from nonmagnetic immobilized systems. Recovery efficiency is evaluated by subjecting the biocatalyst suspension to external magnetic fields of varying strength and measuring the percentage of catalyst recovered within a given time. Important performance parameters include separation time, recovery yield, and stability over repeated reuse cycles. In a previous study, Zhao et al. (2025b) reported that conventional CLEAs retained only about 20% of their initial activity after several cycles of reuse, while mCLEAs maintained more than 70% of their activity even after repeated utilization (Zhao et al., 2025b). Similar observations have been reported for cellulase immobilization, where mCLEAs retained a 74% of their original activity after 6 operational cycles (Jia et al., 2017). In comparison, cellulase immobilized on other magnetic supports showed a relatively 55% after 4 repeated reuse (Gokhale et al., 2013). Efficient and well-designed mCLEAs generally show fast and efficient separation even under low magnetic strength, reflecting adequate MNPs loading and uniform distribution within the aggregates (Noori et al., 2023). Recovery studies are often combined with successive catalytic runs to evaluate enzyme leaching and long-term operational stability. The ease of rapid recovery and reuse with minimal activity loss highlights the suitability of mCLEAs for continuous and bioprocessing applications (Nikpour et al., 2024).

## 6.6 Applications of magnetic cross-linked enzyme aggregates

This section discusses the application of mCLEAs in a range of fields, including chemical synthesis, pharmaceutical manufacturing, food and beverage processing, biofuel production, and environmental bioremediation. These applications are described in detail in the following sections and briefly summarized in Table 6.3.

### 6.6.1 Magnetic cross-linked enzyme aggregates in fine chemical and pharmaceutical synthesis

mCLEAs have become increasingly important in fine chemical and pharmaceutical synthesis due to their high catalytic efficiency, selectivity, and ease of magnetic recovery. Enzymes such as lipase, oxidoreductases, and trans-aminases are commonly employed in mCLEA form for enantioselective reactions. Lipase-based CLEAs, including those prepared from commercial formulations such as Lipozyme 435, are widely used in esterification and transesterification processes and often retain high

**Table 6.3** Practical applications of mCLEAs across different fields.

| Enzyme | Precipitant | Cross-linker | Applications | Reference |
| --- | --- | --- | --- | --- |
| Lipase | Ethanol | Glutaraldehyde | Esterification and transesterification | Miranda et al. (2025) |
| Rhamnopyranosidase | Absolute ethanol, dimethylsulfoxide and 2-propanol | Glutaraldehyde | Glycompounds biosynthesis | Martins et al. (2018) |
| $\beta$-glucuronidase | Saturated $(NH_4)_2\ SO_4$ | Glutaraldehyde | Production of glycyrrhetinic acid (GA) | Wei et al. (2022) |
| $\beta$-galactosidase | | Glutaraldehyde, dextran-aldehyde | Lactose hydrolysis | de Freitas et al. (2023) |
| Lipase | Ethanol, saturated ammonium sulphate | Glutaraldehyde | Lipid hydrolysis, biodiesel synthesis, chemical conversion | Badoei-Dalfard et al. (2019), Sampaio et al. (2022) |
| Laccase | 2–propanol | Glutaraldehyde | Diclofenac degradation, removal of bisphenol | Primožič et al. (2020), Sadeghzadeh et al. (2020) |
| Horseradish peroxidase (HRP) and glucose oxidase (GOx) | $(NH_4)_2\ SO_4$ | Trimesic acid | Phenol degradation | Zhang et al. (2024) |
| Sarcosine oxidase | Ethanol | Glutaraldehyde | Sarcosine detection | Hu et al. (2021) |
| Horseradish peroxidase (HRP) and glucose oxidase (GOx) | | Glutaraldehyde | Glucose detection | Nguyen and Yang (2017) |

catalytic activity over multiple reuse cycles (Miranda et al., 2025). Similarly, oxidoreductase and transaminase mCLEAs have been applied in the asymmetric synthesis of chiral alcohols and amines, showing good operational stability and tolerance to the organic solvents (Martins et al., 2018). $\beta$-Glucuronidase mCLEAs have also been reported in the synthesis of Glycyrrhizinic acid (GA), where immobilization improves enzyme stability and resistance to byproduct inhibition (Wei et al., 2022). The application of mCLEAs in continuous or semi-continuous reactor systems further enhances productivity and process control, while their ability to operate under mild conditions aligns well with green chemistry principles and sustainable pharmaceutical manufacturing (Bennett et al., 2019; Sheldon, 2019).

### 6.6.2 Magnetic cross-linked enzyme aggregates in food and beverage processing

In food biotechnology, mCLEAs are increasingly used to improve enzyme stability while simplifying recovery and reuse. $\beta$-galactosidase mCLEAs have been applied for lactose hydrolysis in dairy products, where they maintain catalytic activity over multiple reaction cycles and can be easily separated from milk or whey using an external magnetic field without affecting product quality (de Freitas et al., 2023). Lipase-based mCLEAs are also employed in the modification of edible oils and fats, enabling the production of structured lipids with enhanced nutritional and physicochemical properties (Guimarães et al., 2018; Sampaio et al., 2022). These examples highlight the potential of mCLEAs as an effective alternative to conventional immobilizing systems in food industries, where mild processing conditions, product safety, and enzyme recyclability are of primary importance (Del Arco et al., 2021).

### 6.6.3 Magnetic cross-linked enzyme aggregates in biofuel production and biomass conversion

mCLEAs have also been widely explored in biofuel production and biomass conversion processes. Lipase-based mCLEAs are commonly used to catalyze the transesterification of vegetable oils and waste cooking oils with methanol or ethanol, providing an eco-friendly route to biodiesel production (Badoei-dalfard et al., 2019). Their magnetic response allows rapid recovery and reuse, which helps lower processing cost and waste generation (Hu et al., 2021). In lignocellulosic biomass conversion, cellulase and xylanase mCLEAs have been reported to efficiently hydrolyze cellulose and hemicellulose into fermentable sugars. These immobilized systems exhibit enhanced resistance to thermal deactivation and inhibitory compounds, making them well-suited for bio-ethanol production. The combination of high catalytic efficiency and magnetic recoverability improves the practicality of mCLEAs in integrated biorefinery operations and supports the development of a sustainable and circular bioeconomy framework (Lucena et al., 2019).

### 6.6.4 Environmental applications of magnetic cross-linked enzyme aggregates

mCLEAs have demonstrated considerable potential in environmental applications, particularly for the removal and degradation of persistent pollutants such as synthetic dyes, phenolic compounds, pesticides, and pharmaceutical residues. Their enhanced stability and recyclability make them well-suited for wastewater treatment and soil remediation applications. Laccase mCLEAs have been widely applied for the decolorization of textile dyes, including Reactive Blue 19 and Congo Red, while maintaining high activity over multiple cycles. In addition, laccase immobilized in the form of mCLEAs has been reported for the sustainable degradation of pharmaceutical contaminants such as diclofenac and the removal of endocrine-disrupting compounds like bisphenol A (BPA) (Primožič et al., 2020; Sadeghzadeh et al., 2020). Peroxidase and oxidase-based mCLEAs show comparable effectiveness in the degradation of aromatic compounds and phenolic compounds present in industrial effluents (Zhang et al., 2024). The magnetic properties of mCLEAs further enable simple and efficient recovery after treatment, minimizing the risk of secondary contamination. Together, these features make mCLEAs a promising bio-catalytic platform for large-scale and environmentally sustainable remediation technologies.

### 6.6.5 Magnetic cross-linked enzyme aggregates in biosensing and analytical systems

Recent advances in nanobiotechnology have expanded the use of mCLEAs beyond traditional catalysis into biosensing and analytical applications. The immobilization of enzymes such as glucose oxidase or horseradish peroxidase on magnetic nanoparticles enhances their electron transfer efficiency, resulting in improved sensitivity and stability of electrochemical sensors (Martins et al., 2018). Similarly, sarcosine oxidase-based mCLEAs have been explored for the detection of sarcosine, an important biomarker associated with prostate cancer, demonstrating good analytical performance along with enhanced stability and reusability (Hu et al., 2021). The use of multienzyme mCLEAs further allows cascade reactions within a confined environment, useful for signal amplification in analytical devices (Nguyen & Yang, 2017). In addition, the magnetic core of mCLEAs allows controlled positioning, easy regeneration, and reuse of the sensing element, facilitating their integration into reusable sensory platforms. These advantages highlight the growing role of mCLEAs in bioanalytical and diagnostic technologies.

## 6.7 Challenges

Despite significant progress in the development and application of mCLEAs, several scientific, technical, and practical challenges remain before their full-scale industrial implementation can be achieved. The following are the key challenges:

### 6.7.1 Nanoparticle toxicity and environmental concerns

The use of MNPs in the mCLEA system has raised important concerns related to toxicity and environmental impacts, particularly when iron oxide-based materials are involved. While materials such as $Fe_3O_4$ and $Fe_2O_3$ are generally considered biocompatible, their surface coating, external chemical functionalization, or leaching of nanoparticles during use can be problematic in food, pharmaceuticals, and biomedical applications. If the nanoparticles are not well stabilized within the enzyme aggregates, it can also complicate product purification and increase the risk of contamination. Moreover, to address these issues, future research should focus on the development of safer magnetic materials, improved encapsulation and immobilization strategies to prevent nanoparticle release, and establish standardized toxicity protocols for evaluating environmental and health impacts.

### 6.7.2 Enzyme leaching and long-term stability

Although mCLEAs are more stable than free enzymes, enzyme leaching remains a significant challenge, especially in challenging solutions like high temperature, extreme pH, or continuous flow operations. A progressive loss of active protein can result from inadequate cross-linking density or weak interactions between enzyme particles, which lowers catalytic efficiency and reproducibility. Furthermore, the aggregate matrix may experience a structural breakdown and mechanical stress as a result of repeated magnetic separation and dispersion cycles. These effects illustrate the necessity of systematic long-term stability and better cross-linking approaches.

### 6.7.3 Limited scalability and product reproducibility

Large-scale production of mCLEAs is still a major challenge, despite their potential in the lab. Batch-to-batch variability in particle size, enzyme loading, and activity retained frequently results from the increasing difficulty of controlling parameters such as mixing efficiency, heat transfer, and cross-linking kinetics in large volumes. The use of organic precipitants and cross-linkers can be more difficult for scale-up because of safety, financial, and environmental concerns.

## 6.8 Future prospectus

### 6.8.1 Hybrid and multifunctional magnetic cross-linked enzyme aggregates

Future research may concentrate on hybrid and multifunctional CLEAs, which combine enzymes with polymers, metal-organic frameworks, or carbon-based nanomaterials. These combinations can expand applications in catalysis and biosensing by enhancing the mechanical strength, improving mass transfer, and introducing multifunctionality.

### 6.8.2 Co-immobilization for multistep catalysis

Immobilizing multiple enzymes within a single mCLEA structure offers an efficient platform for multistep reactions. This approach can reduce diffusion limitations between sequential steps and improve the overall efficiency of multistep biocatalysis processes.

### 6.8.3 Sustainable cross-linking strategies

To advance the synthesis of CLEAs using bio-based and environmentally friendly cross-linkers, compounds such as tannic acid, citric acid, and other polyphenols represent effective alternatives. These cross-linkers offer clear advantages over conventional aldehyde-based cross-linkers by reducing toxicity, operating under mild reaction conditions, and improving overall environmental compatibility.

### 6.8.4 Integration with continuous-flow systems

The inherent magnetic responsiveness of mCLEAs enables their effective integration into an automated reactor system or continuous-flow system. By applying external magnetic fields, the biocatalysts can be retained within the reactor, allowing a prolonged continuous process with minimal enzyme loss and enhanced long-term process stability.

## 6.9 Conclusion

mCLEAs have emerged as a promising biocatalytic system that brings together enzyme immobilization, nanomaterials, and sustainable process design. The combination of carrier-free enzyme aggregates with magnetic recoverability allows mCLEAs to retain high catalytic activity while simplifying separations and reuse, offering a practical solution to many challenges faced by conventional enzyme immobilization approaches. This chapter has outlined the fundamental principles of mCLEA synthesis, key influencing parameters, characterization strategies, major industrial applications, and the associated challenges and future directions. In the future, mCLEA are expected to play an important role in the development of next-generation biocatalytic processes. Their compatibility with continuous flow systems, scope for multifunctional design, and strong alignment with green chemistry principles. Ongoing interdisciplinary research, supported by and closer collaboration between academia and industry, will be crucial for translating the promising features for mCLEAs into sustainable and biomanufacturing solutions.

## AI disclosure

During the preparation of this work, the author(s) used Grammarly in order to correct grammar. After using this tool/service, the author(s) reviewed and edited the content as needed and take(s) full responsibility for the content of the publication.

## References

Abd Rahman, N. H., Jaafar, N. R., Annuar, N. A. S., Rahman, R. A., Abdul Murad, A. M., El-Enshasy, H. A., & Illias, R. M. (2021). Efficient substrate accessibility of cross-linked levanase aggregates using dialdehyde starch as a macromolecular cross-linker. *Carbohydrate Polymers, 267*, 118159. doi:10.1016/j.carbpol.2021.118159.

Abdulhamid, M. B., Hero, J. S., Zamora, M., Gómez, M. I., Navarro, M. C., & Romero, C. M. (2021). Effect of the biological functionalization of nanoparticles on magnetic CLEA preparation. *International Journal of Biological Macromolecules, 191*, 689–698. www.elsevier.com/locate/ijbiomac. doi:10.1016/j.ijbiomac.2021.09.091.

Asgher, M., Shahid, M., Kamal, S., & Iqbal, H. M. N. (2014). Recent trends and valorization of immobilization strategies and ligninolytic enzymes by industrial biotechnology. *Journal of Molecular Catalysis B: Enzymatic, 101*, 56–66. doi:10.1016/j.molcatb.2013.12.016.

Badoei-dalfard, A., Malekabadi, S., Karami, Z., & Sargazi, G. (2019). Magnetic cross-linked enzyme aggregates of Km12 lipase: A stable nanobiocatalyst for biodiesel synthesis from waste cooking oil. *Renewable Energy, 141*, 874–882. http://www.journals.elsevier.com/renewable-and-sustainable-energy-reviews/. doi:10.1016/j.renene.2019.04.061.

Bagus, P. S., Ilton, E. S., & Nelin, C. J. (2013). The interpretation of XPS spectra: Insights into materials properties. *Surface Science Reports, 68*(2), 273–304. doi:10.1016/j.surfrep.2013.03.001.

Bennett, J. A., Campbell, Z. S., & Abolhasani, M. (2019). Role of continuous flow processes in green manufacturing of pharmaceuticals and specialty chemicals. *Current Opinion in Chemical Engineering, 26*, 9–19. http://www.elsevier.com/wps/find/journaldescription.cws_home/725837/description#description. doi:10.1016/j.coche.2019.07.007.

Bhattacharya, A., & Pletschke, B. I. (2014). Magnetic cross-linked enzyme aggregates (CLEAs): A novel concept towards carrier free immobilization of lignocellulolytic enzymes. *Enzyme and Microbial Technology, 61-62*, 17–27. www.elsevier.com/locate/enzmictec. doi:10.1016/j.enzmictec.2014.04.009.

Bilal, M., Mehmood, S., Rasheed, T., & Iqbal, H. M. N. (2019). Bio-catalysis and biomedical perspectives of magnetic nanoparticles as versatile carriers. *Magnetochemistry, 5*(3), 42. https://www.mdpi.com/2312-7481/5/3/42/pdf. doi:10.3390/magnetochemistry5030042.

Bilal, M., Zhao, Y., Rasheed, T., & Iqbal, H. M. N. (2018). Magnetic nanoparticles as versatile carriers for enzymes immobilization: A review. *International Journal of Biological Macromolecules, 120*, 2530–2544. www.elsevier.com/locate/ijbiomac. doi:10.1016/j.ijbiomac.2018.09.025.

Chi, M.-C., Huang, Y.-F., Lu, B.-Y., Lin, M.-G., Wang, T.-F., & Lin, L.-L. (2021). Magnetic cross-linked enzyme aggregates of a transpeptidase-specialized variant (N450D) of *Bacillus licheniformis* γ-glutamyl transpeptidase: An efficient and stable biocatalyst for L-theanine synthesis. *Catalysts, 11*(2), 243. doi:10.3390/catal11020243.

Cruz-Izquierdo, Á., Picó, E. A., López, C., Serra, J. L., Llama, M. J., & Agarwal, P. K. (2014). Magnetic cross-linked enzyme aggregates (mCLEAs) of *Candida antarctica* lipase: An efficient and stable biocatalyst for biodiesel synthesis. *PLoS ONE, 9*(12), e115202. doi:10.1371/journal.pone.0115202.

Cui, J. D., & Jia, S. R. (2013). Optimization protocols and improved strategies of cross-linked enzyme aggregates technology: Current development and future challenges. *Critical Reviews in Biotechnology, 35*(1), 15–28. doi:10.3109/07388551.2013.795516.

Cui, J. D., Cui, L. L., Zhang, S., Zhang, Y. F., Su, Z., Ma, G. H., & Coles, J. A. (2014). Hybrid magnetic cross-linked enzyme aggregates of phenylalanine ammonia lyase from *Rhodotorula glutinis*. *PLoS ONE, 9*(5), e97221. doi:10.1371/journal.pone.0097221.

Danait-Nabar, S., & Singhal, R. S. (2024). Genipin crosslinked L-asparaginase CLEAs effectively mitigate acrylamide in deep-fried sweet potato chips. *Food Bioscience, 60*, 104410. doi:10.1016/j.fbio.2024.104410.

da Rocha, C., Piazza, R., dos Santos, C., Lucena, G., Amantea, B., Jafelicci, M., Jr., de Paula, A., Rodriguez, A. R., Morales, M. A., & Marques, R. F. (2024). Magnetic crosslinked porcine pancreatic lipase aggregates for transesterification process. *Journal of the Brazilian Chemical Society, 35*(6). doi:10.21577/0103-5053.20240002.

de Freitas, L. A., de Sousa, M., Ribeiro, L. B., França, Í. W. L., & Gonçalves, L. R. B. (2023). Magnetic CLEAs of $\beta$-galactosidase from *Aspergillus oryzae* as a potential biocatalyst to produce tagatose from lactose. *Catalysts, 13*(2), 306. https://www.mdpi.com/journal/catalysts. doi:10.3390/catal13020306.

Del Arco, J., Alcántara, A. R., Fernández-Lafuente, R., & Fernández-Lucas, J. (2021). Magnetic micro-macro biocatalysts applied to industrial bioprocesses. *Bioresource Technology, 322*, 124547. doi:10.1016/j.biortech.2020.124547.

Escalante Morales, L. K., Sengar, P., Dorado Baeza, A., Vazquez-Duhalt, R., & Chauhan, K. (2023). Enhanced laccase activity and stability as crosslinked enzyme aggregates on magnetic copper ferrite nanoparticles for biotechnological processes. *ChemCatChem, 15*(22), e202301071. http://onlinelibrary.wiley.com/journal/10.1002/(ISSN)1867-3899. doi:10.1002/cctc.202301071.

Farhan, L. O., Mehdi, W. A., Taha, E. M., Farhan, A. M., Mehde, A. A., & Özacar, M. (2021). Various type immobilizations of isocitrate dehydrogenases enzyme on hyaluronic acid modified magnetic nanoparticles as stable biocatalysts. *International Journal of Biological Macromolecules, 182*, 217–227. www.elsevier.com/locate/ijbiomac. doi:10.1016/j.ijbiomac.2021.04.026.

Gama Cavalcante, A. L., Dari, D. N., da Silva Aires, F. I., de Castro, E. C., dos Santos, K. M., & dos Santos, J. C. S. (2024). Advancements in enzyme immobilization on magnetic nanomaterials: Toward sustainable industrial applications. *RSC Advances, 14*(25), 17946–17988. doi:10.1039/d4ra02939a.

Gokhale, A. A., Lu, J., & Lee, I. (2013). Immobilization of cellulase on magnetoresponsive graphene nano-supports. *Journal of Molecular Catalysis B: Enzymatic, 90*, 76–86. doi:10.1016/j.molcatb.2013.01.025.

Guimarães, J. R., de Lima Camargo Giordano, R., Fernandez-Lafuente, R., & Tardioli, P. W. (2018). Evaluation of strategies to produce highly porous cross-linked aggregates of porcine pancreas lipase with magnetic properties. *Molecules, 23*(11), 2993. doi:10.3390/molecules23112993.

Hojnik Podrepšek, G., Knez, Ž., & Leitgeb, M. (2022). The synthesis of (magnetic) crosslinked enzyme aggregates with laccase, cellulase, $\beta$-galactosidase and transglutaminase. *Frontiers in Bioengineering and Biotechnology, 10*, 813919. doi:10.3389/fbioe.2022.813919.

Hong, J., Jung, D., Park, S., Oh, Y., Oh, K. K., & Lee, S. H. (2021). Immobilization of laccase via cross-linked enzyme aggregates prepared using genipin as a natural cross-linker. *International Journal of Biological Macromolecules, 169*, 541–550. doi:10.1016/j.ijbiomac.2020.12.136.

Hu, Q., Chen, G., Han, J., Wang, L., Cui, X., Wang, P., Chang, C., & Fu, Q. (2021). Determination of sarcosine based on magnetic cross-linked enzyme aggregates for diagnosis of prostate cancer. *Biochemical Engineering Journal, 172*, 108039. doi:10.1016/j.bej.2021.108039.

Ifko, D., Vasić, K., Knez, Ž., & Leitgeb, M. (2023). (Magnetic) cross-linked enzyme aggregates of cellulase from *T. reesei*: A stable and efficient biocatalyst. *Molecules, 28*(3), 1305. doi:10.3390/molecules28031305.

Jia, J., Zhang, W., Yang, Z., Yang, X., Wang, N., & Yu, X. (2017). Novel magnetic cross-linked cellulase aggregates with a potential application in lignocellulosic biomass bioconversion. *Molecules, 22*(2), 269. doi:10.3390/molecules22020269.

Kim, M., Park, J.-M., Um, H.-J., Lee, D.-H., Lee, K.-H., Kobayashi, F., Iwasaka, Y., Hong, C.-S., Min, J., & Kim, Y.-H. (2010). Immobilization of cross-linked lipase aggregates onto magnetic beads for enzymatic degradation of polycaprolactone. *Journal of Basic Microbiology, 50*(3), 218–226. doi:10.1002/jobm.200900099.

Lin, Y., Zhou, M., Tai, X., Li, H., Han, X., & Yu, J. (2021). Analytical transmission electron microscopy for emerging advanced materials. *Matter, 4*(7), 2309–2339. www.cell.com/matter. doi:10.1016/j.matt.2021.05.005.

Liu, D.-M., Chen, J., & Shi, Y.-P. (2018). Advances on methods and easy separated support materials for enzymes immobilization. *TrAC Trends in Analytical Chemistry, 102*, 332–342. doi:10.1016/j.trac.2018.03.011.

Lucena, G. N., Santos, C. C. D., Pinto, G. C., Piazza, R. D., Guedes, W. N., Jafelicci Júnior, M., de Paula, A. V., & Marques, R. F. C. (2020). Synthesis and characterization of magnetic cross-linked enzyme aggregate and its evaluation of the alternating magnetic field (AMF) effects in the catalytic activity. *Journal of Magnetism and Magnetic Materials, 516*, 167326. https://www.journals.elsevier.com/journal-of-magnetism-and-magnetic-materials. doi:10.1016/j.jmmm.2020.167326.

Lucena, G. N., dos Santos, C. C., Pinto, G. C., da Rocha, C. O., Brandt, J. V., de Paula, A. V., Júnior, M. J., & Costa Marques, R. F. (2019). Magnetic cross-linked enzyme aggregates (MCLEAs) applied to biomass conversion. *Journal of Solid State Chemistry, 270*, 58–70. doi:10.1016/j.jssc.2018.11.003.

Martins, S. L., Albuquerque, B. F., Nunes, M. A. P., & Ribeiro, M. H. L. (2018). Exploring magnetic and imprinted cross-linked enzyme aggregates of rhamnopyranosidase in microbioreactors. *Bioresource Technology, 249*, 704–712. www.elsevier.com/locate/biortech. doi:10.1016/j.biortech.2017.10.078.

Mehde, A. A., Mehdi, W. A., Özacar, M., & Özacar, Z. Z. (2018a). Evaluation of different saccharides and chitin as eco-friendly additive to improve the magnetic cross-linked enzyme aggregates (CLEAs) activities. *International Journal of Biological Macromolecules, 118*, 2040–2050. www.elsevier.com/locate/ijbiomac. doi:10.1016/j.ijbiomac.2018.07.075.

Mehde, A. A., Mehdi, W. A., Severgün, O., Çakar, S., & Özacar, M. (2018b). Lipase-based on starch material as a development matrix with magnetite cross-linked enzyme aggregates and its application. *International Journal of Biological Macromolecules, 120*, 1533–1543. www.elsevier.com/locate/ijbiomac. doi:10.1016/j.ijbiomac.2018.09.141.

Melo, R. L. F., Sales, M. B., de Castro Bizerra, V., de Sousa Junior, P. G., Cavalcante, A. L. G., Freire, T. M., Neto, F. S., Bilal, M., Jesionowski, T., Soares, J. M., Fechine, P. B. A., & dos Santos, J. C. S. (2023). Recent applications and future prospects of magnetic biocatalysts. *International Journal of Biological Macromolecules, 253*, 126709. https://www.sciencedirect.com/journal/international-journal-of-biological-macromolecules. doi:10.1016/j.ijbiomac.2023.126709.

Miranda, L. P., Guimarães, J. R., Fernandez-Lafuente, R., & Tardioli, P. W. (2025). Ethanolysis of degummed soybean oil using magnetic CLEAs from Eversa® Transform. *Journal of Biotechnology, 402*, 79–86. https://www.sciencedirect.com/science/journal/01681656. doi:10.1016/j.jbiotec.2025.03.010.

Nadar, S. S., & Rathod, V. K. (2016). Magnetic macromolecular cross linked enzyme aggregates (CLEAs) of glucoamylase. *Enzyme and Microbial Technology, 83*, 78–87. www.elsevier.com/locate/enzmictec. doi:10.1016/j.enzmictec.2015.10.009.

Nguyen, L. T., & Yang, K. L. (2017). Combined cross-linked enzyme aggregates of horseradish peroxidase and glucose oxidase for catalyzing cascade chemical reactions. *Enzyme and Microbial Technology, 100*, 52–59. www.elsevier.com/locate/enzmictec. doi:10.1016/j.enzmictec.2017.02.007.

Nikpour, S., Salehipour, M., Rezaei, S., & Mogharabi-Manzari, M. (2024). Design and fabrication of magnetic cross-linked laccase aggregate using superparamagnetic metal-organic frameworks for phenol removal. *Biochemical Engineering Journal, 209*, 109385. doi:10.1016/j.bej.2024.109385.

Noori, R., Perwez, M., Mazumder, J. A., Ali, J., & Sardar, M. (2023). Bio-imprinted magnetic cross-linked polyphenol oxidase aggregates for enhanced synthesis of L-dopa, a neurodegenerative therapeutic drug. *International Journal of Biological Macromolecules, 227*, 974–985. www.elsevier.com/locate/ijbiomac. doi:10.1016/j.ijbiomac.2022.11.274.

Panwar, D., Kaira, G. S., & Kapoor, M. (2017). Cross-linked enzyme aggregates (CLEAs) and magnetic nanocomposite grafted CLEAs of GH26 endo-$\beta$-1,4-mannanase: Improved activity, stability and reusability. *International Journal of Biological Macromolecules, 105*, 1289–1299. www.elsevier.com/locate/ijbiomac. doi:10.1016/j.ijbiomac.2017.07.154.

Papatola, F., Slimani, S., Peddis, D., & Pellis, A. (2024). Biocatalyst immobilization on magnetic nano-architectures for potential applications in condensation reactions. *Microbial Biotechnology, 17*(6), e14481. doi:10.1111/1751-7915.14481.

Polatoğlu, İ., & Yardım, A. (2022). Determination of effective assay parameters on the activity of magnetite cross-linked invertase aggregates by personal glucose meter. *Biocatalysis and Biotransformation, 40*(1), 1–8. http://www.tandfonline.com/loi/ibab20. doi:10.1080/10242422.2021.1876680.

Primožič, M., Kravanja, G., Knez, Ž., Crnjac, A., & Leitgeb, M. (2020). Immobilized laccase in the form of (magnetic) cross-linked enzyme aggregates for sustainable diclofenac (bio)degradation. *Journal of Cleaner Production, 275*, 124121. https://www.journals.elsevier.com/journal-of-cleaner-production. doi:10.1016/j.jclepro.2020.124121.

Purohit, A., Kumar Rai, S., Chownk, M., Singh Sangwan, R., & Kumar Yadav, S. (2017). Xylanase from *Acinetobacter pittii* MASK 25 and developed magnetic cross-linked xylanase aggregate produce predominantly xylopentose and xylohexose from agro biomass. *Bioresource Technology, 244*, 793–799. doi:10.1016/j.biortech.2017.08.034.

Rai, S. K., Kumar, V., & Yadav, S. K. (2021). Development of recyclable magnetic cross-linked enzyme aggregates for the synthesis of high value rare sugard-tagatose in aqueous phase catalysis. *Catalysis Science and Technology, 11*(6), 2186–2194. http://pubs.rsc.org/en/journals/journal/cy. doi:10.1039/d0cy02397c.

Sadeghzadeh, S., Nejad, Z. G., Ghasemi, S., Khafaji, M., & Borghei, S. M. (2020). Removal of bisphenol A in aqueous solution using magnetic cross-linked laccase aggregates from Trametes hirsuta. *Bioresource Technology, 306*, 123169. doi:10.1016/j.biortech.2020.123169.

Safari, J. B., Ramatsui, L., Nisbet, L., Smith, V. J., Krause, R. W. M., & Pletschke, B. I. (2026). Preparation of magnetic xylanase cross-linked aggregates (m-CLXAs) for the hydrolysis of arabinoxylan by mechano-enzymology. *Enzyme and Microbial Technology, 192*, 110756. https://www.sciencedirect.com/science/journal/01410229. doi:10.1016/j.enzmictec.2025.110756.

Sampaio, C. S., Angelotti, J. A. F., Fernandez-Lafuente, R., & Hirata, D. B. (2022). Lipase immobilization via cross-linked enzyme aggregates: Problems and prospects: A review. *International Journal of Biological Macromolecules, 215*, 434–449. www.elsevier.com/locate/ijbiomac. doi:10.1016/j.ijbiomac.2022.06.139.

Sandler, S. E., Fellows, B., & Mefford, O. T. (2019). Best practices for characterization of magnetic nanoparticles for biomedical applications. *Analytical Chemistry, 91*(22), 14159–14169. doi:10.1021/acs.analchem.9b03518.

Sheldon, R. A. (2019). CLEAs, combi-CLEAs and "smart" magnetic CLEAs: Biocatalysis in a bio-based economy. *Catalysts, 9*(3), 261. doi:10.3390/catal9030261.

Su, E., Meng, Y., Ning, C., Ma, X., & Deng, S. (2018). Magnetic combined cross-linked enzyme aggregates (Combi-CLEAs) for cofactor regeneration in the synthesis of chiral alcohol. *Journal of Biotechnology, 271*, 1–7. www.elsevier.com/locate/jbiotec. doi:10.1016/j.jbiotec.2018.02.007.

Talekar, S., Ghodake, V., Ghotage, T., Rathod, P., Deshmukh, P., Nadar, S., Mulla, M., & Ladole, M. (2012). Novel magnetic cross-linked enzyme aggregates (magnetic CLEAs) of alpha amylase. *Bioresource Technology, 123*, 542–547. doi:10.1016/j.biortech.2012.07.044.

Talekar, S., Joshi, A., Joshi, G., Kamat, P., Haripurkar, R., & Kambale, S. (2013). Parameters in preparation and characterization of cross linked enzyme aggregates (CLEAs). *RSC Advances, 3*(31), 12485–12511. doi:10.1039/c3ra40818c.

Tudorache, M., Nae, A., Coman, S., & Parvulescu, V. I. (2013). Strategy of cross-linked enzyme aggregates onto magnetic particles adapted to the green design of biocatalytic synthesis of glycerol carbonate. *RSC Advances, 3*(12), 4052–4058. http://pubs.rsc.org/en/journals/journalissues. doi:10.1039/c3ra23222k.

Wang, H., Han, S., Wang, J., Yu, S., Li, X., & Lu, L. (2021). Preparation and synthetic dye decolorization ability of magnetic cross-linked enzyme aggregates of laccase from *Bacillus amyloliquefaciens*. *Bioprocess and Biosystems Engineering, 44*(4), 727–735. https://rd.springer.com/journal/449. doi:10.1007/s00449-020-02481-8.

Wang, M., Jia, C., Qi, W., Yu, Q., Peng, X., Su, R., & He, Z. (2011). Porous-CLEAs of papain: Application to enzymatic hydrolysis of macromolecules. *Bioresource Technology, 102*(3), 3541–3545. doi:10.1016/j.biortech.2010.08.120.

Wei, B., Liu, F., Liu, X., Cheng, L., Yuan, Q., Gao, H., & Liang, H. (2022). Enhancing stability and by-product tolerance of $\beta$-glucuronidase based on magnetic cross-linked enzyme aggregates. *Colloids and Surfaces B: Biointerfaces, 210*, 112241. doi:10.1016/j.colsurfb.2021.112241.

Zakharchenko, A., Guz, N., Laradji, A. M., Katz, E., & Minko, S. (2018). Magnetic field remotely controlled selective biocatalysis. *Nature Catalysis, 1*(1), 73–81. https://www.nature.com/natcatal/. doi:10.1038/s41929-017-0003-3.

Zhang, Q., Li, N., Hou, Y., Fan, M., Zhang, Y., & Dang, F. (2024). Co-immobilization of crosslinked enzyme aggregates on lysozyme functionalized magnetic nanoparticles for enhancing stability and activity. *International Journal of Biological Macromolecules, 273*(Pt 2), 133180. doi:10.1016/j.ijbiomac.2024.133180.

Zhang, W., Azizi-Lalabadi, M., Can Karaca, A., Abedi-Firoozjah, R., Assadpour, E., Zhang, F., & Jafari, S. M. (2025). A review of bio-based dialdehyde polysaccharides as multifunctional building blocks for biomedical and food science applications. *International Journal of Biological Macromolecules, 309*(Pt 4), 142964. doi:10.1016/j.ijbiomac.2025.142964.

Zhao, J., Bai, L., Li, S., Cao, Z., Peng, Y., Bai, J., Cai, X., Shi, X., Lin, X., Wei, G., & Zhang, X. (2025a). An overview of advanced instruments for magnetic characterization and measurements. *Frontiers in Electronics, 6*. doi:10.3389/felec.2025.1645594.

Zhao, Y., Zhang, W., Li, Y., Fei, W., & Liang, H. (2025b). Magnetic cross-linked enzyme aggregates of glycoside hydrolase: An efficient and stable biocatalyst for icaritin. *Catalysts, 15*(11), 1034. doi:10.3390/catal15111034.

Zhou, L., Tang, W., Jiang, Y., Ma, L., He, Y., & Gao, J. (2016). Magnetic combined cross-linked enzyme aggregates of horseradish peroxidase and glucose oxidase: An efficient biocatalyst for dye decolourization. *RSC Advances, 6*(93), 90061–90068. http://pubs.rsc.org/en/journals/journalissues. doi:10.1039/c6ra12009a.

Zou, B., Liu, F., Yan, Y., Feng, T., & Dowuona, J. N. N. (2022). Magnetic cross-linked lipase aggregates coupled with ultrasonic pretreatment for efficient synthesis of phytosterol oleate. *Molecular Catalysis, 531*, 112649. doi:10.1016/j.mcat.2022.112649.

CHAPTER

# Methods for the structural characterization in carrier-free immobilization

# 7

**Xingmiao Lu, Feihong Hou, Xiaofei Ren, Yushuo Zhang, Liying Wang and Fufeng Liu**
*College of Biotechnology, Tianjin University of Science and Technology, Tianjin, China*

Carrier-free immobilized enzymes represent a pivotal class of biocatalytic materials, distinguished by a suite of excellent properties including enhanced stability, high activity retention, and unique selectivity. These inherent advantages endow them with substantial application potential across diverse fields, ranging from biocatalysis, biosensing, and environmental remediation. It is well-recognized that the performance and functional behavior of carrier-free immobilized enzymes are predominantly determined by their structural features, such as the spatial configuration of enzyme molecules, aggregation patterns, cross-linking density within the three-dimensional network, particle morphology and size distribution, lattice structure, and internal pore characteristics. A comprehensive understanding of these structural properties is therefore fundamental to unlocking the full potential of carrier-free immobilized enzyme technology.

The field of structural characterization for carrier-free immobilized enzymes has undergone significant development and refinement, evolving from basic morphological observation to sophisticated multidimensional analysis. Early investigations primarily relied on optical microscopy (OM), a technique that provides only a preliminary understanding of the morphology and overall structure of immobilized enzymes due to its limited resolution. Subsequently, scanning electron microscopy (SEM) gradually became the predominant technique for examining the morphology and particle size of carrier-free immobilized enzymes. SEM offers significantly higher resolution compared to OM, enabling clearer visualization of surface morphologies and particle agglomeration states. Nevertheless, the scope of structural characterization using early SEM remained relatively restricted, focusing primarily on surface features rather than internal structures or dynamic conformational changes of enzyme molecules.

With the rapid advancement and interdisciplinary integration of diverse analytical techniques, substantial progress has been achieved in the structural analysis of carrier-free immobilized enzymes. A series of advanced characterization tools have been increasingly adopted, including high-resolution and field-emission electron microscopy (FEEM), confocal laser scanning microscopy (CLSM), advanced X-ray methods (e.g., synchrotron radiation sources), $^{1}H$ and $^{13}C$ nuclear magnetic resonance

**Carrier-Free Enzyme Immobilization: Fundamentals and Innovations in Biocatalysis.**
**DOI: https://doi.org/10.1016/B978-0-443-44139-4.00005-8**

(NMR), as well as in situ/operando characterization and nanomechanical testing. These technological advancements have facilitated multidimensional and multiscale characterization of carrier-free immobilized enzymes, encompassing macroscopic morphology to micro-/nanoscale structures, static to dynamic analysis, and two-dimensional observations to three-dimensional reconstructions.

In recent years, multi-instrument integrated characterization has emerged as a vital and increasingly prevalent approach in the structural analysis of carrier-free immobilized enzymes. This strategy leverages the complementary advantages of different techniques to obtain a comprehensive and holistic understanding of enzyme structures. For instance, a common integrated approach combines SEM, transmission electron microscopy (TEM), and atomic force microscopy (AFM) to investigate the micro- and nano-structural morphologies of immobilized enzymes. SEM provides detailed surface morphology information, TEM offers insights into internal structures and crystal lattices, and AFM enables the characterization of surface topography at the atomic scale as well as the measurement of nanomechanical properties. Meanwhile, fourier transform infrared spectroscopy (FTIR) and circular dichroism (CD) are employed to evaluate the conformational changes. FTIR detects changes in functional groups (e.g., amide bonds in proteins) associated with conformational shifts, while CD is particularly sensitive to the secondary structure of enzymes (e.g., $\alpha$-helices, $\beta$-sheets, and random coils).

In addition to the aforementioned techniques, X-ray diffraction (XRD) is widely used to elucidate the crystalline structure and degree of crystallinity of cross-linked enzyme crystals (CLECs). Dynamic light scattering (DLS) provides rapid nondestructive measurements of the size distribution of enzyme aggregates in solution, which is crucial for understanding their stability and aggregation behavior. Through such multiangle structural characterization, researchers can systematically elucidate the correlations between the structures of carrier-free immobilized enzymes and their functional properties, such as stability, catalytic efficiency, and selectivity. Thereby providing a robust scientific foundation for rational design and performance optimization. The integrated and complementary application of these characterization techniques not only significantly expands the breadth and depth of structural information available, but also facilitates the reliable construction of structure–activity relationships, thereby advancing a deeper understanding of the underlying mechanisms governing enzyme functionality.

The primary objective of this chapter is to provide a comprehensive and systematic introduction to the overarching strategy for the structural characterization of carrier-free immobilized enzymes. Articulate a multitiered research framework that encompasses both microscopic and mesoscopic structural analysis. For each key structural characterization tool employed in the domain of enzyme immobilization, including SEM, TEM, XRD, DLS, FTIR, CD, and NMR, this chapter elaborates on the underlying principles, operational methodologies, and scope of application. Representative case studies are integrated throughout the content to enhance the practical relevance and comprehensibility. These case studies illustrate how each characterization technique can be applied to reveal critical structural parameters, such as enzyme aggregation morphology, cross-linking density, and network architecture. Furthermore, this chapter offers a perspective on emerging trends in structural characterization technologies and explores the potential integration of artificial intelligence (AI) for rational design and molecular-level optimization. AI can facilitate the high-throughput analysis of complex structural data, predict structure–activity relationships, and guide the design of enzyme structures with enhanced performance. Collectively, the insights presented in this chapter aim to provide a theoretical foundation and practical methodological reference for future innovation and efficient application of carrier-free immobilized enzyme technology across various industrial and environmental fields.

## 7.1 Morphological characterization of carrier-free immobilized enzymes

### 7.1.1 Optical microscopy

Optical microscopy (OM) is a fundamental characterization technique that employs visible light transmitted through or reflected off a sample, which is subsequently magnified through a lens system to generate a resolvable image. This technique enables direct observation of macroscopic morphology (e.g., spherical particles, flocculent aggregates, or irregular aggregates), size distribution, and dispersion state of carrier-free immobilization enzymes. For example, the *Candida rugosa* lipase(CRL) was first isolated and purified by ion-exchange chromatography (Lalonde et al., 1995), followed by employing 2-methyl-2,4-pentanediol as the crystallizing agent and subsequent covalent cross-linking with glutaraldehyde (GA) to obtain insoluble CRL-CLECs. Glutaraldehyde is extensively employed as a cross-linking agent in enzyme immobilization due to its ability to facilitate the formation of a cross-linked network that retains a well-defined porous structure. During the synthesis process, the porosity and mechanical properties of the resulting material can be precisely modulated by tailoring reaction parameters, including cross-linker concentration, reaction pH, and temperature (Broun, 1976). OM observations revealed that CRL-CLECs exhibit a uniform plate-like crystalline morphology with an average length of approximately 30 μm and a thickness of 2 μm (Fig. 7.1). The CRL-CLECs are macroporous materials with a solvent content of approximately 50%, featuring interconnected channels that permit the diffusion of solvents, substrates, and reaction products. The uniform and highly ordered crystalline morphology observed via OM macroscopically reflects the well-organized internal lattice structure of CRL-CLECs, which is stabilized by both polar interactions and hydrophobic forces between enzyme molecules. This long-range rigid architecture is further reinforced by the covalent network formed by the GA cross-linker, resulting in a significant synergistic stabilization effect that mitigates thermal denaturation. Concomitantly, GA-mediated aggregation enhanced the stability of enzymes with the enzyme's half-life ($t_{1/2}$) extended to more than 13 days, compared to less than 5 h for its free forms at 40°C. Similarly, under

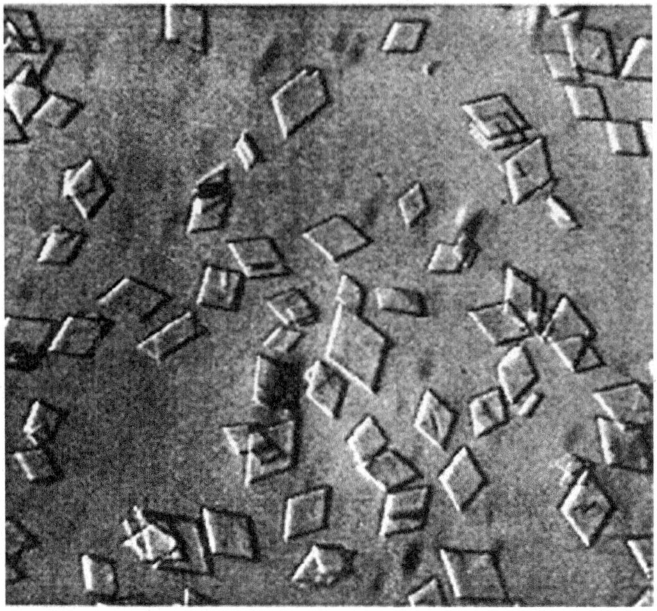

**FIGURE 7.1**

Optical microscopy photograph of cross-linked *C. rugosa lipase* crystals at ´250 magnification.

*From Lalonde, J. J., Govardhan, C., Khalaf, N., Martinez, A. G., Visuri, K., & Margolin, A. L. (1995). Cross-linked crystals of* Candida rugosa *lipase: Highly efficient catalysts for the resolution of chiral esters.* Journal of the American Chemical Society, 117*(26), 6845–6852. https://doi.org/10.1021/ja00131a006. Copyright © 1995, American Chemical Society.*

extreme conditions (such as in the presence of 50% organic solvent), the dense architecture of CRL-CLECs acts as a physical barrier that markedly retarding the disruption of the enzyme's active site by solvent molecules. Consequently, the $t_{1/2}$ increased by over 300-fold relative to the crude enzyme, providing direct macroscopic evidence of the enhanced environmental tolerance conferred by the crystalline morphology. The incorporation of inert protein feeders (e.g., bovine serum albumin), during cross-linking can further improve activity yields and stability. Furthermore, the conformational rigidity of the enzyme molecules within CRL-CLECs minimizes the intrinsic structural fluctuations inherent to free enzymes, enabling the active site to maintain a geometrically constrained and precise chiral microenvironment. This enhanced structural precision ultimately translates to a remarkable improvement in enantioselectivity during the hydrolysis of $\alpha$-aryl propionate esters.

Overall, OM offers several distinct advantages for the characterization of carrier-free immobilized enzymes. Notably, it requires minimal sample preparation, as it eliminates the need for metal coating or vacuum drying, thereby enabling direct observation of samples in a near-native state. Additionally, it is widely accessible and significantly more cost-effective than advanced electron microscopic techniques such as SEM or TEM. Nevertheless, OM possesses inherent limitations. Its spatial resolution is insufficient for visualizing nanoscale structures, such as detailed pore morphology or individual enzyme molecules. Moreover, its penetration depth is limited, precluding the observation of the internal architecture of large aggregates. Thus, to achieve comprehensive structural characterization of carrier-free immobilized enzymes, it is necessary to complement OM with SEM or other advanced characterization techniques.

### 7.1.2 Scanning electron microscopy

Scanning electron microscopy (SEM) is a surface characterization technique based on electron-matter interactions. A high-energy electron beam is employed to perform raster scanning over the sample surface, and the emitted signals are analyzed to reveal the microscopic morphology. It enables direct observation of the microscopic features (e.g., pores and tunnel-like porous structures), size distribution, and crystalline state of carrier-free immobilized enzymes, making it well-suited for surface characterization with highly irregular topographies. Key advantages include a large depth of field, a wide magnification range, and relatively simple sample preparation procedures.

The dialdehyde amylopectin (DAA) was combined with poly (ethylene terephthalate) hydrolase (IsPETase) to form IsPETase/DAA complexes, whose morphological characteristics were characterized by SEM at magnifications of 5000 × and 20,000 ×. Native amylopectin exhibited smooth spherical or oval shapes, whereas DAA presented hollow and wrinkled surface textures, which were attributed to oxidative etching effects (Fig. 7.2 A and B). FEEM images demonstrated that IsPETase/DAA formed porous spherical type I aggregates with an average particle size of approximately 144 μm, and no clustered aggregation was observed (Fig. 7.2 C and D). This porous structure is favorable for reducing substrate diffusion resistance and facilitating efficient mass transfer of solvents, substrates, and products (Lee et al., 2025). Similarly, magnetic nanoparticles (MNPs) ($Fe_3O_4$), enzyme aggregates (Enz-Aggs), cross-linked enzyme aggregates (CLEAs), and nanomagnetic composite cross-linked enzyme aggregates (NM-Combi-CLEAs) were subjected to SEM characterization at magnifications of 10–35 k ×. The $Fe_3O_4$ nanoparticles appeared as uniformly dispersed sphere particles with an average size of ~22.36 nm, Enz-Aggs displayed chain-like morphologies with an average dimension of ~209.96 nm, and CLEAs exhibited star-like structures with an average particle size of ~1.3 μm. The NM-Combi-CLEAs possessed spherical to hemispherical morphologies (average size ~94.38 nm) with filamentous interconnections between individual particles (Fig. 7.3). The small and porous structure of NM-Combi-CLEAs effectively minimized mass-transfer limitations, resulting in a 2-fold enhancement in enzymatic activity compared to free enzymes. Furthermore, the filamentous interconnections further significantly improved structural rigidity, leading to a 3.6-fold increase in carbohydrate catalytic activity at 85°C. The uniform structure feature of $Fe_3O_4$ facilitated the enzyme loading efficiency of up to 90.14%. Through the synergistic effects of composite formation

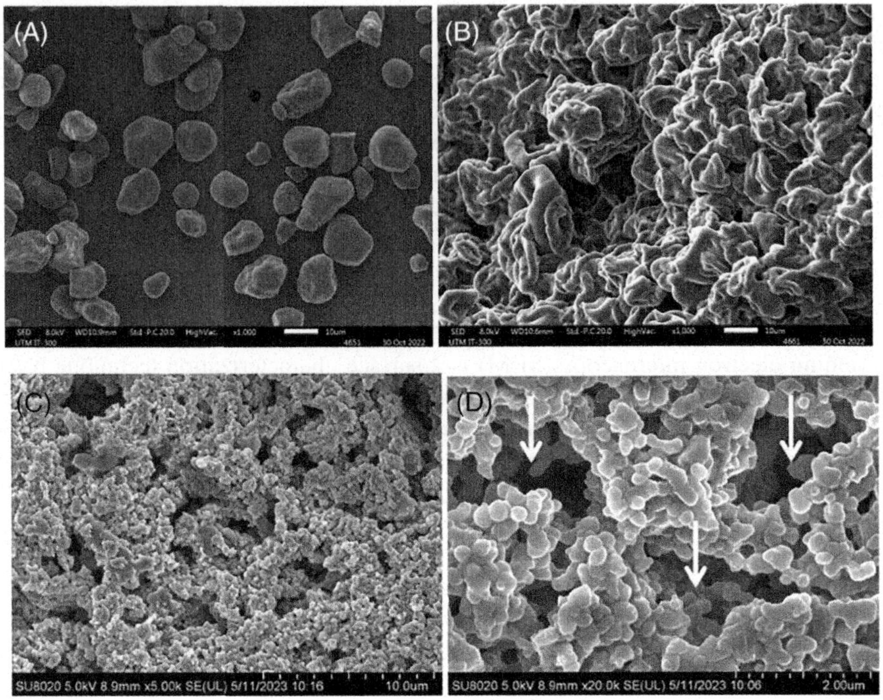


**FIGURE 7.2 The SEM images of amylopectin, DAA and IsPETase/DAA complexes.**

(A) amylopectin at magnification of 1000 ×, (B) DAA at magnification of 1000 ×, (C) IsPETase/DAA at magnification of 5000 ×, and (D) IsPETase/DAA at magnification of 20,000 ×.

*From Lee, Y. L., Jaafar, N. R., Huyop, F., Bakar, F. D. A., Rahman, R. A., & Md Illias, R. (2025). Functionalization of amylopectin as a strategy to improve polyethylene terephthalate hydrolase-cross-linked enzyme aggregate (IsPETase-CLEA) in plastic degradation.* International Journal of Biological Macromolecules, 306, *141492. https://doi.org/10.1016/j.ijbiomac.2025.141492. © 2025 Published by Elsevier B.V.*

and covalent cross-linking, the catalytic efficiency, stability, and reusability of the enzyme were remarkably improved (Torabizadeh et al., 2025). It should be noted that the types of cross-linking agents, enzyme species, cross-linking methodologies, and aggregation modes can all exert significant influences on the morphological characteristics of the resulting aggregates (Table 7.1).

## 7.1.3 Transmission electron microscopy

Transmission electron microscopy (TEM) utilizes the wave–particle duality of high-energy electrons transmitted through ultrathin specimens to achieve high-resolution imaging. It enables direct visualization of the crystal structure, morphology, and dimensional characteristics of carrier-free immobilized enzymes.

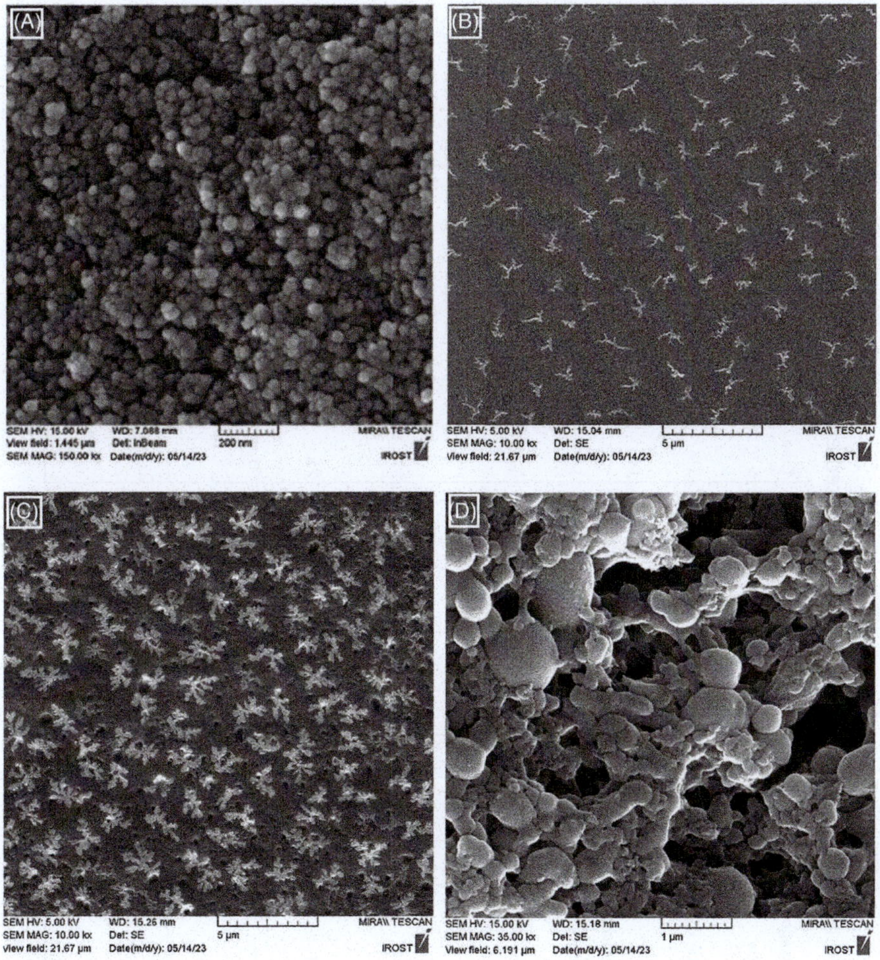


**FIGURE 7.3**

SEM images of (A) magnetic nanoparticles, (B) enzyme aggregates, (C) cross-linked enzyme aggregates (CLEAs), and (D) nanomagnetic composite cross-linked enzyme aggregates (NM-combi-CLEAs).

*From Torabizadeh, H., Abbasi, S., & Tafaghodinia, B. (2025). Multi-immobilization of viscozyme, alcalase and flavourzyme by nanomagnetic combi-CLEAs method for oil and protein hydrolysates extraction from rice bran in aqueous phase.* International Journal of Biological Macromolecules, 311*(Pt 1), 143719. https://doi.org/10.1016/j.ijbiomac.2025.143719.* 

**Table 7.1** The influence of preparation methods on the morphology of cross-linked enzyme aggregates.

| Enzyme species | Precipitant | Crosslinking agent | Additive | pH | Temperature | Time | Morphology | References |
|---|---|---|---|---|---|---|---|---|
| Lipase | Polyethyleneimine | Glutaraldehyde | / | 7.0 | Ambient temperature | 5-10 min | Formed clusters of 5–100 μm as irregular particles | Pan et al. (2011) |
| Mushroom tyrosinase | Ammonium sulfate | Glutaraldehyde | Bovine serum albumin (BSA) | 7.0 | Ambient temperature | 3 h | Amorphous, porous granular, 1–40 μm | Aytar and Bakir (2008) |
| L-asparaginase | Isopropyl alcohol | Sodium tripolyphosphate | BSA | 3.0 | 28°C | 1 h | Spherical microscale aggregates | Danait-Nabar et al. (2025) |
| Penicillin G acylase | Polyethylene glycol | Glutaraldehyde | Dextran sulfate, polyethyleneimine | 7.0 | 2°C | 1 h | Large-sized irregular aggregates | Wilson et al. (2004) |
| Penicillin G acylase | Dimethoxyethane | Glutaraldehyde | BSA | 7.5 | 4°C | 3-4 h | Approximately spherical, with a diameter of around 5 μm | Shah et al. (2006) |
| GDSL-type esterase | Acetone | Pectin | BSA | 8.0 | 25°C | / | Spherical particles with 20–200 nm | Luo et al. (2023) |
| Penicillin G acylase | Ammonium sulfate, tert-Butyl alcohol, Poly(ethylene glycol | Glutaraldehyde | / | 7.0 | 0°C | 2 h | Granular aggregates | Cao et al. (2000) |
| Lipase | Ammonium sulfate | Tripolyphosphate | / | 3.0 | 15°C | 2 h | Porous granular morphology with 249 ± 3.98 nm | Qian et al. (2020) |
| Subtilisin | Ammonium sulphate, polyethylene glycol | Glutaraldehyde | Triton X-100, Tween 20 | 7.0 | 25°C | 20–40 min | Uniform spherical particles with1–2 μm | Sangeetha and Emilia Abraham (2008) |
| Hydroxynitrile lyase | Tert-butanol, $(NH_4)_2SO_4$ | Glutaraldehyde | 2-Butanone | 7.0 | 0°C | 1 h | Approximately 100 μm particles with internal aggregates of about 500 nm, exhibited an irregular granular shape | Cabirol et al. (2008) |

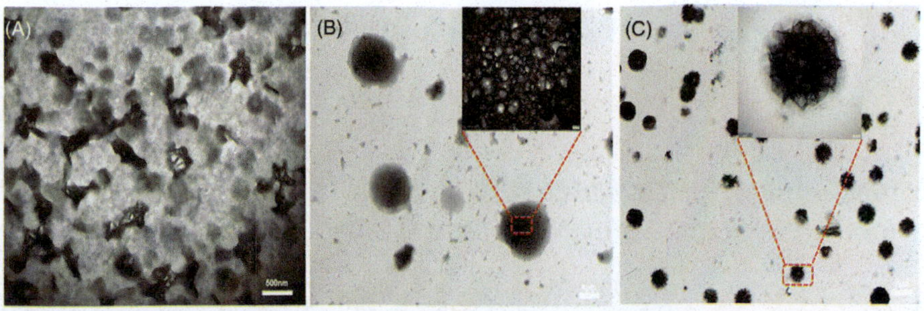


**FIGURE 7.4 The influence of preparation methods on the morphology of cross-linked enzyme aggregates.**

TEM images of (A) free $K_4$-SC, (B) crosslinked SC scaffold, and (C) C-CLEAs.

*Modified from Ma, X., Yang, Y., Wang, M., Zhang, Y., Zhao, B., Hao, S., & Jia, L. (2025). Engineering high-activity crosslinked enzyme aggregates via SpyCatcher/SpyTag-mediated self-assembly.* International Journal of Biological Macromolecules, 295, *139591. https://doi.org/10.1016/j.ijbiomac.2025.139591.* 

Ma et al. (2025) constructed a scaffold via ammonium sulfate-induced precipitation of SpyCatcher (K4-SC), followed by cross-linking with GA. This scaffold was subsequently assembled with SpyTag-labeled cellulase to generate cellulase-based cross-linked enzyme aggregates (C-CLEAs). For TEM analysis, suspensions of cross-linked K4-SC scaffolds, nanomagnetic composite C-CLEAs, and free K4-SC were deposited on carbon-coated copper grids and incubated for 12 h under controlled conditions to facilitate adsorption and air-drying prior to imaging. TEM images revealed that the K4-SC exhibited a uniformly cross-linked network structure. In contrast, the assembled C-CLEAs presented as well-dispersed aggregates (Fig. 7.4). The SpyCatcher-based scaffold served as a stable anchoring matrix for cellulase, minimizing conformational distortion of its active sites. The specific SpyTag–SpyCatcher interaction ensured precise enzyme orientation, thereby enhancing the structural stability and catalytic performance of the C-CLEAs. Consequently, the C-CLEAs maintained high hydrolytic activity, while the porous scaffold architecture reduced mass-transfer resistance and improved operational stability.

In summary, TEM provides unparalleled spatial resolution, with a maximum magnification of approximately 15 million times, enabling the visualization of atomic arrangements and providing profound insights into the structural and crystalline characteristics of carrier-free immobilization enzymes. Despite these advantages, the technique demands precise sample preparation and rigorous operational control to preclude beam-induced damage and structural distortion. Its inherent limitations, including limited penetration depth and incompatibility with liquid samples, further constrain its application scope. Nevertheless, TEM remains an indispensable tool for investigating nanoscale and atomic-level structures, forming a crucial foundation for advanced materials research and characterization.

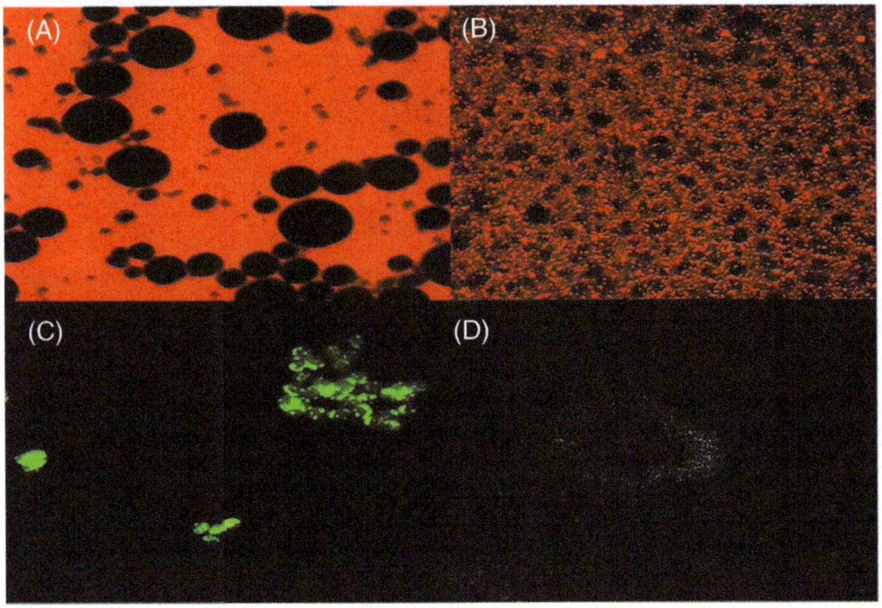


**FIGURE 7.5 Confocal laser scanning microscopy images of different samples.**

(A) The 500 kDa dextran could not diffuse into the CLEAs lipase particles, (B) The 2000 kDa dextran could freely diffuse into the p-CLEAs lipase particles, (C) The 75 nm polystyrene beads could penetrate into the interior of p-CLEAs, and (D) The 100 nm polystyrene beads were stuck to the surface of p-CLEAs.

*Miao, C., Li, H., Zhuang, X., Wang, Z., Yang, L., Lv, P., & Luo, W. (2019). Synthesis and properties of porous CLEAs lipase by the calcium carbonate template method and its application in biodiesel production.* RSC Advances, *9(51), 29665–29675. https://doi.org/10.1039/c9ra04365a.*

## 7.1.4 Confocal laser scanning microscopy

CLSM represents an advanced optical imaging technique that employs laser-mediated excitation of fluorescent labels in conjunction with a confocal pinhole to eliminate out-of-focus stray light. This configuration enables high-resolution optical sectioning and three-dimensional (3D) reconstruction of complex samples. By integrating CLSM with fluorescence labeling strategies, such as tagging enzymes or specific biomolecules, the spatial distribution and localization of target molecules can be precisely visualized and quantitatively analyzed. This technique is particularly effective for visualizing the 3D distribution and uniformity of enzymes within aggregates, as well as for exploring molecular diffusion processes inside these microenvironments. Recently, CLSM has been widely applied to characterize enzyme-related systems, enabling direct visualization of enzyme distribution, structural uniformity, and molecular diffusion within cross-linked and immobilized aggregates. For instance, fluorescein isothiocyanate (FITC) and polyethyleneimine

(PEI) with rhodamine B isothiocyanate were used to visualize their spatial distribution in CLEAs by CLSM. Results revealed that enzyme and PEI were uniformly distributed, forming hemispherical structures, with higher cross-linking degrees resulting in lower porosity (Grajales-Hernández et al., 2025). *Candida antarctica* lipase B was immobilized on silica nanoparticles through physical adsorption, followed by GA-mediated cross-linking. Subsequent labeling of immobilized lipase with FITC and CLSM observation revealed uniform green fluorescence emission across the $SiO_2$ nanoparticle surface, confirming successful immobilization and homogeneous coverage of the lipase (Qian et al., 2023). Magnetic cross-linked enzyme aggregates (CRL-FIL-CLEAs@$Fe_3O_4$) have been developed for the solvent-free synthesis of phytosterol oleate. CLSM imaging of FITC-labeled lipase within this system verified the association of the enzyme with MNPs. After five consecutive reuse cycles, partial detachment of the aggregates was observed, correlating with decreased catalytic activity (Zou et al., 2022). Porous cross-linked enzyme aggregates (p-CLEAs) have also been synthesized using calcium carbonate templates via in situ co-precipitation for biodiesel production. CLSM diffusion studies with fluorescent dextran and polystyrene microspheres showed that 2000 kDa dextran ($\sim$ 56 nm) freely penetrated p-CLEAs, whereas smaller dextran ($<$ 500 kDa) exhibited restricted penetration. Fluorescent polystyrene microspheres of 75 nm penetrated the interior, while 100 nm particles adhered to the surface, revealed that the diffusion behavior was strongly influenced by molecular size and weight, and confirmed the presence of a mesoporous structure ($<$ 100 nm) that facilitates molecular transport. Combined with complementary characterization techniques, these results corroborated that the porous framework of p-CLEAs enhances substrate diffusion and catalytic efficiency (Fig. 7.5) (Miao et al., 2019).

### 7.1.5 Atomic force microscopy

AFM characterizes the surface structure of samples by detecting the intermolecular forces between a nanoscale probe and the sample surface. Distinct from many other imaging techniques, AFM does not require special sample preparation, such as staining or coating. It enables nanoscale resolution imaging under diverse environmental conditions, including ambient air, liquid environments, and even vacuum. Beyond providing detailed surface topographical information, AFM facilitates single-molecule force spectroscopy analysis, which can quantify the mechanical properties of materials. This capability further allows insights into the physical stability of molecular aggregates and the conformational exposure of enzyme active sites.

In recent years, AFM has emerged as a powerful tool for the nanoscale characterization of the surface morphology and structural properties of immobilized enzymes at the nanoscale. Usually, sample pretreatment for AFM characterization is straightforward. The sample suspension is dropped onto a freshly cleaved mica substrate, rinsed with ultrapure water to remove surface-adhered salts, and dried before testing. To explore the potential of laccase in phenol degradation, an immobilization strategy integrating cross-linking with mesoporous silica encapsulation has been developed (Fathali et al., 2019). Various forms of immobilized enzymes have been prepared, including simply adsorbed enzymes, encapsulated cross-linked enzymes, and encapsulated CLEAs. AFM imaging revealed that free enzyme aggregates appeared as dispersed particles with an average height of approximately 120.1 nm. In contrast, CLEAs formed larger assemblies with an increased height of about 387.5 nm, indicating enhanced intermolecular interactions induced by the cross-linking. Furthermore, the 3D porous architecture of the mesoporous silica produced a more uniform particle distribution of the encapsulated CLEAs, thereby alleviating the irregular agglomeration observed in carrier-free CLEAs. Evaluations of storage and operational stability demonstrated that the encapsulated CLEAs exhibited superior thermal and pH stability, as well as enhanced activity retention in both hydrophobic and hydrophilic solvents. Additionally, these encapsulated CLEAs possess a significantly prolonged half-life and improved reusability. Human serum albumin microparticles (HSA-MPs) were prepared via the coprecipitation-cross-linking-dissolution method. AFM was employed to investigate the morphological characteristics of HSA-MPs and to determine their mechanical properties, including stiffness, adhesion force, and energy dissipation. The resulting submicron particles exhibit peanut-shaped morphology with uniform and regular morphology, consisting of two spherical-like protrusions fused along the axis of symmetry (Y-axis). These HSA-MPs also possess excellent mechanical properties, specifically manifested as high stiffness and good colloidal stability (Fig. 7.6 A and B) (Fardelli et al., 2025). Moreover, AFM has been utilized to characterize the microtopography of two lipase-based biocatalysts (Fig. 7.7), namely CLEA and CLEAs onto magnetic particles (CLEMPA). The AFM measurements were performed in tapping mode under ambient air conditions using an etched silicon probe, with a resonance frequency of 300 kHz and a scanning rate of 1.0–1.5 Hz. AFM images revealed that CLEA exhibits a distinct reticular arrangement of enzyme molecules with a regular and ordered morphology. No magnetic particles are involved in the aggregation, resulting in good microstructural uniformity. In contrast, CLEMPA, with -$NH_2$ modified magnetic particles as the carrier, forms larger-sized aggregates with extremely high surface roughness. Notably, CLEMPA exhibited higher catalytic activity compared to CLEAs, which is presumably attributed to the introduction of magnetic particles that modify the microtopography of enzyme aggregates and improve the accessibility of the catalytic structure (Tudorache et al., 2016).

Overall, AFM enables atomic-scale high-resolution imaging. However, its relatively slow scanning speed limits its capability to capture dynamic biological processes. Additionally, the small scanning range of AFM renders it less suitable for visualizing the overall architecture of large-scale samples. Consequently, AFM is often employed in combination with complementary characterization techniques such as SEM and CLSM to obtain more comprehensive surface and structural information about immobilized enzymes.

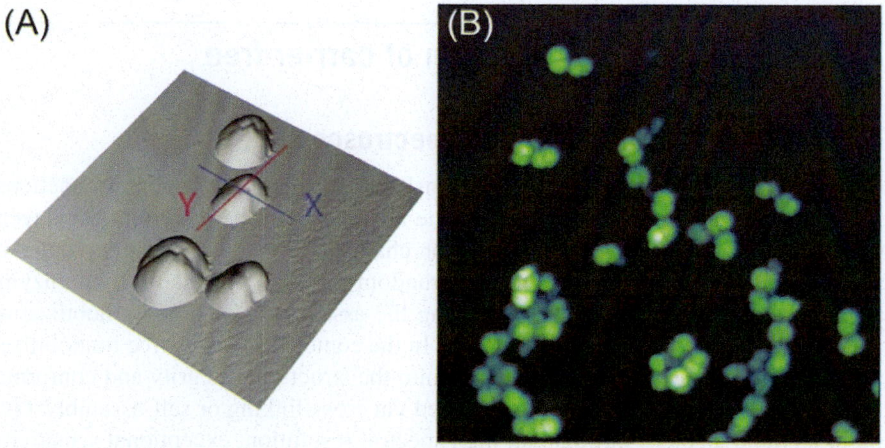


**FIGURE 7.6 Morphology of human serum albumin microparticles.**

Caption (A) 3.5 × 3.5 μm$^2$ 3D rendering and (B) 10 × 10 μm$^2$ AFM image of typical HSA-MPs.

*From Fardelli, E., De Simone, G., Georgieva, R., Xiong, Y., Di Gioacchino, M., Sotgiu, S., Nucara, A., Tavella, A., Baldassarre, L., Sodo, A., Ricci, A., Gasperi, T., Ascenzi, P., Bäumler, H., di Masi, A., & Capellini, G. (2025). Structural instability of human serum albumin during microparticles synthesis.* ACS Applied Bio Materials, 8*(11), 9907–9915. https://doi.org/10.1021/acsabm.5c01228.Copyright © 2025 The Authors. Published by American Chemical Society.*

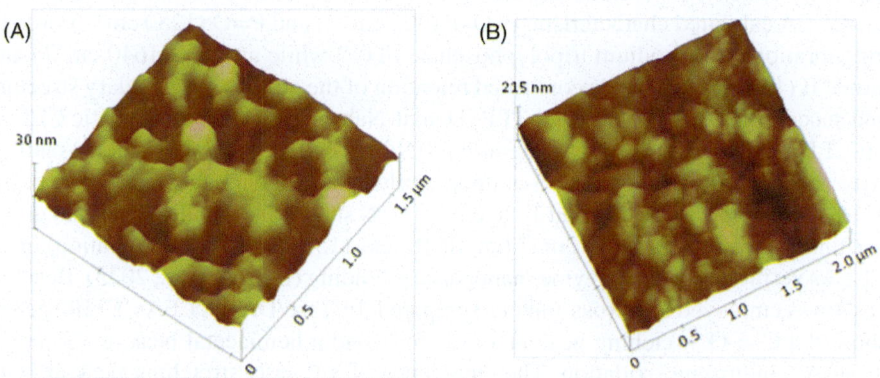


**FIGURE 7.7**

AFM analysis of CLEA (A) and CLEMPA (B).

*From Tudorache, M., Gheorghe, A., Viana, A. S., & Parvulescu, V. I. (2016). Biocatalytic epoxidation of α-pinene to oxy-derivatives over cross-linked lipase aggregates.* Journal of Molecular Catalysis B: Enzymatic, 134, *9–15. https://doi.org/10.1016/j.molcatb.2016.09.009. © 2016 Elsevier B.V. All rights reserved.*

## 7.2 Structural characterization of carrier-free immobilized enzymes

### 7.2.1 Fourier transform infrared spectroscopy

FTIR analyzes the vibrational energy transitions of chemical bonds or functional groups within molecules by quantifying the absorption of infrared light. This technique is extensively employed to monitor changes in the secondary structure of proteins, including $\alpha$-helix, $\beta$-sheet, and random coil that changed during enzyme immobilization processes, thereby enabling the assessment of how immobilization strategies modulate enzyme conformation. In the context of carrier-free immobilized enzymes, FTIR furnishes critical insights into the structural integrity and conformational stability of enzyme aggregates formed via cross-linking or self-assembly. Owing to its rapid analysis throughput, high spectral resolution, exceptional sensitivity, and broad spectral range, FTIR has emerged as an indispensable analytical tool for elucidating structure–function in immobilized enzyme systems.

A multitude of studies have applied FTIR to investigate structural alterations in carrier-free immobilized enzymes. Laccase aggregates were prepared and cross-linked with isopropanol to form CLEAs, and further synthesized hyperbranched polymer-cross-linked laccase aggregates (HPCLEAs) by diamine and acrylate monomers. FTIR analysis revealed an increase in $\beta$-turns and a concomitant decrease in $\beta$-sheets after cross-linking, indicating conformational alterations in the enzyme structure (Peng et al., 2025). L-asparaginase CLEAs were synthesized using sodium tripolyphosphate (TA-CLEA) and GA (GA-CLEA) as cross-linkers. FTIR spectra of TA-CLEA exhibited characteristic P–O–P (906 $cm^{-1}$) and P=O (1250 $cm^{-1}$) stretching vibrations from sodium tripolyphosphate (TPP) while amide I (1640 $cm^{-1}$) and amide II (1525 $cm^{-1}$) bands confirmed retention of the enzyme's secondary structure and successful cross-linking with TPP (Danait-Nabar et al., 2025). Magnetic CLEAs (m-CLEAs) were prepared by integrating MNPs to enhance stability and reusability. After surface modification with 3-aminopropyltriethoxysilane, new absorption peaks at 1625 $cm^{-1}$ (NH bending) and 1050 $cm^{-1}$ (C–N stretching) confirmed amino group grafting, while peaks at 1556 $cm^{-1}$ (amide II) and 1154 $cm^{-1}$ (–$NH_2$ vibration) in m-CLEAs verified covalent enzyme–nanoparticle binding (Morais et al., 2025). Besides, DAA was employed as a cross-linker to prepare IsPETase/DAA CLEAs. FTIR spectra showed a C = O stretching peak at 1730 $cm^{-1}$ and a hemiacetal peak at 880 $cm^{-1}$ in DAA, confirming oxidation. The emergence of a C = N stretching peak at 1643 $cm^{-1}$ and disappearance of the aldehyde peak in IsPETase/DAA indicated successful covalent linkage between enzyme and DAA (Lee et al., 2025). CLEAs-Lip-KV1 was prepared by precipitatinglipase KV1 (a novel lipase isolated from *Acinetobacter haemolyticus*) followed by GA cross-linking. The characteristic C = N band at 1458 $cm^{-1}$, attributed to Schiff base formation, confirmed cross-linking, while the amide I band shifted from 1718 $cm^{-1}$ to 1745 $cm^{-1}$ and broadening of the OH/$NH_2$ peak at 3470 $cm^{-1}$ suggested conformational adjustment and strengthened intermolecular hydrogen bonding (Bouguerra et al., 2025).

Beyond solid-state characterization, FTIR can also be utilized to probe the structural features of immobilized enzymes suspended in solvent systems. For instance, alpha-chymotrypsin was employed as a model enzyme to prepare enzyme precipitated and rinsed with propanol (EPRP), CLEA, protein-coated microcrystals (PCMC), and cross-linked protein-coated microcrystals (CLPCMC). These immobilized formulations were subsequently suspended in organic media and characterized by FTIR. In aqueous solution, the amide I region of alpha-chymotrypsin displayed a strong absorption peak at 1637 $cm^{-1}$, indicating a high abundance of antiparallel beta-sheet structure. In n-propanol, this peak's intensity diminished and shifted to lower wavenumbers, accompanied by the emergence of a new peak at 1657–1659 $cm^{-1}$ corresponding to $\alpha$-helical conformations, suggesting $\beta$-sheet structures. Formulations prepared via reverse addition, such as reverse addition–prepared EPRP and reverse addition–prepared CLEA exhibited a marked reduction in the intensity of the 1637 $cm^{-1}$ peak, which correlated with their transesterification activity hierarchy. These FTIR findings confirmed a positive correlation between the retention of native secondary structure retention and catalytic activity in low-water media (Solanki et al., 2012). These FTIR findings confirmed a positive correlation between the retention of native secondary structure and catalytic activity in low-water environments.

FTIR is a powerful technique for characterizing cross-linking and structural variations in immobilized enzymes. Nevertheless, in aqueous samples, the strong infrared absorption of water frequently obscures target spectral signals, whereas inorganic materials with low infrared activity, such as metals, yield weak or undetectable responses.

### 7.2.2 Circular dichroism

CD quantifies the differential absorption of left- and right-handed circularly polarized light by chiral molecules, offering a highly sensitive approach for the quantitative analysis of secondary structure changes in proteins across both solution and solid states. Consequently, CD serves as a crucial technique for assessing whether enzyme immobilization induces denaturation or conformational perturbation.

Notably, CD has been established as a powerful tool for elucidating the structural alterations of carrier-free immobilized enzymes. A nanomagnetic combi-CLEA system was constructed to co-immobilize three enzymes, including viscozyme, alcalase, and flavourzyme to facilitate the simultaneous rice bran oil extraction and protein hydrolysate production, thereby enabling efficient utilization of rice bran in aqueous media (Torabizadeh et al., 2025). The secondary structures of the mixed free enzymes and nanomagnetic combi-CLEAs (NM-Combi-CLEAs) were characterized using a spectropolarimeter at 25°C within the far-ultraviolet (far-UV) range of 200–260 nm. Both samples exhibited characteristic absorption peaks at 210–220 nm, which correspond to the amide chromophores of peptide bonds (Fig. 7.8). Compared with the free enzymes, NM-Combi-CLEAs displayed distinct alterations in secondary structure. The $\alpha$-helix content increased from 15.3% to 16.8%, $\beta$-sheet content decreased from

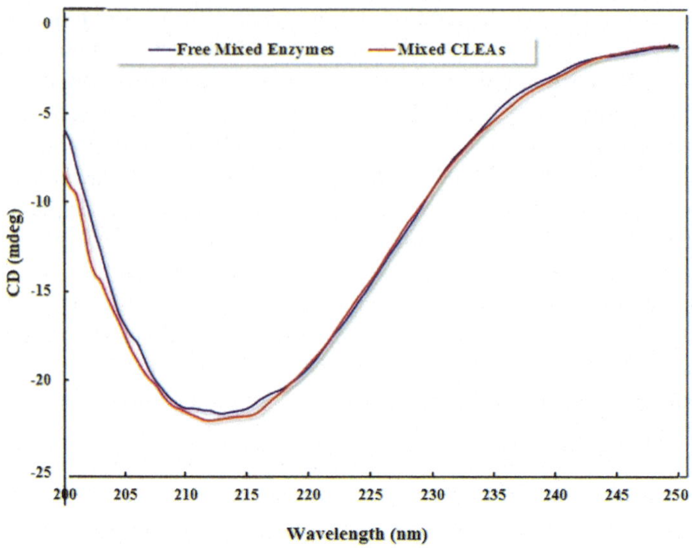


**FIGURE 7.8**

Circular dichroism assessment of free mixed enzymes compared to NM-Combi-CLEAs.

*From Torabizadeh, H., Abbasi, S., & Tafaghodinia, B. (2025). Multi-immobilization of viscozyme, alcalase and flavourzyme by nanomagnetic combi-CLEAs method for oil and protein hydrolysates extraction from rice bran in aqueous phase.* International Journal of Biological Macromolecules, 311*(Pt 1), 143719. https://doi.org/10.1016/j.ijbiomac.2025.143719.*

48.5% to 42.2%, random coils increased from 30.4% to 36.8%, and the turn content decreased from 5.8% to 4.2%. These structural variations reflect conformational rearrangements occurring during the immobilization process, including the formation of new hydrogen bonds and disulfide linkages. Such rearrangements enhance enzyme rigidity, thermostability, and catalytic performance.

The application of a rotating cell accessory has enabled effective structural characterization of cross-linked enzymes suspended in organic phases. Far-UV CD was employed to investigate secondary structural changes in $\alpha$-chymotrypsin and subtilisin following post-cross-linking, revealing that the shifts in characteristic CD peaks correlated directly with changes in catalytic activity. Moreover, near-UV CD provided insights into the integrity of the tertiary structure, demonstrating that formulations with CD peak profiles more closely resembling those of the native enzyme exhibited superior catalytic performance (Solanki et al., 2012). Additionally, CD can be employed to quantify the effects of organic solvents or temperature on enzyme structure (Fig. 7.9).

CD is widely employed for secondary structure analysis of immobilized enzymes, providing a rapid, nondestructive assessment of molecular conformation with

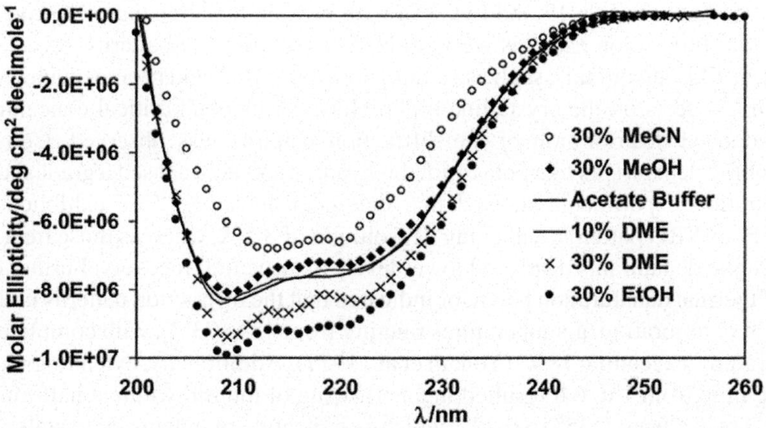


**FIGURE 7.9**

Far-UV CD spectra of the 3-phytase in acetate buffer.

*From Correia, I., Aksu, S., Adão, P., Pessoa, J. C., Sheldon, R. A., & Arends, I. W. C. E. (2008). Vanadate substituted phytase: Immobilization, structural characterization and performance for sulfoxidations.* Journal of Inorganic Biochemistry, 102*(2), 318–329. https://doi.org/10.1016/j.jinorgbio.2007.09.005.* 

minimal sample consumption. However, CD spectra are susceptible to interference from environmental factors, such as buffer composition, and typically yield averaged structural information rather than atomic-level details. Consequently, CD analysis is often complemented by high-resolution techniques, including XRD and NMR for comprehensive structural characterization.

### 7.2.3 X-ray diffraction

XRD stands as a fundamental technique for elucidating the structural characteristics of crystalline materials. It operates by irradiating a crystalline sample with X-rays, which undergo diffraction to produce characteristic patterns when the Bragg condition is satisfied. The positions and relative intensities of the resultant diffraction peaks furnish critical insights into the lattice structure of the sample. Owing to their inherent lack of crystallinity, XRD cannot be used for the structural characterization of CLEAs. However, XRD enables the precise determination of crystal structure, unit cell parameters, and degree of crystallinity in CLECs.

XRD technology was employed as a core characterization tool to investigate the structural changes and renaturation properties of GA-cross-linked CLECs in the presence of denaturants. XRD pattern analysis demonstrated that the CLECs are essentially isomorphous with the native enzyme crystals, both belonging to the $P_1$ space group with only minimal variations in unit cell parameters. Electron density difference maps further facilitated the identification of intermolecular cross-linking sites (e.g., between Lys-13 and Lys-33), with no significant changes in protein conformation. Precession cameras were utilized to record XRD patterns, revealing that detergent-like denaturants, such as sodium dodecyl sulfate and bromoethanol, exerted negligible effects on the crystalline order. However, beyond critical concentrations, a pronounced deterioration of the diffraction patterns was observed. Conversely, hydrophilic denaturants like potassium thiocyanate and urea caused a gradual decline. Renaturation experiments showed that detergent-treated CLECs exhibited partial recovery of XRD patterns, achieving a resolution of 3.2 Å, whereas those treated with hydrophilic denaturants displayed lower levels of structural recovery. Further evaluation of thermal denaturation behavior indicated that the diffraction patterns underwent minimal deterioration at temperatures ranging from 45 to 73°C, with complete loss at temperatures exceeding 78°C (Yonath et al., 1977). Additionally, extending the cross-linking time from 1 to 6 h resulted in a narrowing of the full width at half maximum of XRD peaks from 0.35° to 0.22°, which is indicative of enhanced crystallinity and mechanical stability. Conversely, excessive cross-linking (> 8 h) leads to an increase in lattice defects and a concomitant reduction in mechanical properties. These findings provide XRD-based fundamental guidance for the optimization of the cross-linking process in CLECs preparation (Kubiak et al., 2018).

Fundamentally, XRD offers a precise and nondestructive approach for probing crystalline architecture at the atomic level. Its capacity to yield detailed structural information with minimal alteration to the sample renders it invaluable for the characterization of fragile or high-value enzyme-based materials. Nevertheless, the technique is less effective for amorphous or poorly ordered systems and provides limited spatial resolution, which constrains its ability to reveal local structural heterogeneity. Moreover, prolonged X-ray exposure may induce minor structural perturbations through bond cleavage or free radical generation, subtly affecting the integrity of the crystal lattice.

### 7.2.4 Nuclear magnetic resonance

NMR is a powerful analytical technique based on the resonance phenomenon, wherein atomic nuclei absorb electromagnetic radiation at characteristic frequencies when subjected to a strong external magnetic field. This method facilitates the detailed elucidation of molecular structures, conformational dynamics, and interaction mechanisms of various substances. Owing to its exceptional resolution and versatility, NMR has been widely adopted in the fields of chemistry, biology, and medicine, and plays an increasingly important role in the structural characterization of carrier-free immobilized enzymes.

Solid-state nuclear magnetic resonance (ssNMR) was used to investigate the structure of protein nanocrystals (including ubiquitin and lysozyme) prepared via a rapid bulk crystallization protocol that combines hanging drop screening with centrifugal evaporator concentration (Martin & Zilm, 2003). Results showed that the ssNMR spectral resolution of nanocrystals is comparable to large XRD-quality crystals, with peak linewidths as low as 0.2 ppm, which is significantly superior to that of lyophilized amorphous proteins. Moreover, the nanocrystals maintained local ordered structures and exhibited excellent thermal stability after repeated temperature cycling between liquid nitrogen and room temperature. Concurrently, ssNMR exhibited high sensitivity in distinguishing between crystalline and amorphous states of proteins, thereby serving as a robust analytical tool for assessing crystallization integrity. Additionally, a combined NMR approach integrating $^{29}$Si magic-angle spinning (MAS) NMR and $^{13}$C NMR, provided direct molecular-level evidence for the formation and binding characteristics of $\beta$-glucosidase CLEAs (CLEA-GL) supported on mesoporous silica foams (MCFs) (Reshmi & Sugunan, 2013). $^{29}$Si MAS NMR analysis revealed three distinct characteristic resonance peaks, with notable changes in peak positions and intensities, indicating that enzyme immobilization occurs through covalent bonding or coordination interactions between the enzyme and the geminal silanol groups (Si-OH) on the MCF support surface during CLEA formation. Furthermore, the $^{13}$C NMR further showed reduced peak intensities, reflecting restricted mobility of the carbon environment following aggregation, alongside a peak at 184 ppm that confirmed the chemical cross-linking via Schiff base reactions between GA and the enzyme's amino groups.

NMR offers atomic-level structural resolution and enables comprehensive 3D structural elucidation, supporting multiangle investigations of molecular systems. It also allows for real-time monitoring of dynamic processes and is adaptable with a wide range of physical states and solvent environments. However, the technique is inherently limited by low sensitivity, which hinders the detection of low-concentration samples. In addition, data acquisition and analysis procedures for macromolecules are often time-consuming, leading to relatively high experimental costs and operational complexity.

## 7.3 Physicochemical characterization of carrier-free immobilized enzymes

### 7.3.1 Particle size and dispersibility

As a fundamental technique for characterizing the colloidal properties of carrier-free immobilization systems, including CLEAs and m-CLEAs, dynamic light scattering (DLS) offers precise measurement of particle size distribution, average hydrodynamic diameter, and zeta potential. These parameters provide valuable insights into the assembly behavior of enzyme molecules during the aggregation and cross-linking processes. Notably, the particle size of CLEAs is not a fixed value but dynamically varies depending on factors such as the types of additives, aggregation conditions, and cross-linker parameters (Schoevaart et al., 2004).

Aggregation serves as a prerequisite step in the CLEAs preparation. The type and concentration of precipitant, along with the aggregation temperature, significantly influence the size of the initial aggregates, thereby affecting the dimensions of the final cross-linked product by modulating the precipitation rate and aggregation behavior of enzyme molecules (Torabizadeh & Montazeri, 2020). For instance, in the preparation of CLEAs from lipase, LiCl and NaCl are not observed to induce precipitation, $NH_4Cl$ induced weak precipitation, whereas $Na_2SO_4$ and $(NH_4)_2SO_4$ were highly effective, with $(NH_4)_2SO_4$ being the preferred choice for its capability to disrupt the enzyme's hydration layer and enhance aggregation. Salts with high ionic strength and charge characteristics, such as $Na_2SO_4$ and $(NH_4)_2SO_4$, effectively facilitate rapid protein aggregation by altering the hydration state and electrostatic properties of the solution. In contrast, insufficient ionic strength in LiCl and NaCl prevents the dissolution equilibrium from being disrupted, leading to minimal precipitation (Yu et al., 2006). The mechanisms by which salt-based precipitants (e.g., ammonium sulfate) and organic solvent precipitants (e.g., acetone, tert-butanol) control particle size differ fundamentally. Organic precipitants tend to induce hydrophobic interactions with nonpolar groups of enzymes, often leading to denaturation and hindering the formation of stable large particles, thereby resulting in poor aggregation. Nevertheless, organic solvents can still be employed for enzyme precipitation under specific conditions. For example, 60% acetone had been shown to maximize enzyme activity recovery in m-CLEAs (Safari et al., 2026). Meanwhile, the precipitant concentration is vital. Increasing ammonium sulfate concentration results in a gradual increase in lipase particle size from 1–2 to 16–17 μm, as it enhances the probability of intermolecular interactions (Yu et al., 2006). Similarly, when using polyethylene glycol (PEG) 2000 in concentrations of 2.5% to 45% (w/v), rapid crystallization of streptavidin is observed, with high precipitant concentrations yielding larger crystals within minutes (Martin & Zilm, 2003). The interplay of these factors ultimately determines the efficacy of enzyme immobilization.

The type, concentration, and duration of the cross-linking agent are critical determinants of the final particle size and stability of the aggregates, as they covalently stabilize their architecture through cross-linking bonds. These effects exhibit pronounced dependence on both concentration and time (Gupta et al., 2020). Glutaraldehyde, the most commonly used cross-linker, demonstrates a positive correlation between its concentration and particle size. Higher GA concentrations facilitate covalent cross-linking among more protein molecules, leading to larger aggregates (Yu et al., 2006). However, excessive cross-linkers can interact negatively with amino acid residues in the enzyme's catalytic core, inducing conformational abnormalities and preventing further growth (Joseph et al., 2021). Long-chain cross-linkers, such as 1,11-undecanedicarboxylic acid, can link spatially distant enzyme molecules, forming larger, more loosely structured networks (Ahrari et al., 2022). Additionally, the environmental pH significantly affects aggregate formation by modulating electrostatic interactions between enzyme molecules. In the preparation of lipase-based CLEAs, acidic conditions (pH 2) prevented precipitation due to strong electrostatic repulsion, while alkaline conditions (pH 10) reduced repulsion, facilitating aggregation and resulting in larger particle sizes (Yu et al., 2006).

Additives play a crucial role in regulating the particle size of carrier-free immobilization systems by altering intermolecular interactions among enzymes and shaping the aggregation microenvironment. The effects of different types of additives show significant variations. Bovine serum albumin (BSA), rich in lysine residues, acts as a protein feeder in CLEAs. Its lysine-rich surface helps suppress excessive inter-enzyme cross-linking, thereby alleviating substrate diffusion limitations caused by narrow pores in CLEAs. In the preparation of *Rhizomucor miehei* lipase (RML) CLEs, increased BSA from 0 to 16 mg reduced the particle size from 3908 to 1750 nm, while specific activity improved from 5.0 to 6.6 U/mg (Ahrari et al., 2022). In elastase CLEAs, combined BSA and starch increased the particle size from 1109 to 1497 nm, as the filling effect of starch and spatial support from BSA created a more porous aggregate architecture (Razib et al., 2020). Additionally, lysine can form Schiff base bonds with GA, enhancing covalent linkages and stabilizing the cross-linked network. An optimized enzyme-to-lysine ratio of 1:0.75 during the preparation of nanomagnetic composite CLEAs resulted in high activity that was twice as high as that of an equal enzyme/BSA proportion. This demonstrated lysine's superiority in creating a robust and permeable cross-linked network (Torabizadeh & Montazeri, 2020).

Magnetic nanoparticles can enhance aggregate properties by forming interconnected macroporous structures. Research on m-CLEAs has shown that MNPs' initial particle size of approximately 81.9–88.9 nm increased to about 99.6 nm after immobilization into NM Combi CLEAs. The polydispersity index (PDI) decreased from 0.242 to 0.088, while the zeta potential shifted from –21 to –32 mV, indicating improved colloidal stability through the formation of a protein coating layer and a more uniform dispersion (Torabizadeh & Montazeri, 2020). The particle size of MNPs increased from 115 to 198 nm after silanization and amination (MNP-AP), and further grew to 342 nm after xylanase immobilization (Xyl-MNP-AP). These modifications introduced organic layers and increased the hydrodynamic diameter (Safari et al., 2026). Similarly, in lipase-based CLEAs, magnetic iron oxide nanoparticles (MIONPs) grew from 72.06 to 159.3 nm after silane modification with 3-aminopropyltrimethoxysilane (APTMS), mainly due to the new functional groups (e.g., amino groups) that enhanced hydration thickness and modified the surface charge (zeta potential changed from –28.8 to –20.9 mV). Compared with nonmagnetic CLEAs, magnetic CLEAs exhibit more cavities and interconnected macroporous structures. This results from the altered enzyme aggregation microenvironment due to the incorporation of MIONPs, which creates a porous architecture beneficial for substrate mass transfer, although it may also lead to an apparent increase in particle size due to nanoparticle aggregation (Salgın & Salgın, 2017).

The particle size of enzymes is closely related to their activity, as highlighted by the Michaelis–Menten theory, which states that substrate must first form a complex with the enzyme. Particle size significantly influences the mass transfer process of the substrate, determining access to active sites within CLEAs. Larger particles introduce substantial mass transfer resistance, impeding substrate binding, whereas smaller particles expose enzyme molecules to the reaction medium, potentially jeopardizing structural stability and activity retention. Moreover, the particle size in carrier-free immobilization systems results from a combination of factors, including aggregation conditions, cross-linking parameters, additives, and nanocomposite effects, as evidenced by systematic investigation using techniques such as DLS, laying the theoretical foundation for the rational design and precise control of immobilized enzymes with tailored size, high activity, and enhanced stability.

### 7.3.2 Zeta potential

The zeta potential serves as a crucial indicator for evaluating the dispersion stability of various materials, including carrier-free immobilization enzymes, polymeric nanoparticles, and other colloidal systems. It is determined by measuring the electrophoretic mobility of particles in an electric field, which reflects their surface charge and the properties of the electric double layer surrounding them. Assessing zeta potential provides valuable insight into the electrostatic stabilization mechanisms governing aggregate stability and their interactions with the surrounding environment, including substrates and solvents.

Zeta potential is typically measured by electrophoretic light scattering (ELS) instruments, which can simultaneously provide information on zeta potential, PDI, and particle size distribution. These parameters together offer a comprehensive understanding of the physical characteristics and stability of CLEAs. Arginine deiminase was precipitated and cross-linked with GA to produce novel ADI-CLEAs, which exhibited a zeta potential of −17 mV and an average particle size of 1638 nm. Combined with other characterization results, these findings suggest that CLEAs with higher zeta potential values possess greater porosity, enhanced catalytic activity, and improved mechanical stability (Kawatra et al., 2024). Subsequently, hybrid magnetic arginine deiminase-based CLEAs (mADI-CLEAs) were prepared, showing a zeta potential of −20.1 mV and demonstrating smaller particle size and higher stability compared with the previously reported ADI-CLEAs (Kawatra et al., 2025). Nanomagnetic $\alpha$-amylase and pullulanase CLEAs were synthesized with a PDI of 0.088 and a zeta potential of −32 mV, indicating that immobilization enhanced hydrophilicity, stability, and particle uniformity (Torabizadeh & Montazeri, 2020). In addition, rINANE1–$Fe_3O_4$@PDA CLEAs were prepared with more negative zeta potentials (−27.42 mV) than the free enzyme (−18.33 mV), signifying better colloidal and storage stability (Luo et al., 2023). The heme-modified glucose oxidase CLEAs (HCGOAs-6) was reported with a zeta potential of −25 mV, similar to native GOx, indicating minimal alteration in surface charge during cross-linking (Xiong et al., 2025).

Generally, the zeta potential tends to change, particularly with an increase in its absolute value after the addition of cross-linkers or MNPs during the enzyme immobilization process. These variations indicate that the immobilized nanoparticles exhibited improved dispersibility and enhanced stability in solution, which may contribute to enhancing the catalytic activity and stability of the immobilized enzyme (Safari et al., 2026; Torabizadeh & Montazeri, 2020). Additionally, zeta potential is influenced by various factors, such as the particle size of the aggregates, surface chemical modifications, solution pH, and ionic strength. These factors together determine the surface charge properties of the immobilized enzyme, which affects its behavior in the reaction system. Therefore, monitoring zeta potential changes is crucial for optimizing the enzyme immobilization process. However, zeta potential reflects only an overall average surface charge, without distinguishing between different charge sources or their spatial distribution, and its accuracy decreases when applied to large, irregular, or highly aggregated particles. Moreover, it's worth noting that the zeta potential cannot be directly correlated with enzymatic activity, as catalytic efficiency is also influenced by the internal microstructure, extent of cross-linking, and substrate accessibility.

## 7.4 Functional characterization of cross-linked enzymes

Characterization of enzymatic properties is a key step in evaluating the functional integrity of carrier-free immobilized enzymes, such as CLEAs and CLECs. The primary objectives are to quantitatively assess the impact of immobilization on enzymatic catalytic behavior, to elucidate changes in the structure-activity relationship during the immobilization process, and to provide a scientific basis for optimizing immobilization conditions. This section systematically introduces the core enzymatic parameters (including catalytic kinetic parameters, stability parameters), their determination methods, and the influence of immobilization methods, types of cross-linkers, and other conditions to the enzymatic properties. The changes in the catalytic efficiency, stability, and optimal conditions of the enzyme under different precipitants, cross-linking agents, and additives were systematically summarized (Table 7.2 ).

### 7.4.1 Catalytic kinetic parameters

Catalytic kinetic parameters, including the Michaelis constant ($K_m$), maximum velocity ($V_{max}$), and turnover number ($k_{cat}$), are core indicators for quantitatively assessing enzyme catalytic efficiency, reflecting substrate affinity, maximum catalytic capacity, and intrinsic activity, respectively. By analyzing the kinetic parameters of CLEAs, their catalytic performance can be directly evaluated and compared with that of free enzymes, providing essential evidence for optimizing immobilization strategies and improving enzymatic efficiency.

The catalytic kinetic parameters of immobilized or free enzymes are significantly influenced by multiple factors, including cross-linking agents, the aggregation state of the enzyme, and the presence of additives. For instance, traditional cross-linkers like GA can enhance enzyme stability but usually lead to increased $K_m$ values, indicating reduced substrate accessibility. This was observed with $K_m$ increased from 18.12 mM for the free enzyme to 25.69 mM, resulting in $V_{max}$ decreased from 0.83 to 0.42 μmol•mL$^{-1}$•min$^{-1}$ due to aggregation and cross-linking effects (Kawatra et al., 2024). To overcome this drawback, novel crosslinkers have been developed to construct more favorable microstructures. For example, the use of DAA

**Table 7.2** Influence of immobilization methods, cross-linker types on enzymatic properties.

| Methods | Precipitating agent | Cross-linking agent | Additive | Alterations in kinetic parameters | Stability | Shift in the optimal reaction conditions | References |
|---|---|---|---|---|---|---|---|
| CLEA | Ammonium Sulfate | Glutaraldehyde | / | $K_m$ increased from 18.12 to 25.69 mM, $V_{max}$ decreased from 0.83 to 0.42 μmol/min | ADI-CLEA retained > 50% activity after incubation at 60°C for 120 min or pH 6 for 120 min, retained over 33% activity after 50 days of storage | / | Kawatra et al. (2024) |
| Combi-CLEAs | $CaCl_2$ | Glutaraldehyde | / | $K_m$ increased from 35.21 to 42.11 mM, $V_{max}$ decreased from 12.31 to 10.64 μmol/min | Combi-CLEAs retained > 60% activity after treatment at 45°C for 180 min, retained 69.4% activity after treatment at pH 10 for 180 min | / | Liu et al. (2025) |
| CLEA | Ammonium Sulfate | Dialdehyde amylopectin | / | $V_{max}$ increased by 2.15-fold, $K_m$ decreased to 3.00 mM, $k_{cat}$ increased by 2-fold, $k_{cat}/K_m$ increased by 7-fold | The $t_1/_2$ of IsPETase/DAA increased 1.72–3 times at 35–50°C compared with IsPETase/Amy, it retained > 70% activity after storage at 4°C for 40 days | Increased from 40 to 45°C | Lee et al. (2025) |
| CLEA | Isopropanol | Glutaraldehyde | / | The $K_m$ value increased by 5.19%, the $V_{max}$ decreased by 17% | Thermal stability and pH stability were significantly improved | Increased from 40 to 45°C | Khatik et al. (2022) |
| CLEA | Ammonium sulfate | Tripolyphosphate | / | The $K_m$ of CLEAs increased, but the $V_{max}$ increased significantly | The $t_{1/2}$ was increased to 7.5 times compared with free enzyme at 40°C | / | Qian et al. (2020) |

*(continued on next page)*

| | | | | | | | |
|---|---|---|---|---|---|---|---|
| CLEA | Isopropanol | Tripolyphosphate | BSA | $K_m$ increased by 37%, $V_{max}$ increased by 11% | Its thermal stability was improved, and was stable at pH 3–8 | / | Danait-Nabar et al. (2025) |
| CLEA | Ammonium sulfate | Tripolyphosphate | Chitosan | The $K_m$ value increased from 34.6 to 77.2 g/L, the $V_{max}$ increased from 323 to 857 mmol/min for CLEAs | The $t_{1/2}$ of CLEA was 11.4 times that of the free enzyme at 40°C | / | Yan et al. (2022) |
| CLEA | Ammonium sulfate | Glutaraldehyde | Fructose | The $k_{cat}/K_m$ value was 2-fold and 3-fold that of BSA-CLEA and GA-CLEA, respectively | The $t_1/_2$ of fructose CLEA was higher than that of GA-CLEA and free enzyme at 50°C, fructose CLEA was better than GA-CLEA and free enzyme at pH 3 and 5, fructose CLEA only lost about 14% activity after storage at 4°C for 45 days | Increased from 30 to 40°C | Dey et al. (2023) |
| CLEA | Isopropanol | Glutaraldehyde | Tween 80 | / | Thermal stability and pH stability were enhanced | The optimal temperature increased from 40 to 50°C, the optimal pH increased from 8 to 10 | Bouguerra et al. (2025) |
| CLEA | N-butanol | Glutaraldehyde | Polyethylenemine | $V_{max}$ and $k_{cat}$ increased by 2-fold | Thermal stability was increased by 7.9-fold (60°C), solvent stability was increased by 5.7-fold (in n-hexane/n-butanol/water system) | / | Grajales-Hernández et al. (2025) |

(continued on next page)

**Table 7.2** Influence of immobilization methods, cross-linker types on enzymatic properties—cont'd

| Methods | Precipitating agent | Cross-linking agent | Additive | Alterations in kinetic parameters | Stability | Shift in the optimal reaction conditions | References |
|---|---|---|---|---|---|---|---|
| RP-CLES | / | *Sym-dibenzo-1,5-*21 cyclooctadiene-3,7-diyne | / | $K_m$ decreased from 5.36 to 1.99 mM, $k_{cat}/K_m$ increased by 3.6-fold | $T_1/_2$ was prolonged by about 4.5-fold | / | Li et al. (2020) |
| CLEA mediated by Spy-Catcher/SpyTag | Ammonium sulfate | Glutaraldehyde | / | $K_m$ decreased from 30.74 to 26.61 mg/mL, $V_{max}$ decreased from 1.21 to 0.86 μmol/min• mL | Thermal stability and pH stability were better than those of the free enzyme | / | Ma et al. (2025) |
| Magnetic CLEA | Acetone | Pectin | / | The $K_m$ value increased from 10.86 to 16.55 g/L | It retained over 100% of the initial activity under pH 10, retained >55% of the activity after incubation at 55°C for 12 h, the immobilized enzyme activity remained stable after 45 days | The optimal temperature increased from 35 to 40°C, the optimal pH increased from 8 to 9 | Luo et al. (2023) |
| Magnetic CLEA | Ethanol | Glutaraldehyde | L-lysine | $V_{max}$ increased by about 1.7-fold, $K_m$ decreased by about 1.1–2.0-fold, and $k_{cat}/K_m$ increased by about 2.0-fold | Thermal stability was significantly improved | / | Torabizadeh et al. (2025) |

as a crosslinking agent enables the formation of a macroporous structure, which significantly alleviates diffusion limitations, leading to improved $k_{cat}/K_m$ ratios. Is-PETase/DAA exhibited a 2.15-fold increase in $V_{max}$, alongside a decrease in $K_m$ from 9.74 to 3.00 mM, and a 7-fold improvement in $k_{cat}/K_m$, linked to the macroporous structure formed during immobilization (Lee et al., 2025). To achieve precise cross-linking and avoid undesired covalent bonds, a bifunctional diyne cross-linker (sym-dibenzo-1,5-cyclooctadiene-3,7-diyne) was used, resulting in decreased $K_m$ and a 3.6-fold increase in $k_{cat}/K_m$ due to accelerated cross-linking under microwave irradiation (Li et al., 2020).

The introduction of additives during the crosslinking process is a commonly used strategy to optimize kinetic parameters. For example, CLLAs-GA-BSA demonstrated an increase in $V_{max}$ from 19.23 to 21.97 U $mg^{-1}$ and an enhancement in $k_{cat}/K_m$ from 1.31 to 1.40 mg•$mL^{-1}$ •$s^{-1}$, attributed to BSA's additional amino groups, which react with GA and limit direct cross-linking of the active site (Abd Rahman et al., 2020). In another study, fructose-mediated CLEAs exhibited $k_{cat}/K_m$ values that were approximately double those of conventional BSA-GA-CLEAs and three times those of GA-CLEAs, highlighting the advantages of fructose in enhancing biocatalytic properties (Dey et al., 2023).

Advanced immobilization strategies based on protein engineering and nanomaterials enable more precise control over enzyme orientation and aggregation states. For example, CLEAs formed via SpyCatcher/SpyTag-specific covalent linkage (C-CLEAs) allow for oriented immobilization, yet may lead to the formation of large aggregates. This can create steric hindrance, resulting in both a higher $K_m$ of 26.61 mg/mL and a lower $V_{max}$ of 0.86 μmol/(min•mL) compared to the free enzyme (Ma et al., 2025). In contrast, the nanomagnetic composite CLEAs (NM-Combi-CLEAs), which combine MNPs with co-immobilization techniques, not only facilitate easy recovery but also the thermo robustness was approximately 3.6-fold elevated than free enzymes at 85°C. Specifically, $V_{max}$ was enhanced by 1.7-fold and $k_{cat}/K_m$ improved by 2.0-fold, demonstrating its comprehensive advantages in improving substrate affinity and catalytic performance (Torabizadeh et al., 2025). Furthermore, the physicochemical properties of substrates, environmental conditions (such as temperature and pH), and changes in the spatial structure of the enzyme all significantly affect catalytic performance.

### 7.4.2 Stability evaluation

The primary advantage of carrier-free immobilization is its ability to significantly enhance enzyme stability, essential for achieving industrial-scale biocatalysis. Enzyme stability encompasses multiple aspects, including thermal, pH, operational, and storage stability, each reflecting the enzyme's resistance to structural or functional degradation under different environmental and process conditions, which is essential for its practical applicability and shelf life. Additionally, evaluations of salinity tolerance and resistance to organic solvents or surfactants provide a comprehensive understanding of the enzyme's catalytic robustness under diverse reaction environments.

Enzyme stability is significantly influenced by various factors, including the choice of cross-linkers, additives, and immobilization strategies. The selection of cross-linkers plays a crucial role. For instance, ionic cross-linkers like TPP enhance enzyme stability with milder reactions compared to covalent cross-linkers like GA, which can interfere with active sites and cause activity loss. Incorporating additives such as Tween 80 or calcium ions can improve thermal and pH stability, mitigating environmental fluctuations. Lipase-based CLEAs were prepared with isopropanol as the precipitant, Tween 80 as the additive, and GA as the cross-linking agent. Achieved a relative activity of 19% in a 50% butanol and 50% acetone (v/v) compared to only 6.7% for the free enzyme (Bouguerra et al., 2025). His-tagged enzymes could be precipitated with only 10 mM calcium ions, demonstrating lower reagent consumption and enhanced thermal stability. The combi-CLEAs retained over 60% activity after incubation at 45°C for 180 min, whereas the free enzyme activity dropped below 30%. The pH stability of combi-CLEAs was maintained 54.8% and 69.4% after 180 min at pH 6.0 and 10.0, respectively. The enhanced thermal and pH stability can be attributed to the more rigid structure conferred by the combi-CLEAs (Liu et al., 2025).Qian et al. (2020) pioneered the use of the TPP for the preparation of lipase CLEAs, which significantly improved activity across a broader pH and temperature range, with a thermal half-life 7.5 times longer than that of the free enzyme at optimal temperatures (Qian et al., 2020). Similarly, the L-asparaginase CLEAs using TPP as the cross-linker exhibited higher enzyme activity across the tested pH range than those formed with GA due to reduced interference with the active site (Danait-Nabar et al., 2025). IsPETase/DAA was prepared by cross-linker DAA, which showed a superior thermal stability, pH stability, and 1.72–3-fold increase of $t_1/_2$ within the temperature range of 35–50°C compared with IsPETase/Amy (amylopectin), probably due to the used stronger covalent bonds (Lee et al., 2025). Additionally, the adsorption-precipitation-cross-linking method was applied to prepare (CLEA)–rINANE1–$Fe_3O_4$@PDA, and exhibited improved pH and thermal stability (Luo et al., 2023).

Consequently, CLEAs are formed through covalent crosslinking between enzyme molecules and bifunctional reagents such as PEI or GA. This covalent network reduces the conformational flexibility of the enzyme and enhances its structural rigidity, thereby mitigating activity loss under extreme conditions. As a result, CLEAs exhibit markedly improved thermal, pH, operational, and storage stability compared with their free enzyme counterparts.

### 7.4.3 Variation in optimal reaction conditions

Determining the optimal reaction conditions is critical for maximizing the catalytic performance of immobilized enzymes. Key parameters include the optimal temperature and pH, as well as any dependence on metal ions or cofactors (e.g., $Mg^{2+}$,

$Mn^{2+}$, $Ca^{2+}$, FAD, $NAD(P)^+/NAD(P)H$). The optimal temperature of carrier-free immobilized enzymes commonly shifts to higher values, accompanied by a flatter activity-temperature decay curve. This behavior is generally attributed to the covalent network introduced during crosslinking, which restricts thermally driven conformational fluctuations, increases the energetic barrier to unfolding, and thereby enhances thermal resistance.

Notably, the optimal reaction conditions of immobilized enzymes are influenced by variability, shaped by various factors. For example, lipase-based CLEAs were prepared with isopropanol as the precipitant, Tween 80 as the additive, and GA as the cross-linker. The optimum temperature shifted from 40 to 50°C, which is likely attributable to the extensive formation of covalent imine bonds (C = N) between Lip-KV1 protein molecules. These bonds impart a more rigid and compact structure, reduce conformational flexibility, and thereby enhance thermostability. Concurrently, the optimum pH increased from 8 to 10, indicating that the cross-linking process altered the electrostatic charge of the enzyme and increased the polyanionic character of the CLEAs (Bouguerra et al., 2025). During the preparation of CLEAs for poly (ethylene terephthalate) hydrolase, when the cross-linking agent was switched to DAA, the optimum temperature increased to 45°C. This shift is likely due to the covalent bonds formed between IsPETase and DAA, which altered the conformational flexibility of IsPETase and resulted in higher energy expenditure for the formation of the enzyme-substrate complex (Lee et al., 2025).

In addition, the immobilized enzymes exhibit enhanced tolerance to salinity, organic solvents, and surfactants, which is the key advantage of the carrier-free immobilization strategy. The increased structural rigidity minimizes conformational distortion, while the aggregate architecture limits direct interactions between enzyme and destabilizing agents. Collectively, immobilization modulates the enzyme microenvironment, reinforces structural stability, and optimizes mass transfer, leading to superior functional robustness under diverse conditions.

## 7.5 Conclusions and perspective

Given the inherent structural complexity of carrier-free immobilized enzymes, a comprehensive characterization requires the integration of multiple complementary analytical techniques. For example, SEM and TEM provide direct visualization of morphological features, while FTIR and CD elucidate conformational changes at the molecular level. Such multidimensional structural analyses facilitate the establishment of structure-activity relationships that connect the architectural attributes of carrier-free immobilized enzymes with their superior stability, catalytic activity, and substrate selectivity, thereby laying a robust scientific foundation for the rational design and subsequent optimization. Moreover, the comprehensive characterization of enzymatic properties represents the cornerstone for evaluating the functional performance of these biocatalysts. Through systematic investigation of kinetic parameters, activity recovery, and multidimensional stability indices, the advantages and limitations of various immobilization strategies can be quantitatively assessed. This quantitative evaluation provides invaluable guidance for the development and practical implementation of advanced carrier-free immobilization technologies. In addition, the integration of multidimensional datasets, including substrate specificity and tolerance to different reaction environments, enables a deeper mechanistic understanding of enzyme behavior under industrially relevant conditions, offering critical insights for process optimization and large-scale implementation.

Building upon these comprehensive characterization and analysis strategies, the integration of AI offers transformative potential for the future advancement of carrier-free immobilized enzymes. With the rapid advancement of AI technologies in structural biology and materials science, AI is poised to drive breakthrough innovations in this field, offering an unprecedentedly powerful tool for addressing the complex multiscale structural challenges inherent to such enzyme systems. Leveraging machine learning algorithms, multimodal characterization data, such as average particle size from SEM analyses, $\beta$-sheet content from FTIR spectra, as well as preparation parameters (e.g., crosslinking agent type and concentration, temperature, pH) and performance indicators (activity recovery rate, half-life), can be integrated and analyzed systematically. Through the training of regression models (e.g., random forest, gradient boosting decision tree, eXtreme Gradient Boosting (XGBoost), and neural networks), AI enables the prediction of structural changes and catalytic performance, the deciphering of complex structure-activity relationships, and ultimately the inverse design of enzymes from preparation to functional performance. In this way, AI is propelling the structural characterization of carrier-free immobilized enzymes from traditional "descriptive" and "empirical" approaches toward a new era of "predictive" and "rational" research. Furthermore, AI-driven molecular simulations, virtual screening, and automated characterization platforms will accelerate the intelligent design, directed evolution, and application optimization of these biocatalysts, substantially enhancing research efficiency and innovative capacity. Looking ahead, the deep integration of AI with high-throughput experimentation and advanced characterization techniques is expected to catalyze the industrialization of carrier-free immobilized enzymes, guiding the creation of next-generation systems with tailored structures and superior performance, and providing robust technological support for green catalysis and biomanufacturing innovations.

## References

Abd Rahman, N. H., Jaafar, N. R., Abdul Murad, A. M., Abu Bakar, F. D., Shamsul Annuar, N. A., & Illias, R. M. D. (2020). Novel cross-linked enzyme aggregates of levanase from *Bacillus lehensis* G1 for short-chain fructooligosaccharides synthesis: Developmental, physicochemical, kinetic and thermodynamic properties. *International Journal of Biological Macromolecules, 159*, 577–589. www.elsevier.com/locate/ijbiomac. doi:10.1016/j.ijbiomac.2020.04.262.

Ahrari, F., Yousefi, M., Habibi, Z., & Mohammadi, M. (2022). Application of undecanedicarboxylic acid to prepare cross-linked enzymes (CLEs) of *Rhizomucor miehei* lipase (RML): Selective enrichment of polyunsaturated fatty acids. *Molecular Catalysis, 520*, 112172. doi:10.1016/j.mcat.2022.112172.

Aytar, B. S., & Bakir, U. (2008). Preparation of cross-linked tyrosinase aggregates. *Process Biochemistry, 43*(2), 125–131. doi:10.1016/j.procbio.2007.11.001.

Bouguerra, O. M., Abdul Wahab, R., Huyop, F., Mahat, N. A., & Susanti, E. (2025). Statistically assisted preparation of novel alkaline recombinant crosslinked enzyme aggregates lipase KV1 and their physiochemical attributes. *Arabian Journal for Science and Engineering, 50*(18), 14449–14475. https://www.springer.com/journal/13369. doi:10.1007/s13369-025-09967-1.

Broun, G. B. (1976). Chemically aggregated enzymes. *Methods in Enzymology, 44*(C), 263–280. doi:10.1016/S0076-6879(76)44022-3.

Cabirol, F. L., Pei, L. T., Tay, B., Cheng, S., Hanefeld, U., & Sheldon, R. A. (2008). Linum usitatissimum hydroxynitrile lyase cross-linked enzyme aggregates: A recyclable enantioselective catalyst. *Advanced Synthesis and Catalysis, 350*(14–15), 2329–2338. http://www3.interscience.wiley.com/cgi-bin/fulltext/121427963/PDFSTART. doi:10.1002/adsc.200800309Netherlands.

Cao, L., Van Rantwijk, F., & Sheldon, R. A. (2000). Cross-linked enzyme aggregates: A simple and effective method for the immobilization of penicillin acylase. *Organic Letters, 2*(10), 1361–1364. http://pubs.acs.org/journal/orlef7. doi:10.1021/ol005593x.

Danait-Nabar, S., Gharat, K., & Singhal, R. S. (2025). Sodium tripolyphosphate is a non-toxic and economic alternative to glutaraldehyde for preparation of L-asparaginase CLEAs to reduce acrylamide in potato fries. *Food Chemistry, 472*, 142894. doi:10.1016/j.foodchem.2025.142894.

Dey, B., Panwar, V., & Dutta, T. (2023). Fructose-aided cross-linked enzyme aggregates of laccase: An insight on its chemical and physical properties. *Process Biochemistry, 133*, 220–227. doi:10.1016/j.procbio.2023.09.006.

Fardelli, E., De Simone, G., Georgieva, R., Xiong, Y., Di Gioacchino, M., Sotgiu, S., Nucara, A., Tavella, A., Baldassarre, L., Sodo, A., Ricci, A., Gasperi, T., Ascenzi, P., Bäumler, H., di Masi, A., & Capellini, G. (2025). Structural instability of human serum albumin during microparticles synthesis. *ACS Applied Bio Materials, 8*(11), 9907–9915. doi:10.1021/acsabm.5c01228.

Fathali, Z., Rezaei, S., Faramarzi, M. A., & Habibi-Rezaei, M. (2019). Catalytic phenol removal using entrapped cross-linked laccase aggregates. *International Journal of Biological Macromolecules, 122*, 359–366. www.elsevier.com/locate/ijbiomac. doi:10.1016/j.ijbiomac.2018.10.147.

Grajales-Hernández, D. A., Armendáriz-Ruiz, M. A., Rodríguez, J. A., Velasco-Lozano, S., López-Gallego, F., & Mateos-Díaz, J. C. (2025). Substrate bio-imprinted CLEAs of type B feruloyl esterase from *Aspergillus terreus*: A selective heterogeneous biocatalyst towards butyl caffeate. *Process Biochemistry, 150*, 168–179. https://www.sciencedirect.com/science/journal/13595113. doi:10.1016/j.procbio.2025.01.004.

Gupta, M. N., Perwez, M., & Sardar, M. (2020). Protein crosslinking: Uses in chemistry, biology and biotechnology. *Biocatalysis and Biotransformation, 38*(3), 178–201. http://www.tandfonline.com/loi/ibab20. doi:10.1080/10242422.2020.1733990.

Joseph, J. E., Mary, P. R., Haritha, K. V., Panwar, D., & Kapoor, M. (2021). Soluble and cross-linked aggregated forms of $\alpha$-galactosidase from *Vigna mungo* immobilized on magnetic nanocomposites: Improved stability and reusability. *Applied Biochemistry and Biotechnology, 193*(1), 238–256. http://www.springer.com/humana+press/journal/12010. doi:10.1007/s12010-020-03408-5.

Kawatra, A., Datten, B., Hans, R., & Gulati, P. (2025). Hybrid magnetic nanocomposites of arginine deiminase with improved stability and recyclability for biomedical applications. *Preparative Biochemistry & Biotechnology, 55*(4), 513–520. doi:10.1080/10826068.2024.2430626.

Kawatra, A., Chhabra, D., & Gulati, P. (2024). Cross-linked arginine deiminase aggregates for enhanced production of nutraceutical citrulline. *Process Biochemistry, 136*, 147–155. doi:10.1016/j.procbio.2023.11.015.

Khatik, A. G., Jain, A. K., & Muley, A. B. (2022). Preparation, characterization and stability of cross linked nitrilase aggregates (nitrilase–CLEAs) for hydroxylation of 2-chloroisonicotinonitrile to 2-chloroisonicotinic acid. *Bioprocess and Biosystems Engineering, 45*(9), 1559–1579. https://rd.springer.com/journal/449. doi:10.1007/s00449-022-02766-0.

Kubiak, M., Solarczek, J., Kampen, I., Schallmey, A., Kwade, A., & Schilde, C. (2018). Micromechanics of anisotropic cross-linked enzyme crystals. *Crystal Growth & Design, 18*(10), 5885–5895. doi:10.1021/acs.cgd.8b00637.

Lalonde, J. J., Govardhan, C., Khalaf, N., Martinez, A. G., Visuri, K., & Margolin, A. L. (1995). Cross-linked crystals of *Candida rugosa* lipase: Highly efficient catalysts for the resolution of chiral esters. *Journal of the American Chemical Society, 117*(26), 6845–6852. doi:10.1021/ja00131a006.

Lee, Y. L., Jaafar, N. R., Huyop, F., Bakar, F. D. A., Rahman, R. A., & Illias, R. M. D. (2025). Functionalization of amylopectin as a strategy to improve polyethylene terephthalate hydrolase-cross-linked enzyme aggregate (IsPETase-CLEA) in plastic degradation. *International Journal of Biological Macromolecules, 306*, 141492. https://www.sciencedirect.com/science/journal/01418130. doi:10.1016/j.ijbiomac.2025.141492.

Li, H., Wang, R., Wang, A., Zhang, J., Yin, Y., Pei, X., & Zhang, P. (2020). Rapidly and precisely crosslinked enzymes using bio-orthogonal chemistry from cell lysate for the synthesis of (S)-1-(2,6-dichloro-3-fluorophenyl) ethanol. *ACS Sustainable Chemistry & Engineering, 8*(16), 6466–6478. doi:10.1021/acssuschemeng.0c00987.

Liu, J., Li, R., Miao, J., Sun, H., Chen, Q., Song, H., Peng, H., Chang, Y., & Luo, H. (2025). A cofactor regeneration system for 2-aminobutyric acid production based on combined cross-linked enzyme aggregates: Utilizing His-tagged enzymes with low-concentration calcium ions as precipitant. *Engineering in Life Sciences, 25*(3), e70013. doi:10.1002/elsc.70013.

Luo, C., Hu, Y., Xing, S., Xie, W., Li, C., He, L., Wang, X., & Zeng, X. (2023). Adsorption–precipitation–cross-linking immobilization of GDSL-type esterase from *Aspergillus niger* GZUF36 by polydopamine-modified magnetic clarity tetroxide nanocouplings and its enzymatic characterization. *International Journal of Biological Macromolecules, 245*, 125533. doi:10.1016/j.ijbiomac.2023.125533.

Ma, X., Yang, Y. u, Wang, M., Zhang, Y., Zhao, B., Hao, S., & Jia, L. (2025). Engineering high-activity crosslinked enzyme aggregates via SpyCatcher/SpyTag-mediated self-assembly. *International Journal of Biological Macromolecules, 295*, 139591. doi:10.1016/j.ijbiomac.2025.139591.

Martin, R. W., & Zilm, K. W. (2003). Preparation of protein nanocrystals and their characterization by solid state NMR. *Journal of Magnetic Resonance, 165*(1), 162–174. doi:10.1016/S1090-7807(03)00253-2.

Miao, C., Li, H., Zhuang, X., Wang, Z., Yang, L., Lv, P., & Luo, W. (2019). Synthesis and properties of porous CLEAs lipase by the calcium carbonate template method and its application in biodiesel production. *RSC Advances, 9*(51), 29665–29675. doi:10.1039/c9ra04365a.

Morais, J. R. F., Costa, I. O., Padilha, C. E. A., Rios, N. S., & Santos, E. S. D. (2025). Improving reusability of biocatalysts by exploiting cross-linked enzyme aggregates (CLEAs) with commercial cellulolytic cocktails for hydrolysis of green coconut waste. *Sustainability (Switzerland), 17*(9), 4221. http://www.mdpi.com/journal/sustainability/. doi:10.3390/su17094221.

Pan, J., Kong, X. D., Li, C. X., Ye, Q., Xu, J. H., & Imanaka, T. (2011). Crosslinking of enzyme coaggregate with polyethyleneimine: A simple and promising method for preparing stable biocatalyst of *Serratia marcescens* lipase. *Journal of Molecular Catalysis B: Enzymatic, 68*(3–4), 256–261. doi:10.1016/j.molcatb.2010.11.014.

Peng, Y., Xiong, X., Qiu, M., Lu, Y., Chen, T., & Xu, Z. (2025). Hyperbranched polymer-crosslinked laccase aggregates for efficient aerobic oxidation of alcohols. *Enzyme and Microbial Technology, 189*, 110673. doi:10.1016/j.enzmictec.2025.110673.

Qian, J., Huang, A., Zhu, H., Ding, J., Zhang, W., & Chen, Y. (2023). Immobilization of lipase on silica nanoparticles by adsorption followed by glutaraldehyde cross-linking. *Bioprocess and Biosystems Engineering, 46*(1), 25–38. doi:10.1007/s00449-022-02810-z.

Qian, J. Q., Zhao, C. Y., Ding, J., Chen, Y., & Guo, H. (2020). Preparation of nano-enzyme aggregates by crosslinking lipase with sodium tripolyphosphate. *Process Biochemistry, 97*, 19–26. doi:10.1016/j.procbio.2020.06.026.

Reshmi, R., & Sugunan, S. (2013). Improved biochemical characteristics of crosslinked β-glucosidase on nanoporous silica foams. *Journal of Molecular Catalysis B: Enzymatic, 85-86*, 111–118. doi:10.1016/j.molcatb.2012.08.007.

Razib, M. S. M., Rahman, R. N. Z. R. A., Shariff, F. M., & Ali, M. S. M. (2020). Biochemical and structural characterization of cross-linked enzyme aggregates (Cleas) of organic solvent tolerant protease. *Catalysts, 10*(1), 55. https://www.mdpi.com/2073-4344/10/1/55/pdf. doi:10.3390/catal10010055.

Safari, J. B., Ramatsui, L., Nisbet, L., Smith, V. J., Krause, R. W. M., & Pletschke, B. I. (2026). Preparation of magnetic xylanase cross-linked aggregates (m-CLXAs) for the hydrolysis of arabinoxylan by mechano-enzymology. *Enzyme and Microbial Technology, 192*, 110756. https://www.sciencedirect.com/science/journal/01410229. doi:10.1016/j.enzmictec.2025.110756.

Salgın, S., & Salgın, U. (2017). Bioconjugation of magnetic iron oxide nanoparticles and cross-linked enzyme aggregates: Characterization and comparison of CLEAs and magnetic CLEAs. *Current Nanoscience, 13*(2), 149–158. doi:10.2174/1573413712666161018115439.

Sangeetha, K., & Emilia Abraham, T. (2008). Preparation and characterization of cross-linked enzyme aggregates (CLEA) of subtilisin for controlled release applications. *International Journal of Biological Macromolecules, 43*(3), 314–319. doi:10.1016/j.ijbiomac.2008.07.001.

Schoevaart, R., Wolbers, M. W., Golubovic, M., Ottens, M., Kieboom, A. P. G., van Rantwijk, F., van der Wielen, L. A. M., & Sheldon, R. A. (2004). Preparation, optimization, and structures of cross-linked enzyme aggregates (CLEAs). *Biotechnology and Bioengineering, 87*(6), 754–762. doi:10.1002/bit.20184.

Shah, S., Sharma, A., & Gupta, M. N. (2006). Preparation of cross-linked enzyme aggregates by using bovine serum albumin as a proteic feeder. *Analytical Biochemistry, 351*(2), 207–213. http://www.elsevier.com/inca/publications/store/6/2/2/7/8/1/index.htt. doi:10.1016/j.ab.2006.01.028.

Solanki, K., Gupta, M. N., & Halling, P. J. (2012). Examining structure-activity correlations of some high activity enzyme preparations for low water media. *Bioresource Technology, 115*, 147–151. doi:10.1016/j.biortech.2011.12.066.

Torabizadeh, H., Abbasi, S., & Tafaghodinia, B. (2025). Multi-immobilization of viscozyme, alcalase and flavourzyme by nanomagnetic combi-CLEAs method for oil and protein hydrolysates extraction from rice bran in aqueous phase. *International Journal of Biological Macromolecules, 311*(Pt 1), 143719. doi:10.1016/j.ijbiomac.2025.143719.

Torabizadeh, H., & Montazeri, E. (2020). Nano co-immobilization of α-amylase and maltogenic amylase by nanomagnetic combi-cross-linked enzyme aggregates method for maltose production from corn starch. *Carbohydrate Research, 488*, 107904. doi:10.1016/j.carres.2019.107904.

Tudorache, M., Gheorghe, A., Viana, A. S., & Parvulescu, V. I. (2016). Biocatalytic epoxidation of α-pinene to oxy-derivatives over cross-linked lipase aggregates. *Journal of Molecular Catalysis B: Enzymatic, 134*, 9–15. www.elsevier.com/locate/molcatb. doi:10.1016/j.molcatb.2016.09.009.

Wilson, L., Illanes, A., Abián, O., Pessela, B. C. C., Fernández-Lafuente, R., & Guisán, J. M. (2004). Co-aggregation of penicillin G acylase and polyionic polymers: An easy methodology to prepare enzyme biocatalysts stable in organic media. *Biomacromolecules, 5*(3), 852–857. doi:10.1021/bm0343895.

Xiong, X., Lu, Y., Peng, Y., Qiu, M., Chen, T., & Xu, Z. (2025). Hemin-modified crosslinked glucose oxidase aggregates for efficient degradation of tetracycline and detection of phenol derivatives. *Colloids and Surfaces A: Physicochemical and Engineering Aspects, 713*, 136516. doi:10.1016/j.colsurfa.2025.136516.

Yan, H., Shi, B., Ding, J., Zhu, H., Li, X., & Qian, J. (2022). Preparation and characterization of a nontoxic cross-linked lipase aggregate by using sodium tripolyphosphate and chitosan. *Journal of Applied Polymer Science, 139*(13), 51871. http://onlinelibrary.wiley.com/journal/10.1002/(ISSN)1097-4628. doi:10.1002/app.51871.

Yonath, A., Sielecki, A., Moult, J., Podjarny, A., & Traub, W. (1977). Crystallographic studies of protein denaturation of renaturation. 1. Effects of denaturants on volume and x-ray pattern of crosslinked triclinic lysozyme crystals. *Biochemistry, 16*(7), 1413–1417. doi:10.1021/bi00626a027.

Yu, H. W., Chen, H., Wang, X., Yang, Y. Y., & Ching, C. B. (2006). Cross-linked enzyme aggregates (CLEAs) with controlled particles: Application to *Candida rugosa* lipase. *Journal of Molecular Catalysis B: Enzymatic, 43*(1–4), 124–127. doi:10.1016/j.molcatb.2006.07.001.

Zou, B., Liu, F., Yan, Y., Feng, T., & Dowuona, J. N. N. (2022). Magnetic cross-linked lipase aggregates coupled with ultrasonic pretreatment for efficient synthesis of phytosterol oleate. *Molecular Catalysis, 531*, 112649. doi:10.1016/j.mcat.2022.112649.

SECTION

# Applications and case studies

# III

CHAPTER

# 8 Novel approaches for the production of carrier-free immobilized biocatalysts

**Cesar Mateo**

*Department of Biocatalysis, Institute of Catalysis and Petrochemistry CSIC, Campus UAM, Madrid, Spain*

## 8.1 Introduction

### 8.1.1 Relevance of carrier-free immobilization

Enzyme immobilization has historically been a fundamental tool for expanding the use of biocatalysts in industrial settings, as it enables enzyme reuse, enhances stability under extreme conditions (such as nonphysiological pH, temperature, or the presence of organic solvents), and facilitates enzyme separation from the reaction medium. Traditionally, these systems have relied on solid supports onto which enzymes are immobilized through covalent bonds or physical interactions. In many cases, this approach has led to the development of heterogeneous catalysts with improved properties compared to the soluble enzyme in terms of activity, stability, and selectivity (Mateo et al., 2007).

However, the use of supports can introduce significant drawbacks: dilution of the enzyme's catalytic activity by total mass, diffusional limitations in porous supports, loss of activity during the immobilization process, and increased costs associated with the carrier. These challenges have driven the development of carrier-free immobilization systems, in which the enzyme itself constitutes the main structural component of the biocatalyst, eliminating the need for external materials.

Carrier-free immobilization emerges as a compact, efficient, and potentially more sustainable strategy, as it avoids the use of support materials and allows for higher catalytic activity per total catalyst mass. Such systems can be better suited for processes requiring high enzyme loading, continuous flow operation, or compatibility with nonconventional media. Furthermore, the use of pure or recombinant enzymes opens the possibility of designing biocatalysts that are fully defined at the molecular level, which is particularly attractive in the context of advanced biocatalysis and synthetic biology (Cao et al., 2003).

Among the classical approaches to carrier-free immobilization, the most representative examples are cross-linked enzyme aggregates (CLEAs) and cross-linked enzyme crystals (CLECs), which have been widely described and applied in both industry and academia (Häring & Schreier, 1999; Mateo et al., 2004). However, these

**Carrier-Free Enzyme Immobilization: Fundamentals and Innovations in Biocatalysis.**
**DOI: https://doi.org/10.1016/B978-0-443-44139-4.00006-X**

methods also present limitations that restrict their applicability and have motivated the search for new, more versatile, reproducible, and rationally designable strategies for next-generation biocatalysts.

### 8.1.2 Limitations of cross-linked enzyme aggregates and cross-linked enzyme crystals

CLEAs and CLECs rely on the physical or chemical cross-linking of enzymes to form insoluble structures that retain catalytic activity. Despite their successful use in numerous applications, both systems exhibit limitations that restrict their widespread implementation.

The main limitations of CLEAs include (Cui & Jia, 2015):

1. Inefficient and reproducible precipitation, as aggregate quality and yield can be affected by variations in precipitation conditions.
2. Use of chemical cross-linkers that may react with essential catalytic residues, leading to enzyme inactivation or undesired structural modifications.

Regarding CLECs, although they display a well-ordered structure, they present several disadvantages (Roy & Abraham, 2004):

1. They require enzymes capable of crystallizing under simple, reproducible, and scalable conditions, which excludes many enzymes that cannot be co-crystallized within a single crystal.
2. The cross-linking process is typically more complex, slower, and costlier than in the case of CLEAs, in addition to sharing the adverse effects already mentioned for CLEAs.

Both types of catalysts generally lack fine control over the functional architecture of the biocatalyst. For example, they do not allow the specific orientation of active sites nor the rational integration of multiple enzymes into multienzymatic complexes with direct functional communication.

These limitations have encouraged the exploration and development of alternative strategies that do not rely on precipitation, crystallization, or chemical cross-linking, but instead exploit principles of self-assembly, compartmentalization, or molecular engineering to construct carrier-free biocatalysts that are more stable, versatile, and rationally designable.

### 8.1.3 The need for innovative approaches

The advancement of biocatalysis toward more complex systems involving cascade reactions or continuous-flow processes demands new generations of biocatalysts that can be precisely designed, produced reproducibly at large scale, and operated under stable conditions without loss of activity. These requirements cannot be fully met using the traditional methods described above. This issue will be addressed in more detail in Section 8.2 of this chapter.

In addition, progress in fields such as synthetic biology, protein engineering, nanotechnology, and molecular self-organization has opened the door to new strategies for carrier-free immobilization, overcoming many of the previously mentioned limitations. Emerging approaches include genetically directed assembly mechanisms, the formation of functional compartments, and specific supramolecular interactions. These systems offer several advantages:

- Possibility of controlling spatial organization and enzyme density.
- Integration with recombinant enzyme production technologies for simplified and optimized fabrication.
- Compatibility with labile enzymes or those that are difficult to precipitate or crystallize.
- Potential for designing multifunctional or reconfigurable biocatalysts.

These novel approaches favor greater designability and modularity, facilitating the adaptation of biocatalysts to new processes and products.

In this context, this chapter explores some of the most promising approaches for carrier-free enzyme immobilization, representing a conceptual evolution toward intelligent and functionally integrated biocatalysts.

## 8.2 Microfluidic systems for preparing cross-linked enzymes

To overcome the previously discussed drawbacks of traditional enzyme cross-linking systems based on precipitated or crystallized enzymes (CLEAs, CLECs, and others), one of the most widely adopted solutions has been the use of microfluidic techniques. Microfluidics is the science and technology dedicated to studying and manipulating fluids at a very small scale, typically within channels ranging from 1 to 1000 micrometers in size.

In recent years, the generation of particle emulsions and the formation of microparticles using microfluidic platforms have gained significant traction. In conventional emulsification, precipitation, or polymerization methods, there is very limited control over particle size and distribution, cross-linking density and uniformity, and morphology. As a result, complex and poorly reproducible geometries are often produced, which poses a major challenge for industrial implementation (Shieh et al., 2021).

In contrast, microfluidic devices operate under laminar flow and controlled mixing conditions, allowing for precise regulation of droplet size, cross-linker concentration, and reaction kinetics. In droplet-based microfluidics, aqueous enzyme solutions can be emulsified within an immiscible continuous phase to form uniform microdroplets that act as confined microreactors. Within these compartments, cross-linking agents (e.g., glutaraldehyde, Polyethyleneimine (PEI), or other multifunctional chemicals)

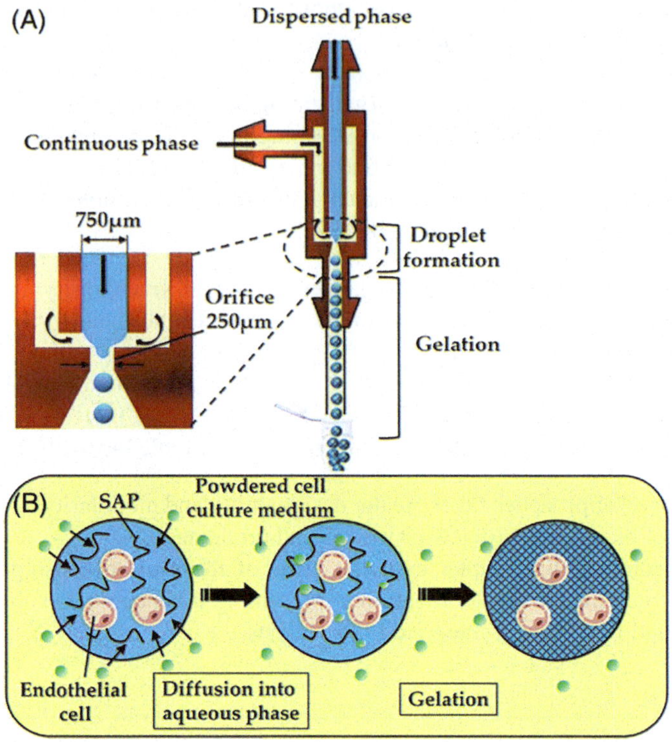


**FIGURE 8.1 Production of self- assembling peptide (SAP) microgels loaded with bovine carotid artery endothelial cells by external gelation using a powdered cell culture medium dispersed in mineral oil.**

(A) Formation of droplets of cell-laden SAP solution and (B) Gelation of droplets due to an increase in the ionic strength caused by diffusion of a powdered cell culture medium into droplets.

*Reproduced with permission from Chen, M., Bolognesi, G., & Vladisavljević, G. T. (2021). Crosslinking strategies for the microfluidic production of microgels.* Molecules, *26(12), 3752. https://doi.org/10.3390/molecules26123752.*

diffuse in a controlled manner, leading to the formation of monodisperse enzymatic microgels or nanogels with tunable physicochemical properties (Fig. 8.1).

The main advantage of microfluidics-assisted cross-linking lies in the reproducibility and uniformity of the resulting biocatalysts. By modulating flow rates and channel geometry, it is possible to precisely tune the size distribution of enzymatic aggregates, typically ranging from a few micrometers to nanometer dimensions. This contrasts with conventional precipitation-based methods, which often yield heterogeneous aggregates with variable activity retention. Moreover, microfluidic platforms inherently minimize reagent consumption, an important consideration when working with costly biocatalysts.

Another major benefit is the potential for continuous operation. Unlike batch preparation of cross-linked enzymes, which requires multiple steps (precipitation, cross-linking, washing, and recovery), microfluidic devices can integrate all these phases into a single automated workflow. This enables process intensification, reduces variability, and enhances scalability. With recent advances in microfluidic chip parallelization and 3D printing, production rates can be significantly increased, paving the way toward industrial-scale biocatalyst manufacturing.

In general, most authors studying this topic classify cross-linking into two main categories (Chen et al., 2021; Farooqi et al., 2023):

1. Ionic methods: These involve the gelation of charged polymers through cross-linking agents. Examples include *external gelation*, where the cross-linker is dissolved or dispersed in the oil phase, and *internal gelation*, where agents are added to the dispersed phase in inactive forms (e.g., chelating agents or controlled-release compounds).
2. Covalent methods: These employ chemical reactions to form covalent bonds between polymer chains. Representative techniques include enzymatic oxidation of modified biopolymers, photopolymerization of monomers or polymers containing cross-linkable groups, and *thiol-ene click* reactions. Other approaches use sol-gel transitions of stimuli-responsive polymers, triggered by pH or temperature changes.

The application of these methods in CLEA production is of great interest since microfluidics enables the precise generation of enzymatic microgels through controlled emulsification and cross-linking. These techniques offer several key advantages:

- Precise control over size and shape: Microgel formation is highly reproducible, which is essential for consistent and scalable applications.
- Efficient enzyme encapsulation: Enzymes can be encapsulated with minimal loss of activity, enhancing their stability and lifetime.
- Reusability in industrial processes: Microfluidically produced CLEAs and CLECs can be reused over multiple reaction cycles, reducing costs and increasing process efficiency.

Numerous examples of the application of fluidic cross-linking have been described. Professor Maeda's group has published several studies developing this type of system. In one such work, a microreactor was developed in which enzymes were immobilized as an enzyme–polymer membrane on the inner walls of the microchannels. A solution of the enzyme (for example, $\alpha$-chymotrypsin) in aqueous buffer was mixed with glutaraldehyde and formaldehyde as cross-linking agents inside a commercial polytetrafluoroethylene (PTFE) tube (internal diameter 500 μm). This resulted in the formation of a CLEA membrane along the internal walls of the tube. This technique has been applied to prepare CLEA-based enzymatic microreactors (CEMs) from a wide variety of enzymes (Honda et al., 2006). In the case of electronegative enzymes (such as aminoacylase), coprecipitation of the enzyme in the presence of poly-L-lysine was used to achieve rapid and efficient CLEA formation. The use of these microchannel reactors with immobilized enzymes demonstrates considerable potential for the design of green and sustainable biotransformations (Honda et al., 2005). In another study, CLEAs of an aminotransferase were produced by separating the acetone-induced precipitation of the enzyme from the cross-linking process. In this way, highly uniform CLEAs (≈100 nm in diameter), stable and with high recovered activity, were obtained, enabling continuous and efficient transamination of (S)-α-methylbenzylamine with pyruvate (Menegatti et al., 2025). Other examples of CLEA production using this technology have been reported for cellulase and laccase enzymes (Nguyen & Yang, 2014; Nguyen et al., 2017), where catalysts with uniform particle size and high catalytic activity were obtained.

## 8.3 Catalytically active inclusion bodies

Various bacteria and other microorganisms produce inclusion bodies (IBs) as a result of the accumulation of misfolded proteins during high-level overexpression processes (Baneyx & Mujacic, 2004). These structures, composed of aggregated proteins, typically display a partially disordered conformation with a significant loss of enzymatic activity, which traditionally required solubilization and refolding protocols to recover the active protein, making them a convenient way to isolate large amounts of pure protein (Singh et al., 2015).

However, studies conducted in recent years have shown that, in addition to misfolded proteins with amyloid-like structures, a substantial fraction of correctly folded proteins is also present within IBs. Evidence suggests that, in cytoplasmic IBs (García-Fruitós et al., 2005), the protein exists in a conformational equilibrium among soluble and correctly folded forms, partially misfolded intermediates, and insoluble aggregates (Jäger et al., 2018). This equilibrium can shift depending on cellular conditions, affecting the relative proportions of soluble, misfolded, aggregated, or protease-degraded forms (Fig. 8.2).

The formation of IBs can be promoted through several strategies: (1) Genetic modification of the enzyme by fusing it with peptides or domains that promote controlled aggregation, (2) Optimization of expression conditions (temperature, induction rate, culture medium, etc.) to favor ordered folding during aggregation, and (3) Co-design with molecular chaperones or co-expression with assembly-assisting proteins that help preserve activity during inclusion formation (Jäger et al., 2020).

These catalytically active IBs (CatIBs) possess a highly compact and mechanically robust nanostructure, making them bionanomaterials. Their strong resistance to denaturing conditions makes them attractive for various biocatalytic, synthetic chemistry, and biomedical applications (Nahalka & Nidetzky, 2007), as heterogeneous biocatalysts that can be used directly in reactions without prior purification or immobilization.

Another major advantage is their direct production in host cells, which reduces intermediate steps and overall process costs. Moreover, their composition, size, and functionality can be modified using genetic engineering tools, providing a degree of control not available in other conventional immobilization methodologies. Altogether,

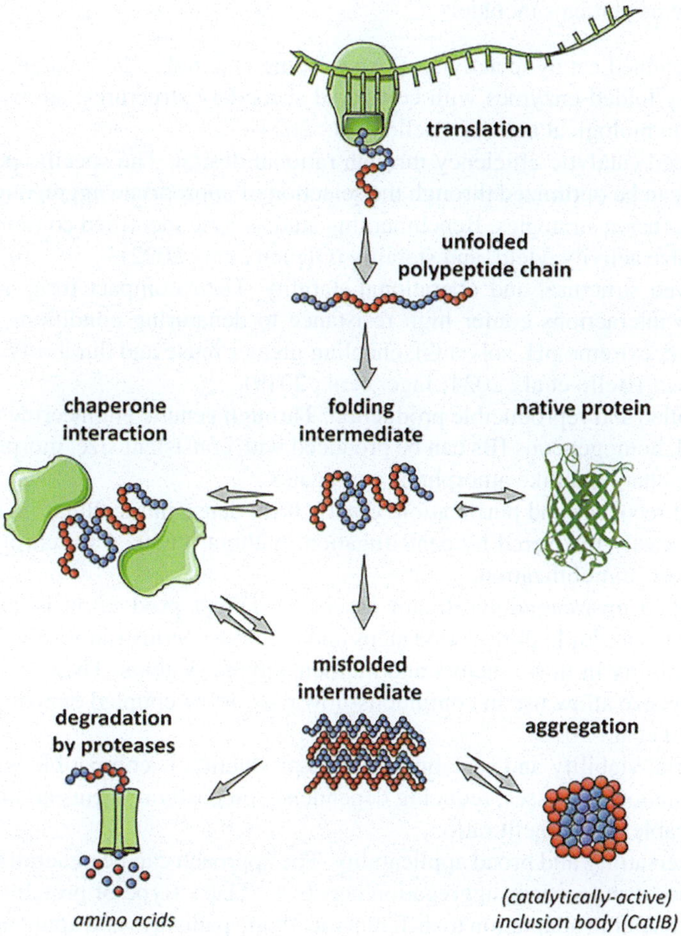


**FIGURE 8.2 Scheme of CatIB formation in bacteria.**

*Modified from Jäger, V. D., Lamm, R., Küsters, K., Ölçücü, G., Oldiges, M., Jaeger, K. E., Büchs, J., & Krauss, U. (2020). Catalytically-active inclusion bodies for biotechnology—general concepts, optimization, and application.* Applied Microbiology and Biotechnology, 104*(17), 7313–7329. https://doi.org/10.1007/s00253-020-10760-3.*

these features make them a powerful emerging technology in the field of carrier-free immobilization.

## 8.3.1 Advantages over amorphous aggregates

One of the key advantages of IBs lies in their structural organization and residual functionality. Whereas amorphous aggregates result from nonspecific and uncontrolled aggregation, IBs exhibit physicochemical properties that make them significantly more attractive as biocatalysts.

Notable advantages include:

- High retained catalytic activity. Despite being insoluble, IBs contain correctly or partially folded enzymes with conserved secondary structures, allowing them to maintain biological activity (Bello et al., 2024).
- Enhanced catalytic efficiency through rational design. The specific performance of IBs can be optimized through the selection of appropriate aggregation tags and genetic fusion strategies. Benchmarking studies have identified combinations that maximize activity, yield, and stability (Ölçücü et al., 2022).
- Improved structural and operational stability. Their compact form and protein-protein interactions confer high resistance to denaturing conditions (high temperature, extreme pH, solvents), enabling greater reuse and durability in catalytic processes (Bello et al., 2024; Jäger et al., 2019).
- Controlled and reproducible production. Through genetic engineering and culture control, homogeneous IBs can be produced with consistent size, morphology, and surface charge, unlike amorphous aggregates.
- Ease of recovery and purification. CatIBs form dense intracellular inclusions that can be easily recovered by centrifugation, without the need for chromatography or further immobilization.
- Simplified upstream/downstream processes. Direct production in the host cell eliminates multiple purification steps and facilitates industrial implementation.
- Applicability in flow reactors and multienzymatic systems. Their robustness and suitable size allow use in continuous-flow reactors or coupled enzymatic systems (Ölçücü et al., 2022).
- Industrial viability and scalability. CatIB production is compatible with scalable fermentation processes, reducing dependence on costly reagents and maintaining a favorable cost-benefit ratio.
- High versatility and broad applicability. The approach can be generalized to many enzymes using various aggregation tags (e.g., TDoT (type of proteins) or coiled-coils), enabling adaptation to different metabolic pathways and applications (Jäger et al., 2020).

## 8.3.2 Drawbacks of catalytically active inclusion bodies compared to cross-linked enzyme aggregates and cross-linked enzyme crystals

- Lower specific activity. Although IBs are catalytically active, their specific activity is generally lower than that of other systems, such as CLEAs or CLECs, particularly when optimal precipitation/crystallization conditions are achieved. This may result from nonoptimal orientations of the active site or diffusional limitations caused by the dense molecular packing within the IB (Ölçücü et al., 2023).
- Protein misfolding and heterogeneity. Since IBs are formed by intracellular aggregation and are designed to be catalytically active, they often contain fractions

of misfolded or inactive protein and exhibit heterogeneous activity distributions among particles (Krauss et al., 2017).

- Dependence on *in vivo* expression. As IBs require controlled intracellular expression of the fusion system in microorganisms, their use is often limited by enzymes that are toxic, poorly expressed, or require specific genetic tools. In contrast, other systems (CLEAs/CLECs) can be prepared from commercial or extracted enzymes without genetic manipulation.
- Lower mechanical robustness. Although insoluble, IBs may display lower mechanical resistance than other systems, affecting both their operational lifetime and reactor design.
- Difficulty in generating rationally co-immobilized multienzymatic systems. Unlike CLEAs, where co-precipitation and cross-linking are straightforward, creating multienzyme CatIB systems in a controlled manner is more complex (Júnior et al., 2021).

## 8.4 Microcompartmentalization and artificial organelles in carrier-free biocatalysis

In recent years, microcompartmentalization has emerged as a strategy inspired by the natural organization of biological systems for the construction of highly efficient and multifunctional biocatalysts. In contrast to traditional immobilization methods on solid supports or with cross-linked protein aggregates (such as CLEAs or CLECs), microcompartmentalization aims to reproduce the spatial confinement and selective protection of cellular organelles, creating controlled environments where enzymes can maintain or even enhance their activity and stability without the need for an external inert support.

Microcompartmentalization is based on the creation of defined microenvironments that allow the physical or dynamic confinement of one or more enzymes within self-assembled supramolecular structures. These structures can be formed from proteins, lipids, polymers, or even bioinspired hybrid systems. In all cases, the goal is to generate catalytic microdomains in which enzymes are concentrated, protected against denaturation, and cooperate more efficiently (Kou et al., 2025; Wang et al., 2023b).

Among the most explored strategies are:

- Complex coacervates are formed through electrostatic association between polymers, peptides, or charged proteins. These liquid compartments can selectively concentrate enzymes and substrates, enhancing catalytic efficiency and offering a degree of dynamic and reversible compartmentalization.
- Microdroplets and protocells, obtained via water-in-oil emulsions or polymeric membrane systems, which allow confinement of complete enzymatic cascades. Their advantage lies in the separation of functional microenvironments that mimic cellular compartmentalization, facilitating sequential reactions that would otherwise be incompatible in a single volume.

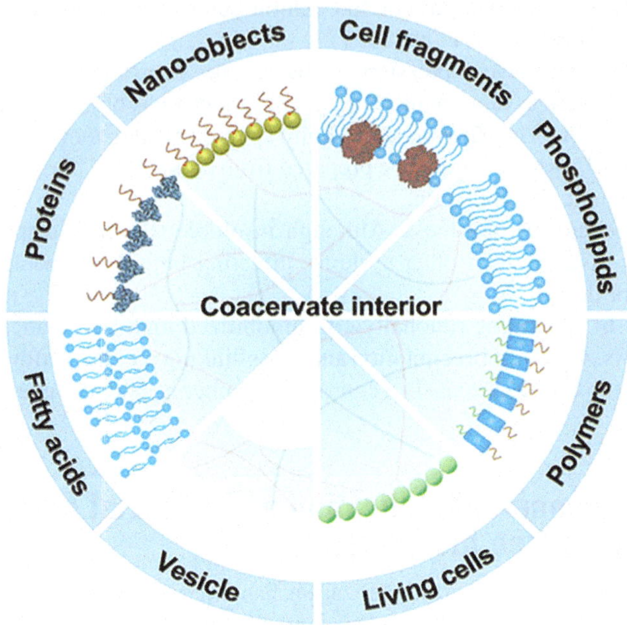


**FIGURE 8.3 Scheme of different systems to produce artificial organelles.**

*Reproduced with permission from Gao, N., & Mann, S. (2023). Membranized coacervate microdroplets: From versatile protocell models to cytomimetic materials.* Accounts of Chemical Research, 56*(3), 297–307. https://doi.org/10.1021/acs.accounts.2c00696.*

- Protein or peptide nanoassemblies, including recombinant viral capsids, modified ferritins, or specific protein-binding assemblies, capable of generating confined nanoscale spaces with high structural stability. These systems provide support-free immobilization, but with a defined architecture that controls the proximity and orientation of the enzymes.

## 8.4.1 Artificial organelles and compartmentalized biocatalysts

The concept of artificial organelles extends microcompartmentalization to more complex functional systems, capable of integrating multiple cascade reactions, regulating selective metabolite exchange, and responding to external stimuli (pH, temperature, light, or chemical signals). These synthetic organelles can be constructed from lipid, polymeric, or protein-based vesicles, incorporating enzymes immobilized in a self-organized manner within their interior or membranes (Fig. 8.3).

For example, polymeric organelles with semipermeable membranes allow the selective entry of small substrates while retaining high-molecular-weight enzymes, creating controlled reaction environments. Similarly, organelles in biomimetic coacervates have demonstrated a high capacity to maintain catalytic activity in nonconventional media (e.g., organic solvents or ionic liquids), which is particularly relevant for synthetic biocatalysis applications.

### 8.4.2 Representative experimental examples

Several studies have demonstrated the potential of these support-free enzymatic systems. For instance:

- Polyelectrolyte coacervates for multienzymatic reactions: Keating's group proposed the use of various polyelectrolytes (polydextrans, polypeptides such as polylysine or polyaspartate, Polyethylene glycol (PEG), etc.) combined with self-assembling proteins and liquid-liquid phase separation to develop new forms of compartmentalization. This approach enabled the creation of multiple distinct liquid compartments, allowing different multienzymatic cascades to take place (Abeysinghe et al., 2024; Mountain & Keating, 2020).
- Polymeric organelles encapsulating enzymatic cascades: van Hest and colleagues developed artificial organelles based on amphiphilic block polymers (polymersomes) encapsulating combinations of redox enzymes. These structures enable sequential reactions in isolated microenvironments, showing significant improvements in thermal stability and selectivity. One example is the use of polystyrene-b-poly(3-(isocyanato-L-alanyl-aminoethyl-thiophene) (PS-b-PIAT) nanoreactors filled with encapsulated enzymes along with free enzymes and substrates inside a larger polybutadiene-b-poly(ethylene oxide) (PB-b-PEO) polymersome, forming a multicompartimentalized structure reminiscent of cells and their organelles. This system enabled a novel three-enzyme cofactor-dependent cascade reaction involving compatible or incompatible enzymes in multiple compartments (Peters et al., 2014).
- Protein nanoassemblies and viral capsids as microreactors: In such systems, biocatalysts and various polymers are encapsulated within viral capsids that act as scaffolds. Encapsulation of enzymes inside protein nanostructures preserves their activity even under chemical or thermal stress. Additionally, genetic or chemical modification of the internal surfaces allows control over enzyme orientation and density, producing self-organized “catalytic nanospheres” (De La Escosura et al., 2009).
- Organelles in biomimetic coacervates and ionic liquids: Complex coacervate microdroplets arise from liquid-liquid phase separation of oppositely charged polyelectrolytes and have applications in synthetic biology, cytomimetic engineering, and microreactor technology. Methods have been developed to produce membrane-bound coacervate microdroplets as a step toward model protocells with high molecular density, increased stability, controllable semipermeability, multilevel organization, and programmable functionality. These systems maintain enzymatic activity in highly nonconventional environments, such as ionic liquids

or biphasic water/organic systems, making them ideal candidates for biotransformations where traditional immobilization systems fail (Gao & Mann, 2023).

### 8.4.3 Advantages and perspectives

Microcompartmentalization offers an innovative alternative to classical immobilization by providing support-free biocatalysts that retain a high degree of spatial organization. These architectures allow:

- Enhanced efficiency of cascade reactions through enzyme colocalization.
- Protection of enzymes against denaturing conditions.
- Introduction of spatiotemporal regulation of catalytic processes.
- Design of smart biocatalysts, responsive to stimuli and capable of self-regulating their activity.

Overall, microcompartmentalized systems and artificial organelles represent one of the most promising frontiers in the development of carrier-free biocatalysts. Their integration with bioassembly techniques, protein engineering, and dynamic materials paves the way for self-organized, reconfigurable, and multifunctional biocatalysts, with tremendous potential for sustainable and industrial biocatalysis in the future.

## 8.5 Tag-mediated assembly

Tag-mediated assembly is an emerging strategy for carrier-free enzyme immobilization that relies on the directed self-organization of proteins through the incorporation of peptide sequences or functional domains that promote highly specific (reversible or irreversible) interactions between enzyme molecules.

In this approach, the genetically modified enzyme itself becomes part of the self-assembling system. The result is the formation of functional protein supramolecular structures, such as fibers, nanogels, granules, networks, or multienzyme complexes, without the need for additional matrices.

Protein tags have traditionally been used for various purposes, including affinity purification, enzyme immobilization and localization, or labeling techniques (Pei et al., 2022).

In carrier-free systems (without external supports), tags allow the direction, ordering, or stabilization of enzymatic self-assembly. Tag-mediated assembly offers multiple advantages that distinguish it from both traditional support-based immobilization methods and other carrier-free strategies such as CLEAs, CLECs, or amorphous IBs. These advantages arise from its modular, programmable, and genetic nature, allowing high precision in biocatalyst design.

Classification of main tags used in carrier-free enzyme immobilization:

- Bioorthogonal covalent tags.

  Various tag systems use genetic fusion technologies for enzyme immobilization. These tags are peptide sequences genetically fused to recombinant proteins

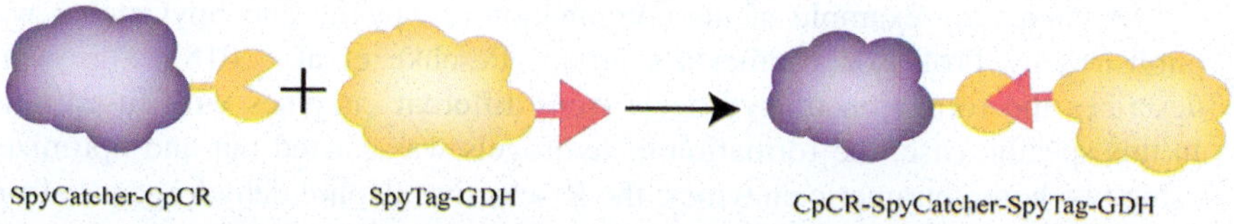


**FIGURE 8.4 Spontaneous reaction between SpyCatcher and SpyTag-fusion enzymes.**

*Modified from Wang, J., Lu, Y., Cheng, P., Zhang, C., Tang, L., Du, L., Li, J., & Ou, Z. (2023). Construction of bi-enzyme self-assembly clusters based on SpyCatcher/SpyTag for the efficient biosynthesis of (R)-ethyl 2-hydroxy-4-phenylbutyrate.* Biomolecules, 13*(1), 91. https://doi.org/10.3390/biom13010091.*

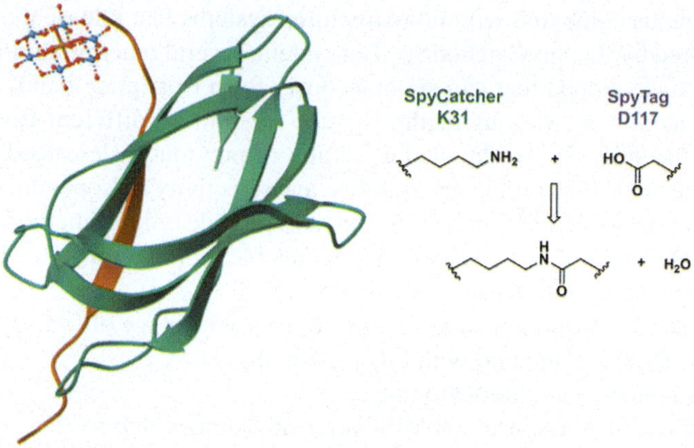


**FIGURE 8.5 Tag- mediated assembly using SpyCatcher system.**

(PDB entry 9OJ3).

for purposes such as enzyme purification via affinity processes, immobilization, or labeling techniques (Ölçücü et al., 2021).

Among these systems, the most widely known and used are Catcher/Tag systems for enzyme immobilization. These systems consist of a peptide and a protein that form a spontaneous and irreversible covalent isopeptide bond (Fig. 8.4) (Essert & Castiglione, 2024; Wang et al., 2023a).

Catcher/Tag systems are protein engineering tools used to create covalent bonds, typically isopeptide bonds, between proteins or protein-peptide pairs (Fan & Aranko, 2024; Keeble & Howarth, 2020; Zakeri et al., 2012). The first and best-characterized system involves the SpyTag peptide, 13 amino acids (aa) long, and the SpyCatcher protein, 116 aa, both derived from the split CnaB2 domain of the fibronectin-binding protein FbaB from *Streptococcus pyogenes*.

When SpyTag and SpyCatcher are mixed, they spontaneously form an irreversible isopeptide bond between a lysine (K31) in SpyCatcher and an aspartic acid (D117) in SpyTag (Patti et al., 2025) (Fig. 8.5).

An interesting example of the combination of SpyTag and SpyCatcher was published by Professor Niemeyer's group (Peschke et al., 2018). This work describes the formation of hydrogels from different enzymes with fused tails. In this specific case, the formation of hydrogels was carried out and optimized using two homotetrameric enzymes: the R-selective alcohol dehydrogenase from *Lactobacillus brevis* fused to a SpyTag, and the redox cofactor-regenerating enzyme glucose 1-dehydrogenase from *Bacillus subtilis*, fused to a SpyCatcher. The optimized hydrogel system was able to catalyze the reaction of different substrates (5-nitrononane-2,8-dione, acetophenone, 4-Cl-acetophenone, and trans-4-phenyl-3-buten-2-one) to convert them into the corresponding R-alcohols with quantitative conversions and enantiomeric excesses greater than 99%. In addition, it was demonstrated that this type of structured system was stable, being advantageous compared with non-structured systems that release the enzymes.

Inspired by the SpyCatcher/SpyTag system, several other protein/peptide pairs have been developed that can spontaneously form isopeptide bonds. These pairs were created using various methods, such as splitting different CnaB domains, altering the split site, or directed evolution of previously described Catcher/Tag pairs. Due to the high efficiency, stability, and selectivity of isopeptide bond formation, currently available Catcher/Tag systems provide reliable methods for creating stable protein conjugates and assemblies, suitable for multienzyme systems as well as hydrogel-type matrices.

Overall, these covalent tag systems offer the advantage of forming permanent, robust structures compared with other methods.

- Modular noncovalent interaction tags.

  Noncovalent protein assemblies offer advantages due to their dynamic and reversible nature, allowing adaptable aggregation/disaggregation and reduced risk of permanent structural damage while enabling functional organization and activity (Hertel et al., 2026). In nature, many noncovalent protein-protein interaction domains, such as PDZ (named after the first three proteins containing this domain: PSD-95, Dlg, ZO-1), GTPase-binding domain (GBD), and Src homology 3 (SH3) orchestrate the assembly of functional complexes for signal transduction and cellular metabolism (Dueber et al., 2009).
- Functional polypeptide-based tags.

  Elastin-like polypeptides (ELPs) and collagen-like polypeptides (CLPs): This technology involves fusing biocompatible polypeptides, such as elastin or collagen, together, and with catalytic proteins. After expression in a host microorganism (e.g., *E. coli*), the fused enzyme can undergo controlled self-assembly to produce the immobilized biocatalyst (Liu et al., 2018).

CLPs offer excellent biocompatibility and low toxicity. For example, collagen fibers modified with Zr(IV) were used to immobilize catalase, showing higher activity than the free enzyme (Song et al., 2011). CLPs can self-assemble into biomaterials with different morphologies (Jiang et al., 2014), facilitating their use as supports for enzyme immobilization (Lu et al., 2007).

Similarly, ELPs are artificial polypeptides (Meyer & Chilkoti, 2002) that exhibit a sharp, reversible phase transition at specific temperatures, a property exploited to separate and purify proteins from fermentation broth (Yeboah et al., 2016).

Hydrophobin Tags: Hydrophobins are small fungal proteins, abundant and cysteine-rich, found in hyphal cell walls and secreted into the culture medium (Scherrer et al., 2000). They participate in the formation of aerial structures, confer surface hydrophobicity, and anchor hyphae to hydrophobic supports (Wösten, 2001).

These proteins self-assemble spontaneously at hydrophilic-hydrophobic interfaces, forming ultrathin (5–12 nm), highly amphipathic films (Wosten et al., 1994). Their properties have enabled diverse applications in medicine, technology, and protein immobilization, particularly for lipases, serving as substitutes for traditional hydrophobic supports (Palomo et al., 2003).

Silk-like polypeptides (SLPs) and Resilins: Silk, produced by insects as a structural material, has remarkable strength, elasticity, and durability, attracting interest in biomaterials. Resilin is an extremely resilient protein critical for insect movement, flight, and sound production (Elvin et al., 2005).

Both resilins and SLPs can polymerize into highly resistant structures. Resilin, which has been synthesized in several host organisms such as E. coli, is composed of three segments:

1. N-terminal hydrophilic segment: highly elastic.
2. Central hydrophobic segment: contains the chitin-binding sequence.
3. C-terminal hydrophilic segment: undergoes reversible conformational changes, storing energy.

This energy storage mechanism, triggered by thermal changes, allows for the synthesis of temperature-responsive cross-linked materials, making resilin-based polypeptides highly attractive for biomaterial applications. While primarily studied for tissue engineering, they also show promise in combination with enzymes and other biomacromolecules (Desai & Lee, 2015).

Advances in this field, together with the progress in computational techniques, polypeptide synthesis methods, and modular protein engineering routines, have enabled the development of new peptides, expanding their applications in tissue engineering, drug delivery, bioimaging, biosensors, catalysis, and bioelectronics (Balu et al., 2021; Manissorn et al., 2025).

Applications:

Tag-mediated assembly has shown great potential in multiple areas of biotechnology, particularly where precise, efficient, and support-free organization of active enzymes is required. Its versatility and modular design capability allow these platforms to be adapted to diverse environments and objectives.

- Industrial biocatalysis: Self-organized enzymatic assemblies can be applied in large-scale chemical reactions, especially those requiring continuous or recyclable processes. Their robustness and thermal stability make them suitable for demanding conditions, such as the synthesis of pharmaceuticals (Rocha et al., 2022), agrochemicals (Kim & Hahn, 2014), and flavor compounds (Gao et al., 2023).

- Multienzyme cascade reactions: By using complementary tags, different enzymes can be assembled within the same supramolecular network, promoting the channeling of intermediates. This design has been applied in the production of modified sugars, organic acids, or bioplastics, improving process efficiency compared to systems with free enzymes (Liu et al., 2023; Luo et al., 2024; Zhu et al., 2020).
- Microreactors and miniaturized platforms: Protein assemblies can be easily integrated into microfluidic systems and lab-on-a-chip devices, facilitating the implementation of miniature bioreactors with high catalytic performance and fine process control. They are also compatible with 3D printing and biofabrication technologies (Bajić et al., 2024; Freitas et al., 2022).
- Stimuli-responsive and controllable systems: By employing tags that respond to stimuli (such as temperature, light, pH, or ligands), it is possible to design smart biocatalysts that activate or deactivate their function depending on the environment. This is useful for biosensors, controlled drug release, or single-use biocatalysts (Gao et al., 2016; Qi & Qiao, 2021).
- Production of functional materials: Assembled enzymatic complexes can form materials with unique structural and catalytic properties, such as bioactive hydrogels, nanofibrillar networks, or functional films, which are valuable in biomedicine, bioplastics, and green technologies (Bitterwolf et al., 2022; Wang et al., 2024).
- *In vivo* expression in recombinant systems: Due to their genetic nature, these assemblies can be produced within modified organisms, such as bacteria or fungi, that express enzymes fused to specific tags. This allows the development of intracellular "nanofactories" for metabolite production or direct bioremediation in the environment (Brockman & Prather, 2015).

Overall, these applications demonstrate the versatility, scalability, and compatibility of this approach across various technological areas. Its ability to integrate biocatalysis, self-assembly, and rational design positions it as a next-generation tool for sustainable and efficient processes.

## 8.6 Comparative analysis of emerging carrier-free enzyme immobilization strategies

Recent advances in carrier-free enzyme immobilization have moved beyond conventional CLEAs and CLECs, opening the door to sophisticated, self-organized, and multifunctional biocatalytic systems. Among the most prominent emerging strategies are catalytically active inclusion bodies (CatIBs), microcompartmentalization and artificial organelles, and tag-mediated enzyme assembly. Each approach offers unique advantages and challenges, which are summarized and analyzed below (Table 8.1).

**Table 8.1** Comparison of the different described strategies.

| Feature/strategy | Catalytically active inclusion bodies (CatIBs) | Microcompartmentalization and artificial organelles | Tag-mediated enzyme assembly |
|---|---|---|---|
| Principle | Formation of insoluble, enzyme-rich aggregates during recombinant expression; aggregates retain catalytic activity | Confinement of enzymes in self-assembled microenvironments (coacervates, vesicles, polymersomes) mimicking cellular organelles | Specific molecular interactions (affinity, covalent, or protein–protein) drive assembly of enzymes into defined complexes or scaffolds |
| Structural control | Limited; morphology and size depend on expression and aggregation conditions | Moderate to high; microenvironment architecture can be tuned via choice of polymers, lipids, or self-assembling components | High; stoichiometry, orientation, and spatial organization can be precisely controlled using tags |
| Activity/substrate access | Good for robust enzymes; diffusion may be limited in dense aggregates | Excellent; confinement enhances local enzyme concentration, substrate channeling, and protection against denaturation | Excellent; site-specific immobilization preserves activity and enables cascade reactions or multienzyme networks |
| Stability | High thermal and chemical stability due to aggregated nature | Moderate to high; stability depends on the compartment material (coacervate, polymer, vesicle) | High, especially for covalent tags; reversible tags allow dynamic regulation |
| Scalability | High; easy to produce in recombinant hosts | Moderate; production of uniform compartments can require microfluidic or advanced assembly techniques | Moderate; requires protein engineering and design, but modularity enables combinatorial assembly |
| Applications | Industrial biocatalysis, harsh conditions, multistep transformations in single-pot | Multienzyme cascades, synthetic biology, confined bioreactors, biosensing | Cascade reactions, artificial metabolons, microreactors, stimuli-responsive biocatalysts |
| Advantages | Simple, genetically encoded, highly concentrated catalytic material | Mimics cellular compartmentalization; enables co-localization and protection; compatible with unconventional media | High precision, modularity, dynamic control, integration with scaffolds or microenvironments |
| Challenges/limitations | Limited control over morphology and particle heterogeneity; diffusion may be restricted | Complexity of design and reproducibility; potential challenges in large-scale production | Requires genetic fusion or chemical modification; scalability and cost may be limiting |

### 8.6.1 Analysis and perspective

- CatIBs: CatIBs represent a genetically encoded, one-step immobilization strategy, making them attractive for industrial biocatalysis. Their robustness and simplicity are unmatched for large-scale processes. However, their structural heterogeneity and potential diffusional limitations restrict fine control over multistep or sensitive reactions.
- Microcompartmentalization and artificial organelles: These strategies mimic cellular organization, providing confined microenvironments that enhance enzyme stability, substrate channeling, and compatibility with harsh or nonconventional media (e.g., ionic liquids). They excel in multienzyme cascades and synthetic biology applications. Challenges include complex assembly protocols, potential reproducibility issues, and the need for microfabrication or microfluidic approaches for precise compartment generation.
- Tag-mediated enzyme assembly: Tag-mediated approaches provide high modularity and structural precision, allowing the construction of artificial metabolons, programmable scaffolds, and stimuli-responsive systems. They integrate seamlessly with other strategies, such as microcompartments, to create dynamic and controllable biocatalysts. Limitations include protein engineering requirements and potentially higher production costs, although the precision they afford can justify the investment for high-value applications.

### 8.6.2 Comparative insights

- Scalability vs. control: CatIBs excel in scalability and simplicity, while tag-mediated assemblies and microcompartments offer greater structural control and functional tunability, albeit with higher design complexity.
- Enzyme protection and stability: Microcompartments provide selective protection and enhanced stability in challenging media. Tag-mediated systems also enable protection, especially when coupled with scaffolds or vesicular environments.
- Multienzyme cascades: Both microcompartments and tag-mediated assemblies favor substrate channeling and sequential reactions, whereas CatIBs are limited to reactions compatible with the dense aggregate environment.
- Integration potential: Tag-mediated systems can bridge CatIBs and microcompartments, combining the robustness of IBs with the spatial organization and dynamic control of compartments or scaffolds.

Based on the previously presented information, it can be concluded that emerging carrier-free immobilization strategies are complementary to one another. CatIBs are ideal for robust, high-density enzyme preparations at scale. Microcompartmentalization and artificial organelles excel in mimicking cellular environments and orchestrating complex, multistep reactions. Tag-mediated enzyme assemblies provide precision, modularity, and dynamic control, enabling the design of next-generation biocatalysts. The integration of these strategies promises multifunctional, programmable, and sustainable carrier-free biocatalysts suitable for industrial and synthetic biology applications.

## 8.7 Conclusions and perspectives

Carrier-free immobilized enzymes have evolved from classical systems based on CLEAs and CLECs to increasingly sophisticated, self-organized, and multifunctional configurations. These strategies overcome many of the intrinsic limitations of conventional immobilization on supports, such as low volumetric activity, diffusional constraints, and high material costs. CLEAs and CLECs represented the first generation of carrier-free biocatalysts, enabling enzyme aggregation and crosslinking to obtain insoluble yet catalytically active materials. However, their structural heterogeneity, diffusional limitations, and limited control over particle morphology continue to restrict their large-scale application in industrial processes.

Recent advances have introduced new avenues to intensify and control the production of these biocatalysts. The integration of microfluidic technologies has enabled more homogeneous particle formation, improved mass and heat transfer, and tunable reaction environments. These approaches have provided unprecedented control over enzyme crosslinking, cofactor retention, and co-immobilization, resulting in more reproducible and scalable preparations.

Beyond traditional aggregates, the field has expanded toward biosynthetic and bioinspired carrier-free systems, such as CatIBs, enzymatic microcompartmentalization, and artificial organelles. These systems leverage principles of self-assembly or biological compartmentalization to generate confined and cooperative catalytic environments without the need for external supports. Concurrently, tag-mediated enzyme assembly strategies have emerged as powerful and modular tools for constructing multienzyme complexes or self-organized scaffolds through specific and reversible interactions. Such assemblies can mimic natural metabolons, promoting substrate channeling and enabling the design of highly efficient cascade reactions.

Looking ahead, the next generation of carrier-free biocatalysts will likely integrate multiple levels of structural and functional control. The combination of synthetic biology, protein engineering, and microfabrication will allow the design of enzymatic assemblies with dynamic regulation, self-repair capacity, and responsiveness to environmental stimuli. The convergence of nanobiotechnology and bioinspired materials science will further enable the creation of hybrid systems that bridge the gap between biological complexity and process robustness.

Ultimately, these advances will lead to truly integrated multicatalytic systems, capable of performing sequential or parallel reactions in continuous flow or confined microenvironments, offering unprecedented efficiency, stability, and sustainability in biocatalytic manufacturing.

## Acknowledgments

This work was supported by the Spanish Ministry of Science, Innovation and Universities under project number PID2024-157256OB-I00. We gratefully acknowledge their financial support.

## References

Abeysinghe, A. A. D. T., Young, E. J., Rowland, A. T., Dunshee, L. C., Urandur, S., Sullivan, M. O., Kerfeld, C. A., & Keating, C. D. (2024). Interfacial assembly of bacterial microcompartment shell proteins in aqueous multiphase systems. *Small, 20*(15), e2308390. http://onlinelibrary.wiley.com/journal/10.1002/(ISSN)1613-6829. doi:10.1002/smll.202308390.

Bajić, M., Khiawjan, S., Hilton, S. T., Lye, G. J., Marques, M. P. C., & Szita, N. (2024). A paradigm shift for biocatalytic microreactors: Decoupling application from reactor design. *Biochemical Engineering Journal, 205*, 109260. https://www.sciencedirect.com/science/journal/1369703X. doi:10.1016/j.bej.2024.109260.

Balu, R., Dutta, N. K., Dutta, A. K., & Choudhury, N. R. (2021). Resilin-mimetics as a smart biomaterial platform for biomedical applications. *Nature Communications, 12*(1), 149. http://www.nature.com/ncomms/index.html. doi:10.1038/s41467-020-20375-x.

Baneyx, F., & Mujacic, M. (2004). Recombinant protein folding and misfolding in *Escherichia coli*. *Nature Biotechnology, 22*(11), 1399–1408. doi:10.1038/nbt1029.

Bello, M. N., Sabri, S., Mohd Yahaya, N., Mohd Shariff, F., & Mohamad Ali, M. S. (2024). Catalytically active inclusion bodies as a potential tool for biotechnology. *Biotechnology and Applied Biochemistry, 71*(6), 1235–1242. http://iubmb.onlinelibrary.wiley.com/hub/journal/10.1002/(ISSN)1470-8744/. doi:10.1002/bab.2624.

Bitterwolf, P., Zoheir, A. E., Hertel, J., Kröll, S., Rabe, K. S., & Niemeyer, C. M. (2022). Intracellular assembly of interacting enzymes yields highly-active nanoparticles for flow biocatalysis. *Chemistry – A European Journal, 28*(66), e202202157. http://onlinelibrary.wiley.com/journal/10.1002/(ISSN)1521-3765. doi:10.1002/chem.202202157.

Brockman, I. M., & Prather, K. L. J. (2015). Dynamic metabolic engineering: New strategies for developing responsive cell factories. *Biotechnology Journal, 10*(9), 1360–1369. http://onlinelibrary.wiley.com/journal/10.1002/(ISSN)1860-7314. doi:10.1002/biot.201400422.

Cao, L., van Langen, L., & Sheldon, R. A. (2003). Immobilised enzymes: Carrier-bound or carrier-free? *Current Opinion in Biotechnology, 14*(4), 387–394. http://www.elsevier.com/locate/copbio. doi:10.1016/S0958-1669(03)00096-X.

Chen, M., Bolognesi, G., & Vladisavljević, G. T. (2021). Crosslinking strategies for the microfluidic production of microgels. *Molecules, 26*(12), 3752. doi:10.3390/molecules26123752.

Cui, J. D., & Jia, S. R. (2015). Optimization protocols and improved strategies of cross-linked enzyme aggregates technology: Current development and future challenges. *Critical Reviews in Biotechnology, 35*(1), 15–28. http://www.tandfonline.com/loi/ibty20. doi:10.3109/07388551.2013.795516.

De La Escosura, A., Nolte, R. J. M., & Cornelissen, J. J. L. M. (2009). Viruses and protein cages as nanocontainers and nanoreactors. *Journal of Materials Chemistry, 19*(16), 2274–2278. doi:10.1039/b815274h.

Desai, M. S., & Lee, S. W. (2015). Protein-based functional nanomaterial design for bioengineering applications. *Wiley Interdisciplinary Reviews: Nanomedicine and Nanobiotechnology, 7*(1), 69–97. http://www3.interscience.wiley.com/journal/121524295/home. doi:10.1002/wnan.1303.

Dueber, J. E., Wu, G. C., Malmirchegini, G. R., Moon, T. S., Petzold, C. J., Ullal, A. V., Prather, K. L. J., & Keasling, J. D. (2009). Synthetic protein scaffolds provide modular control over metabolic flux. *Nature Biotechnology, 27*(8), 753–759. doi:10.1038/nbt.1557.

Elvin, C. M., Carr, A. G., Huson, M. G., Maxwell, J. M., Pearson, R. D., Vuocolo, T., Liyou, N. E., Wong, D. C. C., Merritt, D. J., & Dixon, N. E. (2005). Synthesis and properties of crosslinked recombinant pro-resilin. *Nature, 437*(7061), 999–1002. doi:10.1038/nature04085.

Essert, A., & Castiglione, K. (2024). Dimer stabilization by SpyTag/SpyCatcher coupling of the reductase domains of a chimeric P450 BM3 monooxygenase from *Bacillus spp.* improves its stability, activity, and purification. *ChemBioChem, 25*(3), e202300650. doi:10.1002/cbic.202300650.

Fan, R., & Aranko, A. S. (2024). Catcher/Tag toolbox: Biomolecular click-reactions for protein engineering beyond genetics. *ChemBioChem, 25*(1), e202300600. http://onlinelibrary.wiley.com/journal/10.1002/(ISSN)1439-7633. doi:10.1002/cbic.202300600.

Farooqi, Z. H., Vladisavljević, G. T., Pamme, N., Fatima, A., Begum, R., Irfan, A., & Chen, M. (2023). Microfluidic fabrication and applications of microgels and hybrid microgels. *Critical Reviews in Analytical Chemistry, 54*(7), 2435–2449. http://www.tandf.co.uk/journals/titles/10408347.asp. doi:10.1080/10408347.2023.2177097.

Freitas, A. I., Domingues, L., & Aguiar, T. Q. (2022). Tag-mediated single-step purification and immobilization of recombinant proteins toward protein-engineered advanced materials. *Journal of Advanced Research, 36*, 249–264. http://www.elsevier.com/wps/find/journaldescription.cws_home/722881/description#description. doi:10.1016/j.jare.2021.06.010.

Gao, N., & Mann, S. (2023). Membranized coacervate microdroplets: From versatile protocell models to cytomimetic materials. *Accounts of Chemical Research, 56*(3), 297–307. doi:10.1021/acs.accounts.2c00696.

Gao, X., Wei, C., Qi, H., Li, C., Lu, F., & Qin, H. M. (2023). Directional immobilization of D-allulose 3-epimerase using SpyTag/SpyCatcher strategy as a robust biocatalyst for synthesizing D-allulose. *Food Chemistry, 401*, 134199. doi:10.1016/j.foodchem.2022.134199.

Gao, X., Fang, J., Xue, B., Fu, L., & Li, H. (2016). Engineering protein hydrogels using SpyCatcher-SpyTag chemistry. *Biomacromolecules, 17*(9), 2812–2819. doi:10.1021/acs.biomac.6b00566.

García-Fruitós, E., González-Montalbán, N., Morell, M., Vera, A., Ferraz, R. M., Arís, A., Ventura, S., & Villaverde, A. (2005). Aggregation as bacterial inclusion bodies does not imply inactivation of enzymes and fluorescent proteins. *Microbial Cell Factories, 4*(1), 27. doi:10.1186/1475-2859-4-27.

Häring, D., & Schreier, P. (1999). Cross-linked enzyme crystals. *Current Opinion in Chemical Biology, 3*(1), 35–38. doi:10.1016/s1367-5931(99)80007-6.

Hertel, J. S., Martini, M. A., Stoeckle, M., Weber, A. J., Rabe, K. S., & Niemeyer, C. M. (2026). Harnessing non-covalent protein–protein interaction domains for production of biocatalytic materials systems. *Advanced Functional Materials, 36*(11), e202513931. http://onlinelibrary.wiley.com/journal/10.1002/(ISSN)1616-3028. doi:10.1002/adfm.202513931.

Honda, T., Miyazaki, M., Nakamura, H., & Maeda, H. (2005). Immobilization of enzymes on a microchannel surface through cross-linking polymerization. *Chemical Communications*, (40), 5062–5064. http://pubs.rsc.org/en/journals/journal/cc. doi:10.1039/b510605b.

Honda, T., Miyazaki, M., Nakamura, H., & Maeda, H. (2006). Facile preparation of an enzyme-immobilized microreactor using a cross-linking enzyme membrane on a microchannel surface. *Advanced Synthesis & Catalysis, 348*(15), 2163–2171. doi:10.1002/adsc.200606224.

Jäger, V. D., Kloss, R., Grünberger, A., Seide, S., Hahn, D., Karmainski, T., Piqueray, M., Embruch, J., Longerich, S., Mackfeld, U., Jaeger, K.-E., Wiechert, W., Pohl, M., & Krauss, U. (2019). Tailoring the properties of (catalytically)-active inclusion bodies. *Microbial Cell Factories, 18*(1), 71. doi:10.1186/s12934-019-1081-5.

Jäger, V. D., Lamm, R., Kloß, R., Kaganovitch, E., Grünberger, A., Pohl, M., Büchs, J., Jaeger, K. E., & Krauss, U. (2018). A synthetic reaction cascade implemented by colocalization of two proteins within catalytically active inclusion bodies. *ACS Synthetic Biology, 7*(9), 2282–2295. http://pubs.acs.org/journal/asbcd6. doi:10.1021/acssynbio.8b00274.

Jäger, V. D., Lamm, R., Küsters, K., Ölçücü, G., Oldiges, M., Jaeger, K. E., Büchs, J., & Krauss, U. (2020). Catalytically-active inclusion bodies for biotechnology—general concepts, optimization, and application. *Applied Microbiology and Biotechnology, 104*(17), 7313–7329. link.springer.de/link/service/journals/00253/index.htm. doi:10.1007/s00253-020-10760-3.

Jiang, T., Xu, C., Liu, Y., Liu, Z., Wall, J. S., Zuo, X., Lian, T., Salaita, K., Ni, C., Pochan, D., & Conticello, V. P. (2014). Structurally defined nanoscale sheets from self-assembly of collagen-mimetic peptides. *Journal of the American Chemical Society, 136*(11), 4300–4308. doi:10.1021/ja412867z.

Júnior, A. A. D. T., Ladeira, Y. F. X., França, A. D. S., Souza, R. O. M. A. D., Moraes, A. H., Wojcieszak, R., Itabaiana, I., & Miranda, A. S. D. (2021). Multicatalytic hybrid materials for biocatalytic and chemoenzymatic cascades—strategies for multicatalyst (Enzyme) co-immobilization. *Catalysts, 11*(8), 936. https://www.mdpi.com/2073-4344/11/8/936/pdf. doi:10.3390/catal11080936.

Keeble, A. H., & Howarth, M. (2020). Power to the protein: Enhancing and combining activities using the Spy toolbox. *Chemical Science, 11*(28), 7281–7291. http://pubs.rsc.org/en/journals/journal/sc. doi:10.1039/d0sc01878c.

Kim, S., & Hahn, J. S. (2014). Synthetic scaffold based on a cohesin-dockerin interaction for improved production of 2,3-butanediol in *Saccharomyces cerevisiae*. *Journal of Biotechnology, 192*, 192–196. www.elsevier.com/locate/jbiotec. doi:10.1016/j.jbiotec.2014.10.015.

Kou, Q., Hu, C., Liu, C., Ya, Z., Wang, K., Zheng, J., & Liu, J. (2025). Advances in multicompartment synthetic protocells: From design strategies to biomedical application. *Small, 21*(38), e202507577. doi:10.1002/smll.202507577.

Krauss, U., Jäger, V. D., Diener, M., Pohl, M., & Jaeger, K. E. (2017). Catalytically-active inclusion bodies—Carrier-free protein immobilizates for application in biotechnology and biomedicine. *Journal of Biotechnology, 258*, 136–147. www.elsevier.com/locate/jbiotec. doi:10.1016/j.jbiotec.2017.04.033.

Liu, M., Song, Y., Zhang, Y. H. P. J., & You, C. (2023). Carrier-free immobilization of multi-enzyme complex facilitates *in vitro* synthetic enzymatic biosystem for biomanufacturing. *ChemSusChem, 16*(6), e202202153. http://onlinelibrary.wiley.com/journal/10.1002/(ISSN)1864-564X. doi:10.1002/cssc.202202153.

Liu, D., Du, K., & Feng, W. (2018). Immobilization of enzymes using a multifunctional fusion polypeptide. *Biotechnology Letters, 40*(1), 181–187. doi:10.1007/s10529-017-2458-3.

Lu, A. X., Liao, X. P., Zhou, R. Q., & Shi, B. (2007). Preparation of Fe(III)-immobilized collagen fiber for lysozyme adsorption. *Colloids and Surfaces A: Physicochemical and Engineering Aspects, 301*(1–3), 85–93. www.elsevier.com/locate/colsurfa. doi:10.1016/j.colsurfa.2006.12.027.

Luo, Z., Qiao, L., Chen, H., Mao, Z., Wu, S., Ma, B., Xie, T., Wang, A., Pei, X., & Sheldon, R. A. (2024). Precision engineering of the co-immobilization of enzymes for cascade biocatalysis. *Angewandte Chemie International Edition, 63*(22), e202403539. http://onlinelibrary.wiley.com/journal/10.1002/(ISSN)1521-3773. doi:10.1002/anie.202403539.

Manissorn, J., Promsuk, J., Wangkanont, K., & Thongnuek, P. (2025). Biomimetic peptide conjugates as emerging strategies for controlled release from protein-based materials. *Drug Delivery, 32*(1), 2449703. doi:10.1080/10717544.2025.2449703.

Mateo, C., Palomo, J. M., Van Langen, L. M., Van Rantwijk, F., & Sheldon, R. A. (2004). New and mild cross-linking methodology to prepare cross-linked enzyme aggregates. *Biotechnology and Bioengineering, 86*(3), 273–276. doi:10.1002/bit.20033.

Mateo, C., Palomo, J. M., Fernandez-Lorente, G., Guisan, J. M., & Fernandez-Lafuente, R. (2007). Improvement of enzyme activity, stability and selectivity via immobilization techniques. *Enzyme and Microbial Technology, 40*(6), 1451–1463. doi:10.1016/j.enzmictec.2007.01.018.

Menegatti, T., Lavrič, Ž., Hlebanja, P., & Žnidaršič-Plazl, P. (2025). Microfluidics-based generation and immobilization of nanoscale cross-linked enzyme aggregates for continuous transamination. *Chemical Engineering Journal, 523*, 168865. doi:10.1016/j.cej.2025.168865.

Meyer, D. E., & Chilkoti, A. (2002). Genetically encoded synthesis of protein-based polymers with precisely specified molecular weight and sequence by recursive directional ligation: Examples from the the elastin-like polypeptide system. *Biomacromolecules, 3*(2), 357–367. doi:10.1021/bm015630n.

Mountain, G. A., & Keating, C. D. (2020). Formation of multiphase complex coacervates and partitioning of biomolecules within them. *Biomacromolecules, 21*(2), 630–640. http://pubs.acs.org/journal/bomaf6. doi:10.1021/acs.biomac.9b01354.

Nahalka, J., & Nidetzky, B. (2007). Fusion to a pull-down domain: A novel approach of producing Trigonopsis variabilis D-amino acid oxidase as insoluble enzyme aggregates. *Biotechnology and Bioengineering, 97*(3), 454–461. doi:10.1002/bit.21244.

Nguyen, L. T., Seow, N., & Yang, K. L. (2017). Hollow cross-linked enzyme aggregates (h-CLEA) of laccase with high uniformity and activity. *Colloids and Surfaces B: Biointerfaces, 151*, 88–94. www.elsevier.com/locate/colsurfb. doi:10.1016/j.colsurfb.2016.12.006.

Nguyen, L. T., & Yang, K. L. (2014). Uniform cross-linked cellulase aggregates prepared in millifluidic reactors. *Journal of Colloid and Interface Science, 428*, 146–151. http://www.elsevier.com/inca/publications/store/6/2/2/8/6/1/index.htt. doi:10.1016/j.jcis.2014.04.033.

Ölçücü, G., Krauss, U., Jaeger, K. E., & Pietruszka, J. (2023). Carrier-free enzyme immobilizates for flow chemistry. *Chemie-Ingenieur-Technik, 95*(4), 531–542. http://onlinelibrary.wiley.com/journal/10.1002/(ISSN)1522-2640. doi:10.1002/cite.202200167.

Ölçücü, G., Klaus, O., Jaeger, K. E., Drepper, T., & Krauss, U. (2021). Emerging solutions for *in vivo* biocatalyst immobilization: Tailor-made catalysts for industrial biocatalysis. *ACS Sustainable Chemistry and Engineering, 9*(27), 8919–8945. http://pubs.acs.org/journal/ascecg. doi:10.1021/acssuschemeng.1c02045.

Ölçücü, G., Baumer, B., Küsters, K., Möllenhoff, K., Oldiges, M., Pietruszka, J., Jaeger, K. E., & Krauss, U. (2022). Catalytically active inclusion bodies–benchmarking and application in flow chemistry. *ACS Synthetic Biology, 11*(5), 1881–1896. doi:10.1021/acssynbio.2c00035.

Palomo, J. M., Peñas, M. M., Fernández-Lorente, G., Mateo, C., Pisabarro, A. G., Fernández-Lafuente, R., Ramírez, L., & Guisán, J. M. (2003). Solid-phase handling of hydrophobins:

Immobilized hydrophobins as a new tool to study lipases. *Biomacromolecules, 4*(2), 204–210. doi:10.1021/bm020071l.

Patti, S., Magrini Alunno, I., Pedroni, S., Riva, S., Ferrandi, E. E., & Monti, D. (2025). Advances and challenges in the development of immobilized enzymes for batch and flow biocatalyzed processes. *ChemSusChem, 18*(8), e202402007. http://onlinelibrary.wiley.com/journal/10.1002/(ISSN)1864-564X. doi:10.1002/cssc.202402007.

Pei, X., Luo, Z., Qiao, L., Xiao, Q., Zhang, P., Wang, A., & Sheldon, R. A. (2022). Putting precision and elegance in enzyme immobilisation with bio-orthogonal chemistry. *Chemical Society Reviews, 51*(16), 7281–7304. http://pubs.rsc.org/en/journals/journal/cs. doi:10.1039/d1cs01004b.

Peschke, T., Bitterwolf, P., Gallus, S., Hu, Y., Oelschlaeger, C., Willenbacher, N., Rabe, K. S., & Niemeyer, C. M. (2018). Self-assembling all-enzyme hydrogels for flow biocatalysis. *Angewandte Chemie International Edition, 57*(52), 17028–17032. http://onlinelibrary.wiley.com/journal/10.1002/(ISSN)1521-3773. doi:10.1002/anie.201810331.

Peters, R. J. R. W., Marguet, M., Marais, S., Fraaije, M. W., Van Hest, J. C. M., & Lecommandoux, S. (2014). Cascade reactions in multicompartmentalized polymersomes. *Angewandte Chemie International Edition, 53*(1), 146–150. doi:10.1002/anie.201308141.

Qi, L., & Qiao, J. (2021). Design of switchable enzyme carriers based on stimuli-responsive porous polymer membranes for bioapplications. *ACS Applied Bio Materials, 4*(6), 4706–4719. doi:10.1021/acsabm.1c00338.

Rocha, R. A., Speight, R. E., & Scott, C. (2022). Engineering enzyme properties for improved biocatalytic processes in batch and continuous flow. *Organic Process Research and Development, 26*(7), 1914–1924. http://pubs.acs.org/journal/oprdfk. doi:10.1021/acs.oprd.1c00424.

Roy, J. J., & Abraham, T. E. (2004). Strategies in making cross-linked enzyme crystals. *Chemical Reviews, 104*(9), 3705–3722. doi:10.1021/cr0204707.

Scherrer, S., De Vries, O. M. H., Dudler, R., Wessels, J. G. H., & Honegger, R. (2000). Interfacial self-assembly of fungal hydrophobins of the lichen-forming ascomycetes *Xanthoria parietina* and *X. ectaneoides*. *Fungal Genetics and Biology, 30*(1), 81–93. http://www.journals.elsevier.com/fungal-genetics-and-biology/. doi:10.1006/fgbi.2000.1205.

Shieh, H., Saadatmand, M., Eskandari, M., & Bastani, D. (2021). Microfluidic on-chip production of microgels using combined geometries. *Scientific Reports, 11*(1), 2934. doi:10.1038/s41598-021-81214-7.

Singh, A., Upadhyay, V., & Panda, A. K. (2015). Solubilization and refolding of inclusion body proteins. *Methods in Molecular Biology, 1258*, 283–291. doi:10.1007/978-1-4939-2205-5_15.

Song, N., Chen, S., Huang, X., Liao, X., & Shi, B. (2011). Immobilization of catalase by using Zr(IV)-modified collagen fiber as the supporting matrix. *Process Biochemistry, 46*(11), 2187–2193. doi:10.1016/j.procbio.2011.09.001.

Wang, B. P., Yin, X., Huang, M. Y., Li, T. Y., Long, X. F., Li, Y., & Niu, F. X. (2024). A self-assembling γPFD-SpyCatcher hydrogel scaffold for the coimmobilization of SpyTag-enzymes to facilitate the catalysis of regulated enzymes. *Journal of Agricultural and Food Chemistry, 72*(36), 19940–19947. http://pubs.acs.org/journal/jafcau. doi:10.1021/acs.jafc.4c03403.

Wang, J., Lu, Y., Cheng, P., Zhang, C., Tang, L., Du, L., Li, J., & Ou, Z. (2023a). Construction of bi-enzyme self-assembly clusters based on SpyCatcher/SpyTag for the efficient biosynthesis of (R)-ethyl 2-hydroxy-4-phenylbutyrate. *Biomolecules, 13*(1), 91. doi:10.3390/biom13010091.

Wang, X., Qiao, X., Chen, H., Wang, L., Liu, X., & Huang, X. (2023b). Synthetic-cell-based multi-compartmentalized hierarchical systems. *Small Methods, 7*(12), e2201712. doi:10.1002/smtd.202201712.

Wösten, H. A. B. (2001). Hydrophobins: Multipurpose proteins. *Annual review of microbiology, 55*, 625–646. doi:10.1146/annurev.micro.55.1.625.

Wosten, H. A. B., Wessels, J. G. H., & Schuren, F. H. (1994). Interfacial self-assembly of a hydrophobin into an amphipathic protein membrane mediates fungal attachment to hydrophobic surfaces. *The EMBO Journal, 13*, 5848–5854. https://doi.org/10.1002/j.1460-2075.1994.tb06929.x.

Yeboah, A., Cohen, R. I., Rabolli, C., Yarmush, M. L., & Berthiaume, F. (2016). Elastin-like polypeptides: A strategic fusion partner for biologics. *Biotechnology and Bioengineering, 113*(8), 1617–1627. www.interscience.wiley.com/jpages/0006-3592. doi:10.1002/bit.25998.

Zakeri, B., Fierer, J. O., Celik, E., Chittock, E. C., Schwarz-Linek, U., Moy, V. T., & Howarth, M. (2012). Peptide tag forming a rapid covalent bond to a protein, through engineering a bacterial adhesin. *Proceedings of the National Academy of Sciences of the United States of America, 109*(12), E690–E697. http://www.pnas.org/content/109/12/E690.full.pdf+html. doi:10.1073/pnas.1115485109.

Zhu, Y., Chen, Q., Shao, L., Jia, Y., & Zhang, X. (2020). Microfluidic immobilized enzyme reactors for continuous biocatalysis. *Reaction Chemistry & Engineering, 5*(1), 9–32. doi:10.1039/c9re00217k.

SECTION

# Comparative analysis and future directions

# IV

CHAPTER

# Comparative analysis of immobilization methods

# 9

**Mansour Shahedi[a], Mehdi Mohammadi[b] and Maryam Yousefi[c]**

[a]*Organic and Inorganic Chemistry, Shahid Beheshti University, Tehran, Iran,* [b]*Bioprocess Engineering Department, National Institute for Genetic Engineering and Biotechnology, Tehran, Iran,* [c]*Avicenna Research Institute, ACECR, Nanobiotechnology Research Center, Tehran, Iran*

## 9.1 Introduction

The primary approaches to enzyme immobilization include adsorption, covalent binding, entrapment, encapsulation, and cross-linking (Datta et al., 2013). Each method differs in whether or not a carrier is used, the binding mechanism involved, the strength of enzyme–carrier interactions, the level of retained enzyme activity, and the overall complexity of the process (Garcia-Galan et al., 2011). These differences are crucial, as they directly influence enzyme performance across diverse applications. Furthermore, the choice of immobilization strategy governs essential enzyme properties, including stability, reusability, and tolerance to denaturing conditions (Maghraby et al., 2023).

This chapter primarily addresses enzyme cross-linking as a carrier-free strategy for enzyme immobilization. However, a comprehensive understanding of the full spectrum of immobilization techniques is essential to accurately assess the advantages and limitations of cross-linked enzymes and to support the rational design of biocatalysts with enhanced catalytic performance under industrially relevant conditions. Comparative evaluation further elucidates the contexts in which cross-linking affords superior operational stability and activity, as well as cases where alternative methods may offer greater process efficiency or practical benefits. Hence, the chapter offers a concise comparison of the principal enzyme immobilization methodologies, examining their underlying mechanisms, commonly employed materials, and industrial significance. Through a critical discussion of their respective benefits, limitations, and applicability, it highlights the distinctive role of enzyme cross-linking within the broader framework of immobilization strategies. Such insight is essential for the informed selection and optimization of immobilization systems tailored to the specific requirements of modern biocatalytic processes.

**Carrier-Free Enzyme Immobilization: Fundamentals and Innovations in Biocatalysis.**
**DOI: https://doi.org/10.1016/B978-0-443-44139-4.00008-3**

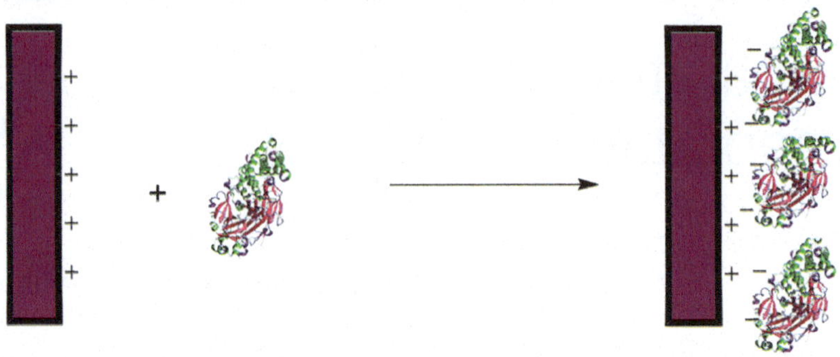

**FIGURE 9.1 Enzyme immobilization by adsorption.**

Enzyme adsorption onto a carrier via ionic interactions.

## 9.2 Immobilization techniques

### 9.2.1 Adsorption

Due to its simplicity, cost-effectiveness, and negligible impact on enzyme structure, adsorption is a fundamental and widely utilized approach for enzyme immobilization (Fig. 9.1). This method employs weak, noncovalent interactions to physically bind enzymes to the surface of a solid support. Examples of these interactions include Van der Waals forces, hydrogen bonding, hydrophobic interactions, and electrostatic attraction (Jesionowski et al., 2014; Nguyen & Kim, 2017).

Electrostatic interactions may occur between ionizable functional groups on the enzyme (such as carboxylate or amino groups) and oppositely charged functional groups on the carrier. For example, this interaction has been utilized for immobilizing *Burkholderia* lipase on magnetic nanoparticles coated with a silica shell functionalized with dimethyl octadecyl [3-(trimethoxysilyl)propyl] ammonium chloride (Tran et al., 2012). Additionally, hydrogen bonds can form between polar amino acid side chains and hydroxylated surfaces such as silica or alumina. This approach has been exploited for the immobilization of *Trametes versicolor* laccase on silica mesoporous materials as a carrier (Bautista et al., 2010). Furthermore, hydrophobic interactions arising from nonpolar residues (such as leucine or valine) interacting with hydrophobic carriers, including butyl Sepabeads or octadecyl Sepabeads, have been employed for the immobilization of *Pyrococcus furiosus* lipase (Branco et al., 2010). These interactions are pH-sensitive and influenced by the isoelectric point (pI) of the enzyme, the ionic strength of the buffer, and the polarity of the environment.

The efficiency of adsorption is significantly affected by both the carrier properties (such as surface charge, porosity, and hydrophobicity) and the enzyme's structural features. Porous materials such as ion-exchange resins, silica gel, alumina, activated carbon, mesoporous silica, zeolites, and polymeric supports like agarose and chitosan are commonly employed due to their high surface area and chemically favorable interaction sites for enzyme adsorption (Chakraborty & Hashmi, 2017). A key advantage of adsorption is that it preserves the native structure and catalytic activity of the enzyme, since no covalent modification is required (Mateo et al., 2000). Furthermore, the gentle and often reversible character of the method allows for straightforward recovery, reuse, or replacement of enzymes. The low cost and procedural simplicity make adsorption particularly attractive for laboratory-scale work, pilot studies, and certain commercial applications where periodic enzyme replacement is acceptable. The broad range of available carriers also provides flexibility for adapting the system to various enzymes and operating conditions (Lima et al., 2021; Tufvesson et al., 2011).

However, adsorption is intrinsically limited by the weak nature of the binding forces. Changes in pH, temperature, ionic strength, or the presence of surfactants and organic solvents can easily disrupt the interactions, leading to enzyme leaching and loss of activity (Hirsh et al., 2010). A further challenge is that enzymes may bind in random orientations or form multiple layers on the support surface, which may reduce the accessibility of the active site or obstruct substrate access (Jiang et al., 2024). In spite of these drawbacks, adsorption remains an effective strategy in numerous settings. It is commonly utilized in batch biotransformation processes (Nguyen et al., 2016), water purification (Wang et al., 2019), biosensor innovation (Sassolas et al., 2012), and analytical biochemistry (Bowers & Carr, 1976).

Physical adsorption has been a cornerstone in the industrial deployment of immobilized enzymes, valued for its simplicity, low cost, and ability to preserve high catalytic activity via noncovalent enzyme–carrier interactions such as hydrophobic forces, ionic linkages, and hydrogen bonding (Jesionowski et al., 2014). This strategy has been widely adopted for commercially important biocatalysts, demonstrating that sufficiently strong adsorption on inorganic or polymeric supports can provide practical operational stability under controlled process conditions (Jesionowski et al., 2014).

Lipases exemplify the strategic value of adsorption-based immobilization. Commercial preparations such as Novozyme 435 (*Candida antarctica* lipase B adsorbed on a macroporous acrylic resin) and Lipozyme RM IM (*Rhizomucor miehei* lipase adsorbed on a macroporous anion-exchange resin) rely on physical adsorption onto hydrophobic or ionic carriers, which enhances stability, solvent tolerance, and reusability for industrial processes (Ortiz et al., 2019).

Glucose isomerase immobilized by adsorption onto silica, alumina, or ion-exchange resins enables the continuous production of high-fructose corn syrup (HFCS), one of the largest industrial applications of immobilized enzymes, providing sufficient retention and stability for extended operation in packed-bed reactors (Datta et al., 2013; Jesionowski et al., 2014).

Oxidoreductases also benefit from adsorption-based immobilization in environmental applications. Laccases and peroxidases immobilized on activated carbon, polymeric matrices, or silica show enhanced operational stability and reusability in wastewater treatment, dye decolorization, and pollutant degradation (Sheldon & van Pelt, 2013; Wang et al., 2019). In the food sector, $\beta$-galactosidase adsorbed onto ion-exchange resins or silica carriers supports repeated lactose hydrolysis under mild processing conditions, enabling the production of lactose-free and prebiotic products (Gaur et al., 2006; Jesionowski et al., 2014).

### 9.2.2 Covalent binding

Covalent binding is a widely utilized method for enzyme immobilization, involving the formation of strong, stable chemical bonds between enzyme molecules and the functional groups on carriers (Fig. 9.2) (Sirisha et al., 2016). This method differs from weaker physical adsorption by chemically anchoring the enzyme through side chains of reactive amino acids such as the ε-amino group of lysine, the thiol group of cysteine, the phenolic hydroxyl of tyrosine, and the carboxyl group of aspartate and glutamate residues.

Covalent enzyme immobilization carriers include both organic and inorganic materials. Inorganic carriers typically include silica, carbon-based nanomaterials such as graphene, and various types of clays, all of which offer structural stability and high surface area for enzyme attachment. Organic carriers, on the other hand, include a range of synthetic polymers such as polyacrylamide, polystyrene, and functionalized poly (aspartic acid), which provide flexibility and functional groups for enzyme binding. Natural polymers like chitosan, agarose, and alginate are also commonly used, offering biocompatibility and biodegradability (Cao, 2006; Zucca & Sanjust, 2014). Composite carriers, such as metal-organic frameworks, are also preferred for their improved mechanical properties and stability (Lian et al., 2017).

Covalent immobilization of enzymes typically employs several strategies, including: (1) epoxy-activated supports (e.g., Eupergit C, Sepabeads EC-EP) which form covalent linkages through nucleophilic attack on epoxide rings (Fig. 10.2A) (Çelem & Önal, 2022); (2) glutaraldehyde-activated amine-functional supports which form imine (Schiff base) with primary amines of enzymes (Fig. 10.2B) (Al-Dhrub et al., 2017); (3) cyanogen bromide (CNBr) activation of hydroxyl-rich supports like agarose, forming moderately stable isourea bonds with amino groups (Fig. 10.2C) (Zucca & Sanjust, 2014); (4) isocyanide based multicomponent approaches, like the Ugi and Passerini reactions, which form covalent bonds between enzymes and various functionalized carriers under mild conditions (Fig. 10.2D) (Mohammadi et al., 2023); and (5) the carbodiimide/N-hydroxysuccinimide (EDC/NHS) coupling strategy, that involves the activation of carboxyl groups on supports using EDCs e.g., EDC in the presence of NHS, which facilitates the formation of amide bonds with primary amines of enzymes (Fig. 10.2E).

The reactivity of these interactions is pH-dependent, governed by the protonation state of the nucleophilic groups. For instance, at pH 9.0, the *N*-terminal amine shows full reactivity, whereas other lysine and tyrosine residues exhibit significantly lower reactivities, helping define a spatially optimal and catalytically competent orientation on the support surface. The difference in the reactivity of various amine groups on the enzyme surface can also be exploited to achieve multipoint covalent immobilization, which generally yields more stable immobilized derivatives (Fig. 9.3). In this approach, several nucleophilic groups on the enzyme surface (typically lysine,

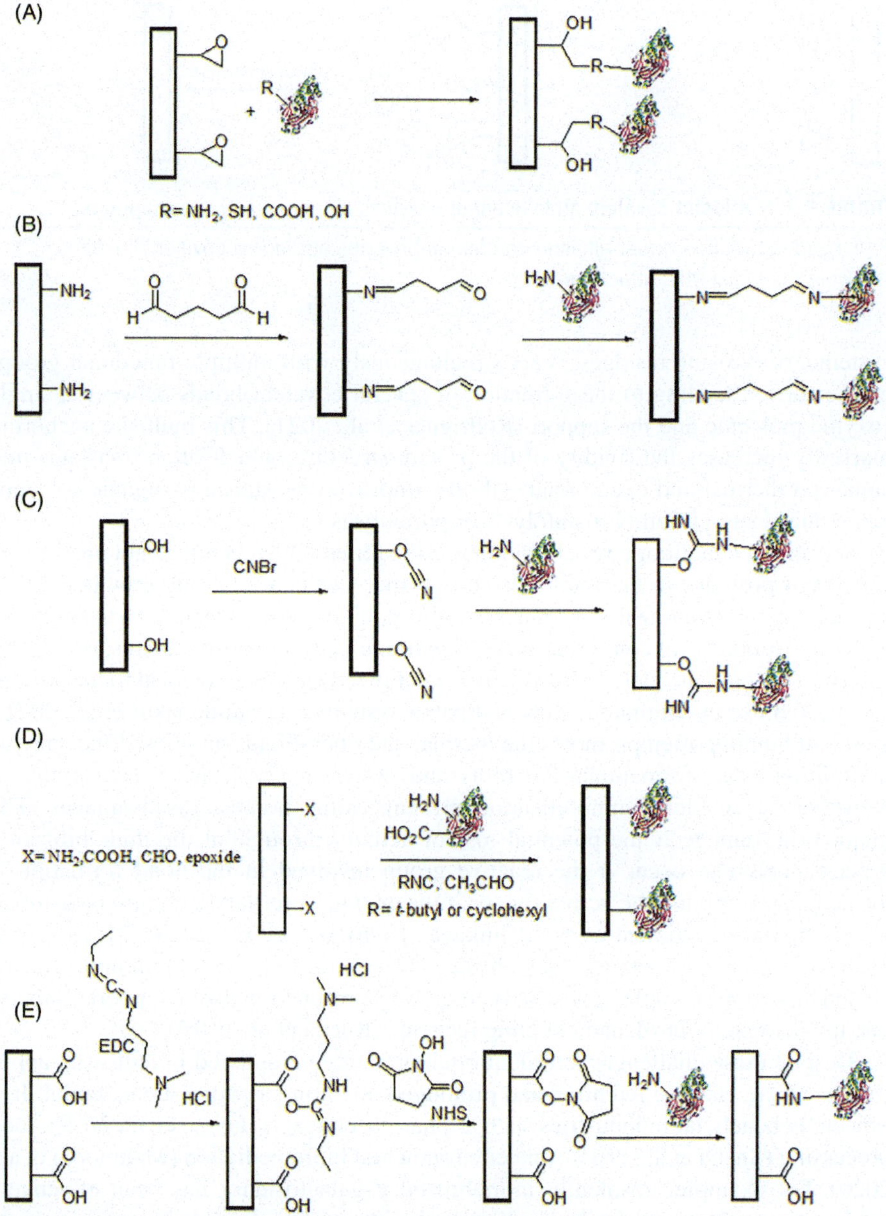


**FIGURE 9.2 Enzyme immobilization by covalent binding.**

Different strategies for covalent attachment of enzymes to carriers.

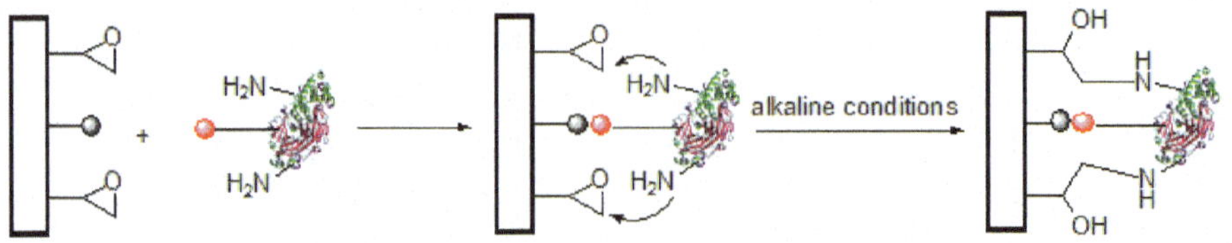


**FIGURE 9.3 Multipoint covalent attachment of enzymes on epoxy-activated supports.**

Prolonged incubation under alkaline conditions provides additional covalent linkages following the initial mild attachment.

cysteine, or tyrosine residues) react simultaneously with multiple functional groups on the carrier, leading to the formation of several covalent bonds between a single enzyme molecule and the support (Rodrigues et al., 2021). This multisite anchoring markedly increases the rigidity of the protein structure, which often translates into enhanced thermal and operational stability, improved resistance to organic solvents, and reduced susceptibility to autolysis or proteolysis.

A primary benefit of covalent binding is the remarkable chemical and operational stability it provides to the immobilized enzyme. Covalent bonding provides strong attachment, maintaining the enzyme securely on the support, preventing leaching even under harsh operating conditions such as extreme pH, high temperatures, or organic solvents (Brena et al., 2013). Covalently immobilized enzymes are ideal for prolonged or repeated use in continuous flow and batch systems (Naramittanakul et al., 2021). Covalent binding often promotes the thermal and operational stability of the enzyme by limiting its conformational flexibility, thus improving resistance to denaturation. Nonetheless, covalent immobilization presents some notable disadvantages. The major disadvantage is the potential loss of activity throughout the immobilization process. This can occur if the reactive group involved in the bond formation is located near or within the active site, or if the enzyme's tertiary structure is distorted by multipoint attachment or rigid linkage (Rodrigues et al., 2013). Moreover, in contrast to simpler methods like physical adsorption, covalent immobilization is chemically more complex and costly, requiring prefunctionalized supports, precise pH, temperature control, and coupling reagents (Rueda et al., 2016).

Despite these challenges, covalent binding is often employed in industrial applications where enzyme retention and prolonged operating stability are essential. It is especially beneficial in industries such as pharmaceuticals (Hassan et al., 2019), food processing (Sneha et al., 2018), and environmental bioremediation (Mahmoodi et al., 2020). For example, covalently immobilized $\beta$-galactosidase has been effectively used in the production of galacto-oligosaccharides (GOS) from lactose (Gaur et al., 2006). The enzymatic degradation of penicillin G to 6-aminopenicillanic acid (6-APA) is efficiently carried out in batch reactors, utilizing a potent immobilized penicillin G acylase, immobilized on hydrophilic resins like epoxy methacrylate. Remarkably, this biocatalyst can be recycled up to 1000 cycles without loss of activity (Parmar et al., 2000). In the healthcare sector, covalently immobilized lipase,

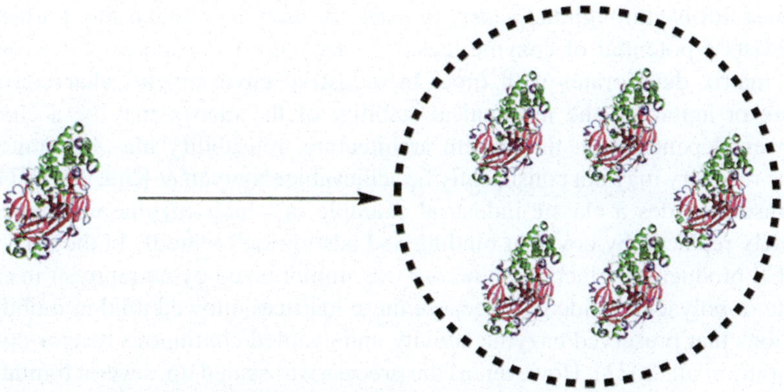

**FIGURE 9.4 Enzyme immobilization by entrapment.**

Enzyme entrapment within a carrier matrix.

derived from *Chromobacterium viscosum, Pseudomonas fluorescens, Burkholderia cepacia*, or *Rhizopus oryzae*, immobilized on polymer beads, has developed as a disposable cylindrical cartridge for enteral feeding systems (Basso & Serban, 2020). The developed system efficiently hydrolyzes up to 90% of the lipids in enteral formulas as they pass through the cartridge.

## 9.2.3 Entrapment

In the entrapment method, enzymes are physically confined within a porous gel or matrix, without forming any direct chemical bonds with the supporting material (Fig. 9.4). The framework acts as a porous cage that facilitates the entry of substrates and the release of products while preventing the enzyme from escaping, thereby maintaining its confinement and catalytic accessibility within the structure (Sassolas et al., 2013). Natural polymers such as alginate (Bhushan et al., 2008), gelatin (Fadnavis et al., 2003), carrageenan (Makshakova et al., 2022), and chitosan (Tavernini et al., 2020), together with synthetic polymers like polyacrylamide (González-Sáiz & Pizarro, 2001) and sol-gel-derived silica (Besanger et al., 2003), are commonly employed for entrapment. The selection of the matrix influences mass transfer characteristics, mechanical strength, and porosity.

Entrapment provides the enzyme with a protected microenvironment, which constitutes one of its principal advantages. The enzyme often retains its native conformation and functionality upon immobilization, as its chemical structure remains unmodified. This approach provides compatibility with various sensitive biological molecules, including multimeric or delicate enzymes, and generally operates under mild reaction conditions (Imam et al., 2021).

Nevertheless, the major limitation of entrapment is mass transfer limitation, which can significantly hinder reaction rates, particularly for large or charged substrates, as they must diffuse through the matrix to reach the enzyme's active site. Furthermore, there exists a potential of enzyme leakage, especially if the pore size is very high or the matrix deteriorates with time. In industrial environments characterized by pressure or agitation, the mechanical stability of the matrix may be a challenge. Moreover, depending on the system architecture, reusability may be limited, and enzyme recovery may not consistently be achievable (Nguyen & Kim, 2017). Glucose isomerase provides a classic industrial example in which enzyme entrapment was ultimately replaced by covalent binding and adsorption methods. In the early stages of HFCS production, glucose isomerase was immobilized by entrapment in calcium alginate or polyacrylamide gels because these matrices allowed mild immobilization conditions that preserved enzyme activity and enabled continuous reactor operation (Maghraby et al., 2023). However, as the process was scaled up, several limitations of entrapment became evident, including significant mass-transfer resistance in highly viscous glucose syrups, gradual enzyme leakage during long-term operation, and poor mechanical stability of gel beads under industrial flow and pressure conditions. These drawbacks led to reduced space–time yields, shorter catalyst lifetimes, and increased operational costs. Consequently, industrial HFCS production transitioned to covalent immobilization or strong adsorption of glucose isomerase onto robust inorganic or polymeric carriers, which provided superior mechanical strength, minimized enzyme loss, improved substrate diffusion, and enabled stable continuous operation over many months, making these methods economically and technically preferable at an industrial scale (Basso & Serban, 2019).

### 9.2.4 Encapsulation

Encapsulation involves enclosing the enzyme within a distinct semipermeable membrane or shell, forming a core-shell structure. In this method, the enzyme is confined in a central core and surrounded by a well-defined membrane that controls the diffusion of substrates and products.

In contrast to entrapment, the enzyme is not dispersed within the pores of a continuous solid or gel matrix; rather, its retention is governed by the presence of a discrete membrane, whereas entrapment relies on physical confinement within the matrix network without a distinct boundary layer (Fig. 9.5) (Brena et al., 2013; Kurzbaum et al., 2020). Encapsulation materials comprise polymeric microcapsules (Parthasarathy & Martin, 1996), liposomes (Yoshimoto, 2016), nanogels (Chauhan, 2014), and sol-gel-derived structures (Pierre, 2004) featuring adjustable porosity. Methods such as interfacial polymerization (Qu et al., 2015), coacervation (Liu et al., 2013), or emulsification (Piacentini et al., 2017) can be employed to fabricate these membrane structures.

The main benefit of encapsulation is the significant protection it provides to the enzyme. The semi-permeable membrane enhances catalytic activity while protecting the enzyme from inhibitors and detrimental environmental factors (Wang et al., 2008). Encapsulation is optimal for biomedical and pharmaceutical applications as it maintains the structural integrity and functionality of sensitive or labile enzymes inside a biocompatible environment (Chang, 2013). Moreover, encapsulation can

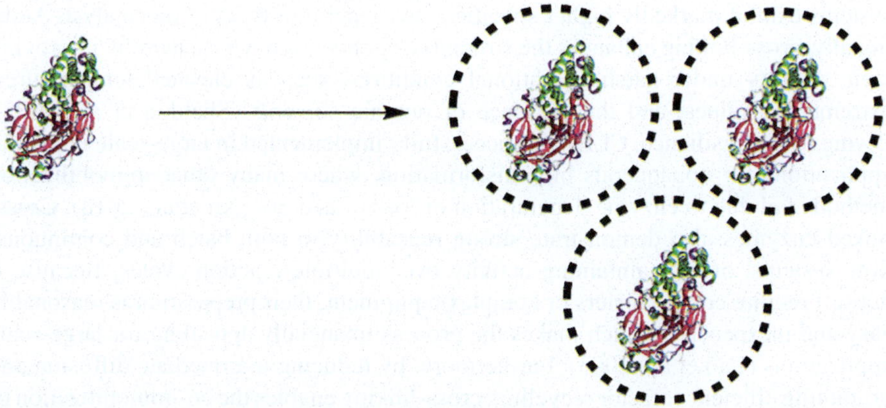

**FIGURE 9.5 Enzyme immobilization by encapsulation.**

Encapsulation of an enzyme within a matrix.

prevent contamination of the final product with the enzyme, which is crucial in food and pharmaceutical production.

However, substantial challenges exist regarding encapsulation. Mass transfer limitation is a considerable disadvantage that can reduce reaction rate, especially for high molecular weight molecules, as substrates and products must diffuse across the membrane. The process of enzyme encapsulation can be complex and requires specialized instruments or conditions to ensure consistency in capsule size, membrane thickness, and permeability. Furthermore, enzyme leakage or diminished activity may occur due to membrane degradation over time or under adverse conditions, such as temperature fluctuations or solvent exposure. The structural fragility of capsules may limit enzyme reusability and hinder their large-scale implementation (Rother & Nidetzky, 2009). Encapsulation is widely applied in drug delivery, controlled-release formulations, biosensing, and bioconversion reactors, offering particular advantages when enzyme stability and separation from the reaction medium are essential (Robescu & Bavaro, 2025). Similarly, encapsulated enzymes are employed in wastewater treatment to inhibit harmful substances from denaturing free enzymes (Vardanyan et al., 2024) and in oral enzyme therapy to protect against gastric acid (Liang et al., 2022).

### 9.2.5 Carrier-free enzyme cross-linking

Carrier-free immobilization of enzymes, particularly in the form of cross-linked enzyme aggregates (CLEAs), offers several significant advantages. Since cross-linked enzyme derivatives typically consist of over 99% active enzyme and do not require inert carrier materials, their exceptionally high volumetric enzyme loading is one of their key benefits (Chen et al., 2023). Unlike carrier-bound systems, where the support can reduce catalytic efficiency and increase production costs, carrier-free immobilized systems exhibit markedly higher specific activity per unit mass of biocatalyst. Additionally, cross-linking enhances the structural rigidity of enzymes, thereby improving their stability under harsh operational conditions such as elevated temperatures, extreme pH values, and the presence of organic solvents (Sheldon et al., 2005). Owing to this resilience, CLEs are successfully implemented in large-scale industrial applications for nonaqueous biotransformation, where many other immobilization methods fail due to enzyme denaturation or loss of activity (Xu et al., 2018). Cross-linked enzymes also demonstrate strong reusability in both batch and continuous-flow systems, often maintaining activity over multiple reaction cycles. Because it doesn't require costly carriers or complex equipment, their preparation is reasonably easy and inexpensive, which makes the process financially appealing for large-scale applications (Cao et al., 2003). Furthermore, by reducing intermediate diffusion and promoting efficient cofactor recycling, cross-linking enables the co-immobilization of multiple enzymes, thereby facilitating cascade reactions within a single particle and enhancing overall process efficiency (Peng et al., 2020). From a broader perspective, cross-linking offers a better balance between catalytic accessibility, mechanical stability, and operational control than methods like entrapment, which frequently result in mass transfer limitations, or adsorption, which suffers from enzyme leaching.

Cross-linking presents certain limitations, including the risk of over-cross-linking, which can cause excessive structural rigidity or impede access to the enzyme's active site, ultimately resulting in partial or complete loss of catalytic activity (Do et al., 2018). On the other hand, inadequate cross-linking may yield fragile aggregates that lose structural integrity and deteriorate during operation. The difficulty of accurately regulating particle size and homogeneity, which can impact mass transfer, reproducibility, and process scalability, is another critical issue with CLEs (Cui & Jia, 2015). Furthermore, diffusion within dense cross-linked particles may be restricted, particularly in low-porosity formulations or for substrates with a high molecular weight. Additionally, cross-linking requires careful optimization of several parameters, including temperature, pH, reaction time, the precipitation or crystallization process, and cross-linker concentration. Inadequate control of these factors can introduce significant variability in the resulting immobilized biocatalyst (Chen et al., 2023). Despite these limitations, enzyme cross-linking, along with its recently developed subtypes, remains one of the most attractive and rapidly expanding immobilization techniques in industrial biotechnology (Sheldon, 2011).

## 9.3 Comparative performance

Each immobilization strategy, adsorption, covalent binding, entrapment, encapsulation, and cross-linking, exhibits a distinct balance of advantages and limitations, whose suitability depends on the specific biocatalytic application, operational parameters, and environmental conditions. A comparative analysis of these strategies based on essential performance metrics offers a practical framework for identifying the most suitable technique.

### 9.3.1 Thermal and chemical stability

Both cross-linking and covalent binding confer superior thermal and chemical stability to enzymes by forming strong covalent bonds that preserve the structural integrity of the immobilized enzyme (Robescu & Bavaro, 2025). Covalent binding on solid carriers combines this robustness with greater versatility, making it particularly advantageous when a carrier-based approach is desired.

Entrapment and encapsulation provide moderate stability by physically shielding enzymes from external conditions. Because enzyme leaching may occur due to the unsuitable pore size of the matrix or its degradation over time. On the other hand, adsorption offers the least stability, making enzymes more susceptible to leaching when exposed to changes in pH, hydrophobicity, or ionic strength (Mohidem et al., 2023).

### 9.3.2 Enzyme reusability and operational stability

Reusability is a critical factor for economically viable enzyme immobilization. Enzymes that are immobilized for recurrent use across numerous cycles must have significant mechanical and operational stability (Maghraby et al., 2023). In contrast to adsorption or encapsulation, covalent bonding and cross-linking provide superior reusability, frequently surpassing 20 to 100 cycles, contingent upon the method employed. Cross-linked enzymes can be readily isolated through filtration or centrifugation owing to their particle characteristics (Hassan et al., 2025; Patti et al., 2025). In regulated industries such as food and pharmaceuticals, the leaching of immobilized enzymes can lead to product contamination (Sheldon & van Pelt, 2013). Carrier-free CLEs are beneficial in this context, since they diminish the incorporation of extraneous materials and minimize the possibility of enzyme leakage in the final product.

Table 9.1 presents examples of enzyme immobilization via cross-linking, highlighting their number of reuses and residual activity. Different enzymes, such as *Trametes versicolor* laccase, have been successfully immobilized using genipin as a cross-linker and tested for oxidation of 2,2′-azinobis(3-ethylbenzothiazoline-6-sulfonate) (ABTS) in 10 cycles. Remarkably, after 10 reuse cycles, the residual activity of the laccase enzyme retained 85% of its original activity (Hong et al., 2021). *Thermomyces lanuginosus* and *Rhizomucor miehei* lipases have been immobilized through isocyanide-based multicomponent reactions using cysteamine as the cross-linking agent. The resulting immobilized biocatalysts exhibited good operational stability during repeated use. After four consecutive cycles of fish oil hydrolysis, the immobilized enzymes retained 79% and 77% of their initial activities, respectively.

Ahrari et al. (2023). Similarly, *Thermomyces lanuginosus* lipase has been immobilized for biodiesel production using Tween 80 and amino-functionalized magnetite nanoparticles, retaining 70% of its activity after 10 reaction cycles (Zhang et al., 2015). In addition to lipases, α-amylase has also been immobilized for starch hydrolysis, employing glutaraldehyde as a cross-linker, with an activity retention of 63% after 15 reaction cycles (Torabizadeh et al., 2014).

**Table 9.1** Operational performance of cross-linked enzymes in repeated catalytic cycles.

| Enzyme | Cross-linking method | Reaction | Immobilization conditions | Number of reuses | % Activity retention (after reuse) | Time of each reuse (h) | References |
|---|---|---|---|---|---|---|---|
| *Thermomyces Lanuginosus* lipase | CLEAs | Biodiesel production | Tween 80 and amino functionalized magnetite nanoparticles | 10 | 70 | 48 | Zhang et al. (2015) |
| *Trametes versicolor* laccase | CLEAs | ABTS oxidation reaction | Genipin as a cross linker/BSA as an additive | 10 | 85 | – | Hong et al. (2021) |
| α-amylase | CLEAs | starch hydrolysis | Glutaraldehyde as a cross linker/ BSA as an additive | 15 | 63 | 2 | Torabizadeh et al. (2014) |
| *Rhizomucor miehei* lipase and *Candida antarctica* lipase | co-CLEs | Hydrolysis of fish oil | Glutaraldehyde as a cross linker/cyclohexyl isocyanide and BSA as additives | 5 | 72 | 8 | Ahrari and Mohammadi (2024) |
| *Rhizomucor miehei* lipase or *Thermomyces lanuginosa* lipase | CLEs | Hydrolysis of fish oil | Cysteamine as a cross linker/cyclohexyl isocyanide and BSA as additives | 4 | 79% for *Thermomyces lanuginosa* lipase and 77% for *Rhizomucor miehei* lipase | 8 | Ahrari et al. (2023) |

(continued on next page)

| | | | | | | | |
|---|---|---|---|---|---|---|---|
| *Thermomyces lanuginosa* lipase | CLEs | Hydrolysis of fish oil | 1,11-undecanedicarboxylic acid as across linker/cyclohexyl isocyanide as an additive | 5 | 32% | 8 | Ahrari et al. (2024) |
| *Rhizomucor miehei* lipase | CLEs | Hydrolysis of fish oil | 1,11-undecanedicarboxylic acid as across linker/cyclohexyl isocyanide and BSA as additives | 4 | 25% | 8 | Ahrari et al. (2022) |
| galactitol dehydrogenase and NADH oxidase | co-CLEAs | L-tagatose synthesis | glutaraldehyde | 8 | 51.5% | 24 | Li et al. (2022) |
| *Candida Antarctica* lipase B | CLEAs | Transesterification of 5-hydroxymethylfurfural | Glutaraldehyde as a cross linker/BSA and starch as additives | 8 | 44 | 12 | Saikia et al. (2021) |

### 9.3.3 Economic considerations

A general comparison of the process costs associated with common enzyme immobilization techniques helps establish an overall understanding of their economic characteristics. Nevertheless, the final selection of an appropriate immobilization method should be based on a detailed evaluation of the specific enzyme, support material, and process requirements (Homaei et al., 2013). Physical adsorption is the most cost-effective immobilization technology owing to its simplicity and minimal material requirements, as it does not require any reagents or complex procedures during immobilization (Rodrigues et al., 2013). Enzyme cross-linking technology, particularly the one-step formation of CLEs and the preparation of CLEAs, offers significant economic advantages for large-scale applications due to its simplicity, carrier-free nature, and reliance on inexpensive reagents. (Ahrari & Mohammadi, 2024; Ahrari et al., 2022, 2023; Ashjari et al., 2020). Covalent bonding and encapsulation are generally more costly, as they require specialized supports and chemical activation of either the enzyme's functional residues or the support (Garcia-Galan et al., 2011). Entrapment is comparatively economical because it generally requires inexpensive polymeric materials and mild operating conditions; however, supplementary expenses may arise from managing polymer synthesis or gelation processes and from specialized apparatus needed to control pore structure and mass transfer limitations (Imam et al., 2021).

### 9.3.4 Mass transfer efficiency

Mass transfer efficiency plays a critical role in the overall performance of enzyme immobilization systems, particularly in biocatalytic applications where substrate access to the enzyme's active site is crucial for catalytic efficiency (Bolivar et al., 2022; Malar et al., 2020). However, diffusion limitations in immobilized enzyme systems are not determined solely by whether the enzyme is located on the surface of the carrier; they are also influenced by factors such as the structural characteristics of the support (e.g., porous versus nonporous materials), the intrinsic catalytic properties of the enzyme (e.g., turnover number), and the enzyme loading within the immobilized catalyst (Eldin & Mita, 2014; Liese & Hilterhaus, 2013; Sheldon & van Pelt, 2013). Adsorption and covalent binding techniques typically offer the best mass transfer properties since enzymes are generally immobilized on the surface of the carrier, making them readily accessible to the substrate. This surface availability facilitates efficient diffusion of the substrate to the active sites of the enzyme, ensuring that the reaction proceeds at optimal rates. In these methods, the enzyme remains exposed on the surface, minimizing diffusion barriers and enhancing the rate of interaction between the substrate and enzyme (Sigurdardóttir et al., 2018).

In contrast, entrapment and encapsulation methods often face limitations related to mass transfer due to the presence of diffusion barriers. In these systems, enzymes are enclosed within a matrix or polymeric structure, which can impede the movement of substrate molecules to the enzyme's active sites. The size and structure of the encapsulating material create a physical barrier that may slow down substrate diffusion, reducing the efficiency of the enzymatic reaction. These methods, while effective in preventing enzyme leaching and providing stability, often result in slower reaction kinetics due to the restricted access of the substrate to the enzyme's active site (Tadesse & Liu, 2025).

Cross-linking techniques offer a compromise between surface exposure and diffusion control, providing moderate to good mass transfer properties. The degree of accessibility depends largely on factors such as the particle size of the cross-linked aggregates and the porosity of the immobilization matrix. Larger particles and highly porous structures tend to allow better substrate penetration, thereby improving mass transfer efficiency. However, cross-linking may still face diffusional restrictions, particularly when the enzyme is immobilized within a dense network or when the cross-linking density is too high, which could limit substrate access (Sheldon, 2011). Since the cross-linked biocatalysts are composed predominantly of protein, they may exhibit significant internal diffusion limitations; therefore, strategies such as the incorporation of inert proteins (e.g., bovine serum albumin) or the use of porogenic agents are often employed to increase particle size, generate a more open structure, and improve mass transfer (Ahrari et al., 2023). In addition, the extent to which diffusion affects catalytic performance strongly depends on the intrinsic activity of the enzyme toward the target substrate (Sheldon, 2018). Highly active enzymes may become diffusion-limited because the rate of catalytic turnover exceeds the rate at which substrate molecules can diffuse to the active sites. In contrast, for enzymes with lower catalytic rates, the reaction itself rather than mass transfer may remain the rate-limiting step (Berendsen et al., 2006).

To optimize mass transfer in cross-linking systems, careful consideration must be given to factors such as the enzyme's loading density, the cross-linker's molecular size, and the overall structure of the resulting enzyme matrix (Xu et al., 2018). In some cases, modifications to the cross-linking procedure may be required to improve the porosity and particle size, thereby reducing any diffusional restrictions and enhancing the system's overall catalytic efficiency (Yamaguchi et al., 2018). Therefore, while cross-linking offers a balanced approach with good mass transfer properties, its success depends on optimizing these parameters to prevent the formation of diffusion-limiting barriers and ensure the best performance in biocatalytic applications.

### 9.3.5 Retention of enzyme activity

Retention of enzyme activity is a critical factor in the successful use of immobilized enzymes, particularly in industrial and biotechnological applications where consistent performance over time is essential. Different immobilization techniques have varying impacts on enzyme activity, each with its strengths and potential challenges (Prabhakar et al., 2025). Adsorption generally preserves enzyme activity effectively, as the enzyme's active site remains largely intact. However, one of the significant drawbacks of adsorption is the possibility of desorption, where the enzyme detaches from the support during the reaction process, leading to a gradual loss of activity over time. To minimize desorption and maximize retention, factors such as the nature of the support material, the enzyme's affinity for the support, and the surrounding conditions (e.g., pH, temperature, and ionic strength) must be optimized (Jesionowski et al., 2014).

Entrapment and encapsulation techniques offer the advantage of protecting the enzyme from external stressors. As the enzyme remains isolated from the external environment, the conditions within the matrix or capsule help maintain its structure and activity over time. These methods also have the advantage of not altering the enzyme's active site, allowing for continued substrate binding and catalysis (Imam et al., 2021).

In contrast, covalent binding and cross-linking involve the formation of strong, stable covalent bonds between the enzyme and the support or between enzymes, respectively. These techniques can alter the enzyme's conformation or active site due to the chemical reactions involved, potentially leading to a loss of catalytic activity (Chen et al., 2023; Zucca & Sanjust, 2014). If the covalent bond forms near or within the enzyme's active site, it may hinder the binding of substrates or disrupt the enzyme's overall structure. However, with careful optimization, such as selecting milder cross-linking agents or controlling the extent of covalent binding, cross-linking can be effective in retaining enzyme activity while providing enhanced stability and resistance to denaturation over time (Ahrari et al., 2024). Therefore, a combination of techniques or a tailored approach to optimize enzyme retention and performance is often employed to achieve the best results in industrial and research applications. Ultimately, the goal is to strike a balance that maximizes enzyme activity, stability, and efficiency while minimizing the risk of deactivation or loss of function.

Table 9.2 presents various enzymes that have been successfully immobilized using cross-linking methods, showcasing the effects of different cross-linkers and additives on enzyme activity. While cross-linking can sometimes reduce enzyme activity due to the formation of covalent bonds between enzyme molecules, recent studies indicate that enzyme activity is often well-preserved or even enhanced in certain cases. In particular, lipases appear highly compatible with cross-linking methods, likely because their interfacial activation mechanism allows the enzyme to adopt a more active open conformation upon contact with hydrophobic surfaces, enhancing substrate accessibility (Fernández-Lorente et al., 2007).

Application of isocyanide-based multicomponent reactions for enzyme cross-linking, where linkers like glutaraldehyde and dicarboxylic acids are used alongside isocyanides, has shown positive effects not only in maintaining enzyme activity but also in significantly boosting the residual activity of the enzymes (Ashjari et al., 2020). For instance, when *Rhizomucor miehei* lipase was cross-linked with 1,11-undecanedicarboxylic acid as the cross-linker and cyclohexyl isocyanide, a remarkable 255% increase in residual activity was observed (Ahrari et al., 2022). This boost in activity is largely attributed to the increased stability and structural rigidity provided by the cross-linking process.

Moreover, a simultaneous cross-linking strategy involving both *Rhizomucor miehei* lipase and *Candida antarctica* lipase B has led to a substantial 404% increase

**Table 9.2** Catalytic activity of enzymes immobilized via cross-linking methods.

| Enzyme type | Cross-linking method | Reaction performed | Immobilization conditions | Activity of immobilized enzyme (%) | References |
|---|---|---|---|---|---|
| *Rhizomucor miehei* lipase and *Candida antarctica* lipase | co-CLEs | Hydrolysis of 4-nitrophenyl butyrate | Glutaraldehyde as a cross linker/cyclohexyl isocyanide and BSA as additives | 404 | Ahrari and Mohammadi (2024) |
| *Rhizomucor miehei* lipase | CLEs | Hydrolysis of 4-nitrophenyl butyrate | 1,11-undecanedicarboxylic acid as a cross linker/cyclohexyl isocyanide and BSA as additives | 255 | Ahrari et al. (2022) |
| *Thermomyces lanuginosa* lipase | CLEs | Hydrolysis of 4-nitrophenyl butyrate | 1,11-undecanedicarboxylic acid as across linker/cyclohexyl isocyanide as a additive | 152 | Ahrari et al. (2024) |
| endolevanase from *Bacillus lehensis* G1 | CLEAs | levan-type-fructooligosaccharides production | Dextran as a cross linker | 90.5 | Makki et al. (2024) |
| *Aspergillus niger* GZUF36 lipase | CLEAs | Triolein hydrolysis | Glutaraldehyde as a cross linker | 99.7 | Zhu et al. (2021) |
| laccase from *Trametes versicolor* IBL-04 | CELAs | Degradation of dye | Glutaraldehyde as a cross linker | 72.85 | Bilal et al. (2021) |
| $\beta$-glucosidase from *Aspergillus niger* | CLEAs | Hydrolyzing $\rho$-nitrophenyl-$\beta$-D-glucopyranoside | Glutaraldehyde as a cross linker | 33 | da Cunha et al. (2023) |

in residual activity (Ahrari & Mohammadi, 2024). This synergistic effect between the two lipases demonstrates that enzyme mixtures, when cross-linked properly, can offer superior activity, making them highly beneficial for industrial applications.

### 9.3.6 Scalability and industrial applications

In terms of scalability, all five immobilization approaches demonstrate versatility across different scales. However, adsorption, covalent binding, and cross-linking are predominantly employed in industrial systems owing to their reproducibility, robustness, and easy integration into continuous processes. Entrapment and encapsulation, although scalable, are predominantly utilized in specialized applications such as biomedical and environmental sectors, where gentle immobilization and superior biocompatibility are essential (Maghraby et al., 2023).

### 9.3.7 Comparative summary

In summary, no single enzyme immobilization method is universally superior; however, covalent binding is recognized for its exceptional stability and reusability, making it ideal for industrial applications that require long-term enzyme function. In contrast, cross-linking, though stable relative to other methods, does not offer the same level of durability as covalent binding. It is generally used when moderate stability is sufficient, without the need for permanent attachment. Adsorption is optimal for cost-sensitive, short-term applications, while entrapment and encapsulation are preferable for mild, protective immobilization where substrate dispersion is not significantly constrained. Choosing the most suitable technique necessitates balancing these aspects in relation to the enzyme, process, and performance objectives. Table 9.3 provides a summary of the comparison of these methods.

## 9.4 Decision for method selection

The selection of a suitable enzyme immobilization technique is a multifactorial decision-making process that must integrate biochemical, engineering, and economic considerations. The selected strategy directly affects immobilization yield, catalytic performance, and enzyme stability, as well as process integration, cost efficiency, regulatory compliance, and product quality. A rational, data-driven approach is therefore essential to align the immobilization strategy with the enzyme's molecular properties and the specific requirements of the intended bioprocess.

### 9.4.1 Enzyme structure and surface chemistry

The structural characteristics of the enzyme are a critical factor in method selection. Enzymes possessing a high concentration of accessible amino groups (e.g., lysine residues) are ideal candidates for covalent binding or cross-linking, necessitating

**Table 9.3** Comparative evaluation of enzyme immobilization methods.

| Parameter | Adsorption | Covalent binding | Entrapment | Encapsulation | Cross-linking (CLEAs/CLECs) |
|---|---|---|---|---|---|
| Stability | Low | High | Moderate | Moderate | High |
| Reusability | Low | High | Moderate | Low–Moderate | High |
| Cost | Very Low | Medium–High | Low–Medium | Medium–High | Low–Medium |
| Mass transfer | Excellent | Good | Limited | Limited | Moderate–Good |
| Activity retention | High (if no leaching) | Medium (risk of active site modification) | High | High | Medium–High |
| Scalability | High | High | Medium | Low–Medium | High |
| Process complexity | Very Simple | Moderate to Complex | Simple | Complex | Moderate |
| Carrier requirement | Yes | Yes | Yes | Yes | No |
| Typical applications | Lab-scale bioconversions, biosensors | Industrial biocatalysis, biosensors | Food processing, low-cost systems | Drug delivery, biosensing | Industrial synthesis, bioremediation |

reactive surface groups for attachment. Enzymes characterized by limited solubility, low stability, a pronounced tendency for aggregation, or a scarcity of reactive functional groups can be more effectively preserved through physical entrapment or encapsulation to prevent deactivation (Barbosa et al., 2013). Enzymes sensitive to oxidative stress, pH shifts, or temperature extremes may benefit more from gentle physical methods such as encapsulation or adsorption on biocompatible supports. In contrast, enzymes already known for high thermal and chemical resilience can tolerate more aggressive chemical modifications as required in covalent immobilization or cross-linking. Post-translational modifications (PTMs), such as glycosylation, may also interfere with or enhance certain immobilization strategies by influencing the accessibility and reactivity of the modified residues.

The position and accessibility of the enzyme's active site are critical parameters in selecting an appropriate immobilization method. Covalent attachment occurring near the catalytic region, either during covalent immobilization or cross-linking, can cause steric hindrance or even direct deactivation of the active site. In such cases, the use of spacer arms or site-specific covalent attachment may be necessary to maintain catalytic accessibility and prevent structural interference. For multimeric enzymes, immobilization strategies must also ensure the preservation of quaternary structure and native subunit interactions (Tran & Balkus, 2011).

### 9.4.2 Process conditions and operational environment

Process-specific characteristics, including reaction temperature, pH, ionic strength, the presence of organic solvents, and mechanical stress, determine the necessary robustness of the immobilized system. In high-temperature bioreactors, organic media, or strongly alkaline/acidic conditions, only highly stable immobilization strategies such as covalent attachment of enzymes to rigid supports and, to a lesser extent, cross-linked enzyme systems can ensure long-term enzyme durability (Xu et al., 2018). In contrast, when the application operates under near-physiological conditions (e.g., in biosensors or diagnostic systems), milder approaches such as adsorption or encapsulation are often sufficient to maintain enzyme function.

In processes requiring continuous operation, such as packed-bed reactors or membrane bioreactors, mechanical stability and resistance to washout or shear stress are essential. Covalent immobilization on structural supports (such as resins, membranes, and monoliths) or stabilized CLEAs is frequently preferred in these configurations. In batch or semi-continuous systems operating at low flow rates, the use of more flexible carriers, such as hydrogels for entrapment, may be acceptable (Bolivar & López-Gallego, 2020).

### 9.4.3 Substrate and product characteristics

The physicochemical characteristics of the substrate, particularly molecule size, hydrophobicity, and diffusion, should be assessed prior to choosing an immobilization technique. Most immobilization techniques ensure enough mass transfer for small, water-soluble compounds. In circumstances involving bulky, hydrophobic, or charged substrates, diffusional resistance emerges as a significant impediment in systems such as entrapped or encapsulated enzymes. Methods that immobilize the enzyme on accessible surfaces, such as adsorption, covalent binding, or porous CLEAs, are favored to reduce substrate transport restrictions.

## 9.5 Conclusion

The comparative analysis of enzyme immobilization strategies, including adsorption, covalent binding, entrapment, encapsulation, and cross-linking, highlights the necessity of a context-specific, multidisciplinary approach in selecting the optimal technique for industrial, biomedical, and environmental applications. Each method offers a distinct set of strengths and limitations: adsorption is simple and cost-effective but suffers from weak binding and limited stability; covalent binding ensures strong enzyme-support interaction and high reusability, though it risks activity loss and involves higher process complexity. Entrapment and encapsulation are gentle methods that preserve enzyme conformation but often face mass transfer limitations and structural instability. Cross-linking, especially in the form of CLEAs or cross-linked enzyme crystals (CLECs), emerges as a highly promising approach, providing excellent stability, high catalytic loading, and operational resilience without the need for carriers.

Experimental studies demonstrate that covalent binding and cross-linking offer superior thermal and pH stability, reusability, and activity retention, making them ideal for continuous, high-throughput industrial bioprocesses. In contrast, adsorption, entrapment, and encapsulation are more appropriate for short-term, low-cost, or biocompatibility-focused applications such as biosensors, food production, or therapeutic systems.

The selection of an immobilization strategy must therefore be tailored to the molecular characteristics of the enzyme, the physicochemical nature of the substrates and products, the environmental demands of the process, and the economic constraints of scaling. The integration of materials science, bioengineering, and computational optimization tools, including artificial intelligence (AI)-assisted parameter tuning, is expected to increasingly shape future innovations in this field. Furthermore, the emergence of nanostructured carriers, multienzyme co-immobilization platforms, and biodegradable materials paves the way for more efficient, sustainable, and scalable immobilization technologies.

## References

Ahrari, F., Yousefi, M., & Mohammadi, M. (2024). The effect of carbon chain length of cross-linking agent on the functionality of carrier- free immobilized *Thermomyces lanuginosa* lipase particles. *International Journal of Biological Macromolecules, 270*, 132076. doi:10.1016/j.ijbiomac.2024.132076.

Ahrari, F., Yousefi, M., Habibi, Z., & Mohammadi, M. (2022). Application of undecanedicarboxylic acid to prepare cross-linked enzymes (CLEs) of *Rhizomucor miehei* lipase (RML): Selective enrichment of polyunsaturated fatty acids. *Molecular Catalysis, 520*, 112172. doi:10.1016/j.mcat.2022.112172.

Ahrari, F., Yousefi, M., Habibi, Z., & Mohammadi, M. (2023). Cross-linked lipase particles with improved activity: Application of a non-toxic linker for cross-linking. *LWT, 173*, 114371. doi:10.1016/j.lwt.2022.114371.

Ahrari, F., & Mohammadi, M. (2024). Combined cross-linking of *Rhizomucor miehei* lipase and *Candida antarctica* lipase B for the effective enrichment of omega-3 fatty acids in fish oil. *International Journal of Biological Macromolecules, 260*(Pt 2), 129362. doi:10.1016/j.ijbiomac.2024.129362.

Al-Dhrub, A. H. A., Sahin, S., Ozmen, I., Tunca, E., & Bulbul, M. (2017). Immobilization and characterization of human carbonic anhydrase I on amine functionalized magnetic nanoparticles. *Process Biochemistry, 57*, 95–104. doi:10.1016/j.procbio.2017.03.025.

Ashjari, M., Garmroodi, M., Ahrari, F., Yousefi, M., & Mohammadi, M. (2020). Soluble enzyme cross-linking via multi-component reactions: A new generation of cross-linked enzymes. *Chemical Communications, 56*(67), 9683–9686. doi:10.1039/d0cc03429k.

Barbosa, O., Torres, R., Ortiz, C., Berenguer-Murcia, Á., Rodrigues, R. C., & Fernandez-Lafuente, R. (2013). Heterofunctional supports in enzyme immobilization: From traditional immobilization protocols to opportunities in tuning enzyme properties. *Biomacromolecules, 14*(8), 2433–2462. doi:10.1021/bm400762h.

Basso, A., & Serban, S. (2020). Overview of immobilized enzymes' applications in pharmaceutical, chemical, and food industry. In *Methods in molecular biology: 2100* (pp. 27–63). Springer. doi:10.1007/978-1-0716-0215-7_2.

Basso, A., & Serban, S. (2019). Industrial applications of immobilized enzymes—A review. *Molecular Catalysis, 479*, 110607. doi:10.1016/j.mcat.2019.110607.

Berendsen, W. E., Lapin, A., & Reuss, M. (2006). Investigations of reaction kinetics for immobilized enzymes – Identification of parameters in the presence of diffusion limitation. *Biotechnology Progress, 22*(5), 1305–1312. doi:10.1021/bp060062e.

Besanger, T. R., Chen, Y., Deisingh, A. K., Hodgson, R., Jin, W., Mayer, S., Brook, M. A., & Brennan, J. D. (2003). Screening of inhibitors using enzymes entrapped in sol-gel-derived materials. *Analytical Chemistry, 75*(10), 2382–2391. doi:10.1021/ac026370i.

Bhushan, I., Parshad, R., Qazi, G. N., & Gupta, V. K. (2008). Immobilization of lipase by entrapment in Ca-alginate beads. *Journal of Bioactive and Compatible Polymers, 23*(6), 552–562. doi:10.1177/0883911508097866.

Bilal, M., Noreen, S., Asgher, M., & Parveen, S. (2021). Development and characterization of cross-linked laccase aggregates (Lac-CLEAs) from *Trametes versicolor* IBL-04 as ecofriendly biocatalyst for degradation of dye-based environmental pollutants. *Environmental Technology & Innovation, 21*, 101364. doi:10.1016/j.eti.2021.101364.

Bolivar, J. M., Woodley, J. M., & Fernandez-Lafuente, R. (2022). Is enzyme immobilization a mature discipline? Some critical considerations to capitalize on the benefits of immobilization. *Chemical Society Reviews, 51*(15), 6251–6290. doi:10.1039/d2cs00083k.

Bolivar, J. M., & López-Gallego, F. (2020). Characterization and evaluation of immobilized enzymes for applications in flow reactors. *Current Opinion in Green and Sustainable Chemistry, 25*, 100349. doi:10.1016/j.cogsc.2020.04.010.

Bowers, L. D., & Carr, P. W. (1976). Applications of immobilized enzymes in analytical chemistry. *Analytical Chemistry, 48*(7), 544A–559a. doi:10.1021/ac60371a033.

Branco, R. V., Gutarra, M. L. E., Freire, D. M. G., & Almeida, R. V. (2010). Immobilization and characterization of a recombinant thermostable lipase (Pf2001) from pyrococcus furiosus on supports with different degrees of hydrophobicity. *Enzyme Research, 2010*, 180418. doi:10.4061/2010/180418.

Brena, B., González-Pombo, P., & Batista-Viera, F. (2013). Immobilization of enzymes: A literature survey. *Methods in Molecular Biology, 1051*, 15–31. doi:10.1007/978-1-62703-550-7_2.

Cao, L. (2006). *Carrier-bound immobilized enzymes: Principles, application and design* (pp. 1–563). Wiley-VCH. doi:10.1002/3527607668.

Cao, L., van Langen, L., & Sheldon, R. A. (2003). Immobilised enzymes: Carrier-bound or carrier-free? *Current Opinion in Biotechnology, 14*(4), 387–394. doi:10.1016/S0958-1669(03)00096-X.

Çelem, E. B., & Önal, S. (2022). Sepabeads EC-EP immobilized $\alpha$-galactosidase: Immobilization, characterization and application in the degradation of raffinose-type oligosaccharides. *Process Biochemistry, 116*, 136–147. doi:10.1016/j.procbio.2022.02.020.

Chakraborty, M., & Hashmi, M. S. J. (2017). *An overview of biosensors and devices* (pp. 1–23). Elsevier BV. doi:10.1016/b978-0-12-803581-8.10316-9.

Chang, T. M. S. (2013). *Biomedical applications of immobilized enzymes and proteins*: 1. Springer Science & Business Media.

Chauhan, G. S. (2014). Evaluation of nanogels as supports for enzyme immobilization. *Polymer International, 63*(11), 1889–1894. doi:10.1002/pi.4734.

Chen, N., Chang, B., Shi, N., Yan, W., Lu, F., & Liu, F. (2023). Cross-linked enzyme aggregates immobilization: Preparation, characterization, and applications. *Critical Reviews in Biotechnology, 43*(3), 369–383. doi:10.1080/07388551.2022.2038073.

Cui, J. D., & Jia, S. R. (2015). Optimization protocols and improved strategies of cross-linked enzyme aggregates technology: Current development and future challenges. *Critical Reviews in Biotechnology, 35*(1), 15–28. doi:10.3109/07388551.2013.795516.

da Cunha, T. M., Mendes, A. A., Hirata, D. B., & Angelotti, J. A. F. (2023). Optimized conditions for preparing a heterogeneous biocatalyst via cross-linked enzyme aggregates (CLEAs) of $\beta$-glucosidase from *Aspergillus niger*. *Catalysts, 13*(1), 62. doi:10.3390/catal13010062.

Datta, S., Christena, L. R., & Rajaram, Y. R. S. (2013). Enzyme immobilization: An overview on techniques and support materials. *3 Biotech, 3*(1), 1–9. doi:10.1007/s13205-012-0071-7.

Do, M. Q., Henry, E., Kato, M., & Cheruzel, L. (2018). Cross-linked cytochrome P450 BM3 aggregates promoted by Ru(II)-diimine complexes bearing aldehyde groups. *Journal of Inorganic Biochemistry, 186*, 130–134. doi:10.1016/j.jinorgbio.2018.06.001.

Eldin, M. S. M., & Mita, D. G. (2014). Immobilized enzymes: Strategies for overcoming the substrate diffusion-limitation problem. *Current Biotechnology, 3*(3), 207–217.

Fadnavis, N. W., Sheelu, G., Kumar, B. M., Bhalerao, M. U., & Deshpande, A. A. (2003). Gelatin blends with alginate: Gels for lipase immobilization and purification. *Biotechnology Progress, 19*(2), 557–564. doi:10.1021/bp010172f.

Fernández-Lorente, G., Palomo, J. M., Cabrera, Z., Guisán, J. M., & Fernández-Lafuente, R. (2007). Specificity enhancement towards hydrophobic substrates by immobilization of lipases by interfacial activation on hydrophobic supports. *Enzyme and Microbial Technology, 41*(5), 565–569. doi:10.1016/j.enzmictec.2007.05.004.

Bautista, F., Morales, G., & Sanz, R. (2010). Immobilization strategies for laccase from *Trametes versicolor* on mesostructured silica materials and the application to the degradation of naphthalene. *Bioresource Technology, 101*(22), 8541–8548. doi:10.1016/j.biortech.2010.06.042.

Garcia-Galan, C., Berenguer-Murcia, Á., Fernandez-Lafuente, R., & Rodrigues, R. C. (2011). Potential of different enzyme immobilization strategies to improve enzyme performance. *Advanced Synthesis & Catalysis, 353*(16), 2885–2904. doi:10.1002/adsc.201100534.

Gaur, R., Pant, H., Jain, R., & Khare, S. (2006). Galacto-oligosaccharide synthesis by immobilized *Aspergillus oryzae* β-galactosidase. *Food Chemistry, 97*(3), 426–430. doi:10.1016/j.foodchem.2005.05.020.

González-Sáiz, J. M., & Pizarro, C. (2001). Polyacrylamide gels as support for enzyme immobilization by entrapment. Effect of polyelectrolyte carrier, pH and temperature on enzyme action and kinetics parameters. *European Polymer Journal, 37*(3), 435–444. doi:10.1016/s0014-3057(00)00151-8.

Hassan, M. E., Zhu, X., de Souza, E. F., Elnashar, M. M., & Lu, F. (2025). Enzyme immobilization advances: A key to unlocking renewable bioenergy potential. *Green Chemistry, 27*(37), 11289–11311. doi:10.1039/d5gc03388h.

Hassan, M. E., Yang, Q., Xiao, Z., Liu, L., Wang, N., Cui, X., & Yang, L. (2019). Impact of immobilization technology in industrial and pharmaceutical applications. *3 Biotech, 9*(12), 439. doi:10.1007/s13205-019-1969-0.

Hirsh, S. L., Bilek, M. M. M., Nosworthy, N. J., Kondyurin, A., dos Remedios, C. G., & McKenzie, D. R. (2010). A comparison of covalent immobilization and physical adsorption of a cellulase enzyme mixture. *Langmuir, 26*(17), 14380–14388. doi:10.1021/la1019845.

Homaei, A. A., Sariri, R., Vianello, F., & Stevanato, R. (2013). Enzyme immobilization: An update. *Journal of Chemical Biology, 6*(4), 185–205. doi:10.1007/s12154-013-0102-9.

Hong, J., Jung, D., Park, S., Oh, Y., Oh, K. K., & Lee, S. H. (2021). Immobilization of laccase via cross-linked enzyme aggregates prepared using genipin as a natural cross-linker. *International Journal of Biological Macromolecules, 169*, 541–550. doi:10.1016/j.ijbiomac.2020.12.136.

Imam, H. T., Marr, P. C., & Marr, A. C. (2021). Enzyme entrapment, biocatalyst immobilization without covalent attachment. *Green Chemistry, 23*(14), 4980–5005. doi:10.1039/d1gc01852c.

Jesionowski, T., Zdarta, J., & Krajewska, B. (2014). Enzyme immobilization by adsorption: A review. *Adsorption, 20*(5-6), 801–821. doi:10.1007/s10450-014-9623-y.

Jiang, Y., Zheng, J., Wang, M., Xu, W., Wang, Y., Wen, L., & Dong, J. (2024). Pros and cons in various immobilization techniques and carriers for enzymes. *Applied Biochemistry and Biotechnology, 196*(9), 5633–5655. doi:10.1007/s12010-023-04838-7.

Kurzbaum, E., Raizner, Y., Kuc, M. E., Kulikov, A., Hakimi, B., Iasur Kruh, L., & Menashe, O. (2020). Phenol biodegradation by bacterial cultures encapsulated in 3D microfiltration-membrane capsules. *Environmental Technology, 41*(22), 2875–2883. doi:10.1080/09593330.2019.1587005.

Li, X. Y., Xu, M. Q., Liu, H., Zhou, Q., Gao, J., & Zhang, Y. W. (2022). Preparation of combined cross-linked enzyme aggregates containing galactitol dehydrogenase and NADH oxidase for l-tagatose synthesis via in situ cofactor regeneration. *Bioprocess and Biosystems Engineering, 45*(2), 353–364. doi:10.1007/s00449-021-02665-w.

Lian, X., Fang, Y. u, Joseph, E., Wang, Q. i, Li, J., Banerjee, S., Lollar, C., Wang, X., & Zhou, H.-C. (2017). Enzyme–MOF (metal–organic framework) composites. *Chemical Society Reviews, 46*(11), 3386–3401. doi:10.1039/C7CS00058H.

Liang, X., Wen, K., Chen, Y., Fang, G., Yang, S., & Li, Q. (2022). Oral administration of therapeutic enzyme capsule for the management of inflammatory bowel disease. *International Journal of Nanomedicine, 17*, 4843–4860. doi:10.2147/IJN.S378073.

Liese, A., & Hilterhaus, L. (2013). Evaluation of immobilized enzymes for industrial applications. *Chemical Society Reviews, 42*(15), 6236–6249. doi:10.1039/C3CS35511J.

Liu, H., Cui, Y., Li, P., Zhou, Y., Chen, Y. u, Tang, Y., & Lu, T. (2013). Polyphosphonate induced coacervation of chitosan: Encapsulation of proteins/enzymes and their biosensing. *Analytica Chimica Acta, 776*, 24–30. doi:10.1016/j.aca.2013.03.040.

Maghraby, Y. R., El-Shabasy, R. M., Ibrahim, A. H., & Azzazy, H. M. E. S. (2023). Enzyme immobilization technologies and industrial applications. *ACS Omega, 8*(6), 5184–5196. doi:10.1021/acsomega.2c07560.

Mahmoodi, N. M., Dastgerdi, H., & Hayati, B. (2020). Environmentally friendly novel covalently immobilized enzyme bionanocomposite: From synthesis to the destruction of pollutant. *Composites Part B: Engineering, 184*, 107666. doi:10.1016/j.compositesb.2019.107666.

Makki, H. H., Rizwana Jaafar, N., Jailani, N., Alqasem, A. A., Rahmat, Z., & Illias, R. M. (2024). Strategy in synthesizing longer-chain levan-type fructooligosaccharides by selective dextran macromolecular cross-linked *Bacillus lehensis* G1 endolevanase aggregate immobilization. *Catalysts, 14*(9), 584. doi:10.3390/catal14090584.

Makshakova, O. N., Bogdanova, L. R., Makarova, A. O., Kusova, A. M., Ermakova, E. A., Kazantseva, M. A., & Zuev, Y. F. (2022). κ-Carrageenan hydrogel as a matrix for therapeutic enzyme immobilization. *Polymers, 14*(19), 4071. doi:10.3390/polym14194071.

Malar, C. G., Seenuvasan, M., Kumar, K. S., Kumar, A., & Parthiban, R. (2020). Review on surface modification of nanocarriers to overcome diffusion limitations: An enzyme immobilization aspect. *Biochemical Engineering Journal, 158*, 107574. doi:10.1016/j.bej.2020.107574.

Mateo, C., Abian, O., Fernandez-Lafuente, R., & Guisan, J. M. (2000). Reversible enzyme immobilization via a very strong and nondistorting ionic adsorption on support-polyethylenimine composites. *Biotechnology and Bioengineering, 68*(1), 98–105. doi:10.1002/(sici)1097-0290(20000405)68:1⟨98::aid-bit12⟩3.0.co;2-t.

Mohammadi, M., Shahedi, M., Ahrari, F., Mostafavi, M., Habibi, Z., & Yousefi, M. (2023). Isocyanide-based multi-component reactions for carrier-free and carrier-bound covalent immobilization of enzymes. *Nature Protocols, 18*(5), 1641–1657. doi:10.1038/s41596-023-00812-z.

Mohidem, N. A., Mohamad, M., Rashid, M. U., Norizan, M. N., Hamzah, F., & Hanapi, M. (2023). Recent advances in enzyme immobilisation strategies: An overview of techniques and composite carriers. *Journal of Composites Science, 7*(12), 488. doi:10.3390/jcs7120488.

Naramittanakul, A., Buttranon, S., Petchsuk, A., Chaiyen, P., & Weeranoppanant, N. (2021). Development of a continuous-flow system with immobilized biocatalysts towards sustainable bioprocessing. *Reaction Chemistry & Engineering, 6*(10), 1771–1790. doi:10.1039/d1re00189b.

Nguyen, H. H., & Kim, M. (2017). An overview of techniques in enzyme immobilization. *Applied Science and Convergence Technology, 26*(6), 157–163. doi:10.5757/asct.2017.26.6.157.

Nguyen, L. N., Hai, F. I., Dosseto, A., Richardson, C., Price, W. E., & Nghiem, L. D. (2016). Continuous adsorption and biotransformation of micropollutants by granular activated carbon-bound laccase in a packed-bed enzyme reactor. *Bioresource Technology, 210*, 108–116. doi:10.1016/j.biortech.2016.01.014.

Ortiz, C., Ferreira, M. L., Barbosa, O., dos Santos, J. C. S., Rodrigues, R. C., Berenguer-Murcia, Á., Briand, L. E., & Fernandez-Lafuente, R. (2019). Novozym 435: The "perfect" lipase immobilized biocatalyst? *Catalysis Science & Technology, 9*(10), 2380–2420. doi:10.1039/c9cy00415g.

Parmar, A., Kumar, H., Marwaha, S. S., & Kennedy, J. F. (2000). Advances in enzymatic transformation of penicillins to 6-aminopenicillanic acid (6-APA). *Biotechnology Advances, 18*(4), 289–301. doi:10.1016/s0734-9750(00)00039-2.

Parthasarathy, R. V., & Martin, C. R. (1996). Enzyme and chemical encapsulation in polymeric microcapsules. *Journal of Applied Polymer Science, 62*(6), 875–886. doi:10.1002/(SICI)1097-4628(19961107)62:6⟨875::AID-APP3⟩3.0.CO;2-M.

Patti, S., Alunno, I. M., Pedroni, S., Riva, S., Ferrandi, E. E., & Monti, D. (2025). Advances and challenges in the development of immobilized enzymes for batch and flow biocatalyzed processes. *ChemSusChem., 18*(8), e202402007. doi:10.1002/cssc.202402007.

Peng, F., Ou, X.-Y., Guo, Z.-W., Zeng, Y.-J., Zong, M.-H., & Lou, W.-Y. (2020). Co-immobilization of multiple enzymes by self-assembly and chemical crosslinking for co-factor regeneration and robust biocatalysis. *International Journal of Biological Macromolecules, 162*, 445–453. doi:10.1016/j.ijbiomac.2020.06.141.

Piacentini, E., Yan, M., & Giorno, L. (2017). Development of enzyme-loaded PVA microspheres by membrane emulsification. *Journal of Membrane Science, 524*, 79–86. doi:10.1016/j.memsci.2016.11.008.

Pierre, A. C. (2004). The sol-gel encapsulation of enzymes. *Biocatalysis and Biotransformation, 22*(3), 145–170. doi:10.1080/10242420412331283314.

Prabhakar, T., Giaretta, J., Zulli, R., Rath, R. J., Farajikhah, S., Talebian, S., & Dehghani, F. (2025). Covalent immobilization: A review from an enzyme perspective. *Chemical Engineering Journal, 503*, 158054. doi:10.1016/j.cej.2024.158054.

Qu, Y., Huang, R., Qi, W., Su, R., & He, Z. (2015). Interfacial polymerization of dopamine in a Pickering emulsion: Synthesis of cross-linkable colloidosomes and enzyme immobilization at oil/water interfaces. *ACS Applied Materials & Interfaces, 7*(27), 14954–14964. doi:10.1021/acsami.5b03787.

Robescu, M. S., & Bavaro, T. (2025). A comprehensive guide to enzyme immobilization: All you need to know. *Molecules, 30*(4), 939. doi:10.3390/molecules30040939.

Rodrigues, R. C., Ortiz, C., Berenguer-Murcia, Á., Torres, R., & Fernández-Lafuente, R. (2013). Modifying enzyme activity and selectivity by immobilization. *Chemical Society Reviews, 42*(15), 6290–6307. doi:10.1039/c2cs35231a.

Rodrigues, R. C., Berenguer-Murcia, Á., Carballares, D., Morellon-Sterling, R., & Fernandez-Lafuente, R. (2021). Stabilization of enzymes via immobilization: Multipoint covalent attachment and other stabilization strategies. *Biotechnology Advances, 52*, 107821. doi:10.1016/j.biotechadv.2021.107821.

Rother, C., & Nidetzky, B. (2009). Enzyme immobilization by microencapsulation: Methods, materials, and technological applications. In M. C Flickinger (Ed.), *Encyclopedia of Industrial Biotechnology* (pp. 1–21). Wiley. doi:10.1002/9780470054581.eib275.

Rueda, N., dos Santos, J. C. S., Ortiz, C., Torres, R., Barbosa, O., Rodrigues, R. C., Berenguer-Murcia, Á., & Fernandez-Lafuente, R. (2016). Chemical modification in the design of immobilized enzyme biocatalysts: Drawbacks and opportunities. *The Chemical Record, 16*(3), 1436–1455. doi:10.1002/tcr.201600007.

Saikia, K., Rathankumar, A. K., Kumar Vaithyanathan, V., Cabana, H., & Vaidyanathan, V. K. (2021). Preparation of highly diffusible porous cross-linked lipase B from *Candida antarctica* conjugates: Advances in mass transfer and application in transesterification of 5-Hydroxymethylfurfural. *International Journal of Biological Macromolecules, 170*, 583–592. doi:10.1016/j.ijbiomac.2020.12.178.

Sassolas, A., Blum, L. J., & Leca-Bouvier, B. D. (2012). Immobilization strategies to develop enzymatic biosensors. *Biotechnology Advances, 30*(3), 489–511. doi:10.1016/j.biotechadv.2011.09.003.

Sassolas, A., Hayat, A., & Marty, J.-L. (2013). Enzyme immobilization by entrapment within a gel network. In J. M Guisan (Ed.), *Immobilization of enzymes and cells* (pp. 229–239). Springer.

Sheldon, R. A. (2011). Cross-linked enzyme aggregates as industrial biocatalysts. *Organic Process Research and Development, 15*(1), 213–223. doi:10.1021/op100289f.

Sheldon, R. A., & van Pelt, S. (2013). Enzyme immobilisation in biocatalysis: Why, what and how. *Chemical Society Reviews, 42*(15), 6223–6235. doi:10.1039/c3cs60075k.

Sheldon, R. A., Schoevaart, R., & Van Langen, L. M. (2005). Cross-linked enzyme aggregates (CLEAs): A novel and versatile method for enzyme immobilization (a review). *Biocatalysis and Biotransformation, 23*(3-4), 141–147. doi:10.1080/10242420500183378.

Sheldon, R. A. (2018). Cross-linked enzyme aggregates (CLEAs): From concept to industrial biocatalyst. *Biocatalysis for green chemistry and chemical process development*, 363–396 RSC Catalysis Series No. 29. doi:10. 1039/9781782629993-00363.

Sigurdardóttir, S. B., Lehmann, J., Ovtar, S., Grivel, J. C., Negra, M. D., Kaiser, A., & Pinelo, M. (2018). Enzyme immobilization on inorganic surfaces for membrane reactor applications: Mass transfer challenges, enzyme leakage and reuse of materials. *Advanced Synthesis and Catalysis, 360*(14), 2578–2607. doi:10.1002/adsc.201800307.

Sirisha, V. L., Jain, A., & Jain, A. (2016). Enzyme immobilization: An overview on methods, support material, and applications of immobilized enzymes. *Advances in Food and Nutrition Research, 79*, 179–211. doi:10.1016/bs.afnr.2016.07.004.

Sneha, H. P., Beulah, K. C., & Murthy, P. S. (2018). Enzyme immobilization methods and applications in the food industry. In M Kuddus (Ed.), *Enzymes in food biotechnology: Production, applications, and future prospects* (pp. 645–658). Elsevier. doi:10.1016/B978-0-12-813280-7.00037-2.

Lima, J. S., Boemo, A. P. S. I., de Araújo, P. H. H., & de Oliveira, D. (2021). Immobilization of endoglucanase on kaolin by adsorption and covalent bonding. *Bioprocess and Biosystems Engineering, 44*(8), 1627–1637. doi:10.1007/s00449-021-02545-3.

Tadesse, M., & Liu, Y. (2025). Recent advances in enzyme immobilization: The role of artificial intelligence, novel nanomaterials, and dynamic carrier systems. *Catalysts, 15*(6), 571. doi:10.3390/catal15060571.

Tavernini, L., Ottone, C., Illanes, A., & Wilson, L. (2020). Entrapment of enzyme aggregates in chitosan beads for aroma release in white wines. *International Journal of Biological Macromolecules, 154*, 1082–1090. doi:10.1016/j.ijbiomac.2020.03.031.

Torabizadeh, H., Tavakoli, M., & Safari, M. (2014). Immobilization of thermostable $\alpha$-amylase from *Bacillus licheniformis* by cross-linked enzyme aggregates method using calcium and sodium ions as additives. *Journal of Molecular Catalysis B: Enzymatic, 108*, 13–20. doi:10.1016/j.molcatb.2014.06.005.

Tran, D. N., & Balkus, K. J. (2011). Perspective of recent progress in immobilization of enzymes. *ACS Catalysis, 1*(8), 956–968. doi:10.1021/cs200124a.

Tran, D.-T., Chen, C.-L., & Chang, J.-S. (2012). Immobilization of *Burkholderia* sp. lipase on a ferric silica nanocomposite for biodiesel production. *Journal of Biotechnology, 158*(3), 112–119. doi:10.1016/j.jbiotec.2012.01.018.

Tufvesson, P., Törnvall, U., Carvalho, J., Karlsson, A. J., & Hatti-Kaul, R. (2011). Towards a cost-effective immobilized lipase for the synthesis of specialty chemicals. *Journal of Molecular Catalysis B: Enzymatic, 68*(2), 200–205. doi:10.1016/j.molcatb.2010.11.004.

Vardanyan, A., Agback, T., Golovko, O., Diétre, Q., & Seisenbaeva, G. A. (2024). Natural silicates encapsulated enzymes as green biocatalysts for degradation of pharmaceuticals. *ACS ES&T Water, 4*(2), 751–760. doi:10.1021/acsestwater.3c00811.

Wang, M., Mohanty, S. K., & Mahendra, S. (2019). Nanomaterial-supported enzymes for water purification and monitoring in point-of-use water supply systems. *Accounts of Chemical Research, 52*(4), 876–885. doi:10.1021/acs.accounts.8b00613.

Wang, R., Xia, B., Li, B.-J., Peng, S.-L., Ding, L.-S., & Zhang, S. (2008). Semi-permeable nanocapsules of konjac glucomannan–chitosan for enzyme immobilization. *International Journal of Pharmaceutics, 364*(1), 102–107. doi:10.1016/j.ijpharm.2008.07.026.

Xu, M.-Q., Wang, S.-S., Li, L.-N., Gao, J., & Zhang, Y.-W. (2018). Combined cross-linked enzyme aggregates as biocatalysts. *Catalysts, 8*(10), 460. doi:10.3390/catal8100460.

Yamaguchi, H., Kiyota, Y., & Miyazaki, M. (2018). Techniques for preparation of cross-linked enzyme aggregates and their applications in bioconversions. *Catalysts, 8*(5), 174. doi:10.3390/catal8050174.

Yoshimoto, M. (2016). Stabilization of enzymes through encapsulation in liposomes. In H. Y. Kim, & J. M Kim (Eds.), *Enzyme stabilization and immobilization: Methods and protocols* (pp. 9–18). Springer.

Zhang, W. W., Yang, X. L., Jia, J. Q., Wang, N., Hu, C. L., & Yu, X. Q. (2015). Surfactant-activated magnetic cross-linked enzyme aggregates (magnetic CLEAs) of *Thermomyces lanuginosus* lipase for biodiesel production. *Journal of Molecular Catalysis B: Enzymatic, 115*, 83–89. doi:10.1016/j.molcatb.2015.02.003.

Zhu, R., Li, C., Chen, C., Xing, S., Cai, Y., Zeng, X., & He, L. (2021). Effect of cross-linked enzyme aggregate strategy on characterization of sn-1,3 extracellular lipase from Aspergillus niger GZUF36. *Applied Microbiology and Biotechnology, 105*(5), 1925–1941. doi:10.1007/s00253-021-11160-x.

Zucca, P., & Sanjust, E. (2014). Inorganic materials as supports for covalent enzyme immobilization: Methods and mechanisms. *Molecules, 19*(9), 14139–14194. doi:10.3390/molecules190914139.

CHAPTER

# Future perspectives and challenges in carrier-free immobilization

10

**Lixue Zhao[a,b], Yingjia Li[a], Manman Shi[a], Yingjie Du[a] and Jiandong Cui[a,*]**

[a] *Laboratory of Industrial Fermentation Microbiology, Tianjin University of Science and Technology, Tianjin, China,* [b] *College of Biotechnology, Tianjin University of Science and Technology, Tianjin, China*

* *Corresponding author.*

## 10.1 Introduction

Carrier-free immobilized enzyme technology is a frontier in biocatalysis. It allows enzyme molecules to form stable catalytic assemblies via direct cross-linking or self-assembly, thus avoiding the use of traditional external carriers (Bilal et al., 2018; Lopez et al., 2017; Margolin & Navia, 2001; Sheldon & Woodley, 2018). This strategy provides a fundamentally different route to enzyme stabilization. In recent years, research has advanced from proof-of-concept to practical application, giving rise to typical methods including cross-linked enzyme aggregates (CLEAs) (Brem et al., 2012), enzyme self-assembled hydrogels (Bitterwolf et al., 2019; Chen et al., 2023), and polymer-enzyme nanocomposites (Li et al., 2015). Carrier-free immobilized enzyme technology overcomes major drawbacks of conventional immobilization, such as activity loss caused by masking effects, mass transfer resistance, and high costs (Liu et al., 2025; Patti et al., 2025), and promotes the development of efficient, large-scale biocatalytic processes. Driven by progress in synthetic biology, nanotechnology, and materials science, carrier-free immobilization is developing rapidly (Hudson et al., 2008). This chapter addresses several key challenges: balancing enzyme stability and catalytic activity, establishing reproducible and scalable preparation methods, and expanding universal applicability while maintaining task-specific performance. We systematically analyze the underlying mechanisms and discuss emerging trends, including flexible "immobilized but not rigid" enzyme architectures, stimuli-responsive systems, and ordered multienzyme co-immobilization. This review aims to provide a clear theoretical basis and practical reference for future research and industrial applications.

**Carrier-Free Enzyme Immobilization: Fundamentals and Innovations in Biocatalysis.**
**DOI: https://doi.org/10.1016/B978-0-443-44139-4.00011-3**

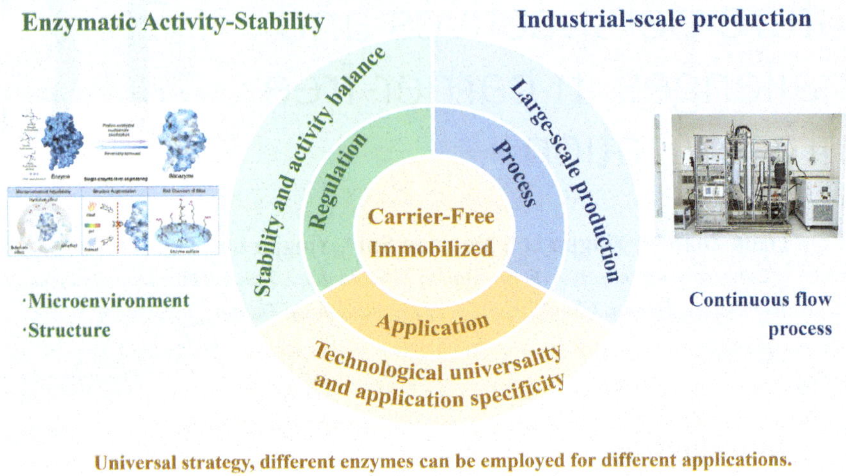


**FIGURE 10.1**

Challenges with carrier-free immobilization of enzymes. Current challenges facing carrier-free immobilized enzymes.

## 10.2 Current technical challenges and bottlenecks

Although carrier-free immobilization technology demonstrates significant advantages, its transition from laboratory to industrial application still faces a series of critical challenges, primarily concentrated in the following three areas: regulation of the stability and activity balance, process challenges in large-scale production, and technological universality and application specificity (Fig. 10.1).

### 10.2.1 Regulation of the stability and activity balance

One of the central difficulties in carrier-free immobilization is striking a balance between the stability and catalytic efficacy of enzyme molecules (Barbosa et al., 2015; Cao et al., 2003; Roessl et al., 2010; Sheldon et al., 2005). Although carrier-free immobilization usually improves the thermal stability and operational sturdiness of enzymes, the cross-linking procedure might still cause irreversible alterations in the conformation of enzyme molecules, resulting in a loss of activity (Fig. 10.2) (Chauhan et al., 2022; Chen et al., 2023; Hou et al., 2023). This method exposed significant drawbacks. Chemical cross-linking made enzyme structures more rigid, and using too much glutaraldehyde (GA) (Akkas et al., 2020) for cross-linking damaged the active sites (Akkas et al., 2020; Cao et al., 2003). To address the balance between

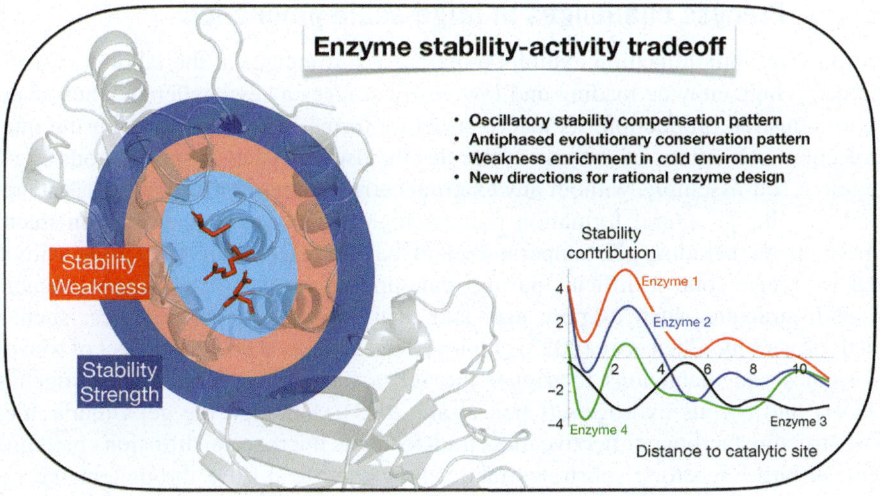


**FIGURE 10.2**

Enzyme stability-activity trade-off. The balance between enzyme stability and catalytic activity.

*From Hou, Q., Rooman, M., & Pucci, F. (2023). Enzyme stability-activity trade-off: New insights from protein stability weaknesses and evolutionary conservation.* Journal of Chemical Theory and Computation, 19 *(12), 3664–3671. https://doi.org/10.1021/acs.jctc.3c00036.*

enzyme stability and activity, researchers tried to introduce sacrificial templates like bovine serum albumin (BSA), polyethylene glycol (PEG) polymers, or crystallizing salts as pore-forming agents, which were later eluted after cross-linking, successfully creating macropores inside the aggregate, which made mass transfer easier (Bose et al., 2013). For example, adding an appropriate quantity of BSA as a co-cross-linking framework during the preparation of lipase CLEAs enhanced mechanical stability (Jung et al., 2013). The pores remaining after elution markedly increased the specific surface area, facilitating diffusion of the hydrophobic triglyceride substrate toward the enzyme's active site; however, this macroporous structure also rendered the enzyme more susceptible to leakage (Alves et al., 2021). Wang et al. (2011) proposed a new method for preparing porous CLEAs (p-CLEAs) by incorporating starch as a porogen, which was subsequently removed through $\alpha$-amylase digestion to create a porous structure. This approach significantly improved CLEA catalytic efficiency for macromolecular substrates; nevertheless, poor mechanical stability still remained a tenacious obstacle (Wang et al., 2011). How to accurately control cross-linking density while maintaining the flexibility of active sites and at the same time guarantee the long-term stability of enzymes in complex industrial settings remained a crucial matter in urgent need of a solution.

### 10.2.2 Process challenges in large-scale production

Carrier-free immobilization exhibits remarkable advantages at the laboratory scale, including high enzyme loading and low cost, but faces a key challenge when scaled up to industrial production. Its macro-structure formation relies solely on the intermolecular interactions of enzyme molecules themselves, such as spontaneous aggregation or self-assembly, without any external carrier support or formation assistance, rendering the structural formation process highly sensitive to the reaction microenvironment, including pH, temperature, and mass transfer efficiency. This sensitivity makes precise control difficult during scale-up and leads to batch inconsistencies in both structure, such as pore size and distribution, and performance, such as catalytic activity. Chen et al. (2023) employed the combined bridging effect of triazole molecules and metal ions to trigger the self-assembly of alcohol dehydrogenase (ADH) into porous hydrogels (Chen et al., 2023). Although the gel's hierarchical pore structure facilitates effective mass transfer at the microscale, diffusion challenges arise in larger reactors: substrates take a long time to diffuse throughout the gel, and products cannot be extracted promptly. These diffusion constraints lower the overall efficiency of the reactor and hinder the complete exploitation of the system's enzymatic capabilities. Guauque Torres et al. (2014) prepared layered cross-linked CLEAs of *Candida antarctica* lipase B (CALB) using BSA as a core component (Guauque Torres et al., 2014). The hierarchical structural design effectively prevented activity loss from excessive enzyme aggregation and significantly improved catalytic activity and stability, although activity retention declined to 51.96% after 10 cycles.

At present, carrier-free immobilized enzymes encounter difficulties in controllability and process scaling during large-scale production. However, these challenges can be addressed through intelligent regulatory strategies, optimized reactor designs to enhance mass and heat transfer, and low-cost inducible molecules. These improvements would enable uniform structure, stable performance, and controllable costs, gradually transforming laboratory-scale potential into industrial competitiveness (Shen et al., 2022).

### 10.2.3 Technological universality and application specificity

Another drawback of carrier-free immobilization method is that they lack enough universality because there are substantial differences in the structural features, the distribution of surface functional groups, and the stability of various enzyme molecules which indicates that the immobilization approaches designed for particular enzymes can't be easily utilized for other enzyme systems (Cao et al., 2000; Noori et al., 2019; Ottone et al., 2020; Reis et al., 2019). While food-grade enzymes require completely nontoxic cross-linking agents, environmental remediation enzymes must withstand high concentrations of organic solvents (Ruiz et al., 2000). And these special requirements make higher demands on immobilization procedures so in order to improve universality, researchers tried to boost cross-linking efficiency by modifying the enzyme's surface. Abdul Wahab et al. (2019) proposed a method combining enzyme surface modification to increase the number of free amino groups with CLEA preparation (Abdul Wahab et al., 2019). By engineering the protein surface, specific sites on the xylanase were altered to lysine, enhancing its cross-linking efficiency with GA.

Studies have demonstrated that porous CLEAs used for the kinetic resolution of chiral pharmaceutical intermediates exhibit much higher catalytic efficiency and reusability than conventional carrier-immobilized enzymes. This advantage is due to their open pore structure, which allows rapid exchange of substrates and products while maintaining the enzyme's active conformation (Sheldon, 2011). However, the technology continues to face challenges related to limitations in the broad applicability of its preparation process and the lack of reliability in its mass production. Bounegru and Apetrei (2023) pointed out that the cross-linking immobilization of tyrosinase is marked by its very high usage rate, and the biosensors developed had excellent performance (Bounegru & Apetrei, 2023). Moreover, recent research has centered on new cross-linking-based immobilization approaches like magnetically cross-linked enzyme aggregates (mCLEAs). Nevertheless, this approach remains limited to specific enzymes and cannot be used for systems involving multiple enzymes, so the tension between the broad applicability of the technology and the particular requirements of its use restricts the widespread adoption of carrier-free immobilization techniques. Therefore, future efforts should focus on developing a universal enzyme-catalyzed system to establish it as a versatile platform for enzyme-catalyzed applications (Chauhan et al., 2022).

## 10.3 Future prospects for carrier-free immobilization

With a range of pioneering accomplishments, carrier-free immobilization has shown tremendous promise. Its future possibilities go well beyond the current state, driving the whole industry toward a new phase of enhanced efficiency and cost-effectiveness. The future outlook for carrier-free immobilization will be explored below through the following aspects (Fig. 10.3).

### 10.3.1 Immobilized but not rigid state

The advantage of carrier-free immobilization lies in breaking away from the traditional approach of confining enzymes to rigid carriers. However, excessive cross-linking agents may also induce conformational rigidity in enzymes; thus, maintaining enzyme flexibility during cross-linking remains a significant challenge. Our research team proposed the concept of "Immobilized but not rigid" state to explore whether enzymes can maintain a free-like state similar to unbound molecules when immobilized (Du et al., 2022). Zhang et al. (2020) designed and synthesized nitrogen-coordinated cyclic boron ester (NCB) bonds, which improve the hydrothermal stability of boron esters via N→B coordination while retaining their ability to undergo dynamic exchange (Zhang et al., 2020). Cross-linking enzymes using boron ester or imine bonds

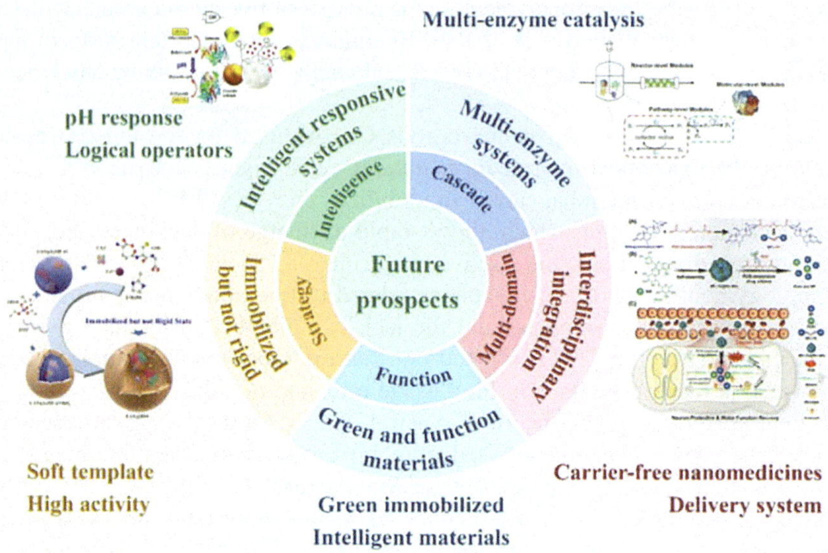


**FIGURE 10.3**

Future prospects of carrier-free immobilization. Emerging opportunities for carrier-free immobilization in biocatalysis.

allows for reversible cleavage and reformation within the catalytic microenvironment, thereby preserving the conformational dynamic equilibrium of the enzyme's active site (Fig. 10.4).

Similarly, surface modification strategies have also been widely employed to address activity loss in immobilized enzymes. Li et al. (2018) introduced flexible PEG chains as a "molecular brush" coating on the surface of magnetic beads, which not only preserved the advantages of enzyme immobilization but also restored the enzyme's catalytic activity by leveraging the hydrophilicity and spatial flexibility of PEG to prevent enzyme rigidity (Li et al., 2018). This approach was successfully applied to next-generation sequencing library construction. Therefore, applying the concept of "Immobilized but not rigid state" to carrier-free immobilization can more effectively address the rigidity issues caused by enzyme cross-linking. Carrier-free immobilization, which does not rely on rigid external supports, provides natural technical benefits for achieving an "immobilized but not rigid" state. Current approaches include gentle cross-linking methods (such as enzymatic cross-linking and metal-coordinated weak cross-linking), the use of conformational stabilizers (like BSA and glycoprotein chaperones), and targeted modifications of enzyme molecules (including engineering flexible sites on their surfaces) (Cao et al., 2000). For instance, sacrificial template pore regulation technology can be deeply integrated with the carrier-free immobilization preparation process. These approaches optimize carrier-free systems like CLEAs, achieving enzyme immobilization while mitigating conformation rigidity

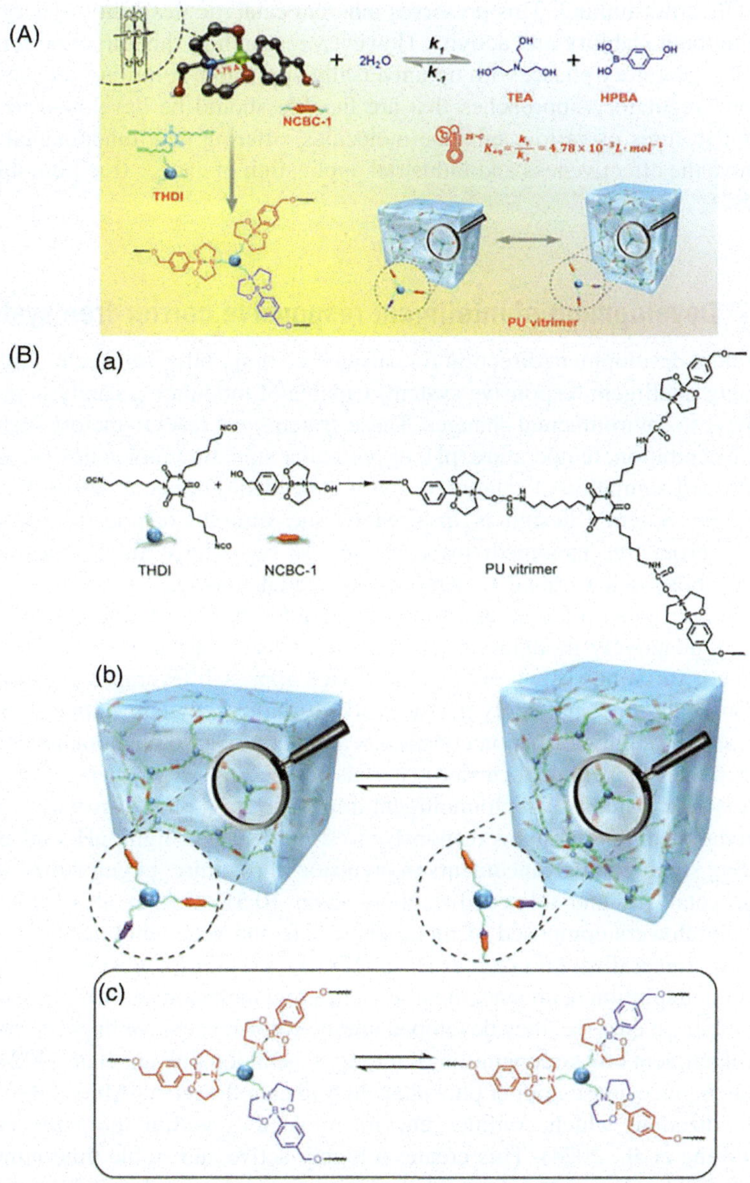


**FIGURE 10.4**

Schematic illustration of dynamic reversible cross-linking based on nitrogen-coordinated cyclic boronate bonds. (A) Design concept for nitrogen-coordinated cyclic boronic diesters (NCB) bonds and (B) The specific chemical pathways and reactions of the design concept.

*From Zhang, X., Wang, S., Jiang, Z., Li, Y., & Jing, X. (2020). Boronic ester based vitrimers with enhanced stability via internal boron-nitrogen coordination.* Journal of the American Chemical Society, 142 *(52), 21852–21860. https://doi.org/10.1021/jacs.0c10244.*

induced by cross-linking. This preserves inherent catalytic flexibility, synergistically enhancing both stability and activity. However, research in this direction remains in its initial exploratory phase, with no universally applicable solutions yet established. Therefore, regulatory approaches that are flexible should be developed to suit the structural features of various enzyme molecules, offering new fundamental support to improve the effectiveness and industrial application of carrier-free immobilization methods.

### 10.3.2 Development of intelligent responsive carrier-free systems

A key future development direction for carrier-free immobilization technology lies in developing intelligent responsive systems capable of modulating catalytic properties in response to environmental changes. These systems can react to factors such as substrate concentration, temperature, pH, or particular signaling molecules, allowing for the controlled regulation of catalytic activity as needed (Díez et al., 2014; Shen et al., 2022). This system's design is inspired by the stimulus-responsive mechanisms found in carrier-free nanomedicines, and several types have already been created, including those that respond to temperature, metal ions, and thermal stimulation. Shen et al. designed a temperature-responsive intelligent pesticide delivery system capable of adaptively modulating the pesticide release rate in response to ambient temperature variations (Shen et al., 2022). This approach leverages conformational alterations within a smart polymer network to realize "dynamic immobilization," thereby addressing the diffusion constraints and burst release phenomena commonly associated with conventional carrier materials embedding active agents. Looking forward, it is anticipated that such intelligent immobilization platforms will evolve into multimodal regulatory systems responsive to stimuli such as light, pH, and enzymes, facilitating significant advancements in precision agriculture, targeted drug delivery, dynamic catalysis, and related disciplines (Fig. 10.5). Zhai et al. (2022) created hybrid nanoflowers composed of metal phosphate and acetylcholinesterase through biomimetic mineralization (Zhai et al., 2022). By utilizing the distinct allosteric effects (either inhibition or enhancement) that different metal ions have on the activity of the immobilized enzyme, they developed unique color-responsive fingerprints that allow for intelligent cross-recognition of multiple metal ions. Wang et al. (2024) developed starch glucosidase-copper phosphate hybrid nanoflowers (AMG@NFs) through biomineralization, which confines enzyme molecules within inorganic petal-like layers (Wang et al., 2024). This creates a highly active and stable time-temperature indicator (TTI). The system uses cumulative thermal stimulation to activate the immobilized enzyme's catalyzed hydrolysis reaction, resulting in an intelligent color change from burgundy to colorless. The activation energy (Ea) of this reaction precisely matches the kinetics of quality degradation in cold-chain pork, allowing for intelligent closed-loop monitoring of the "immobilized enzyme-ambient temperature-quality signal" relationship. The development of intelligent responsive carrier-free immobilization will transform this field from static immobilization to dynamic

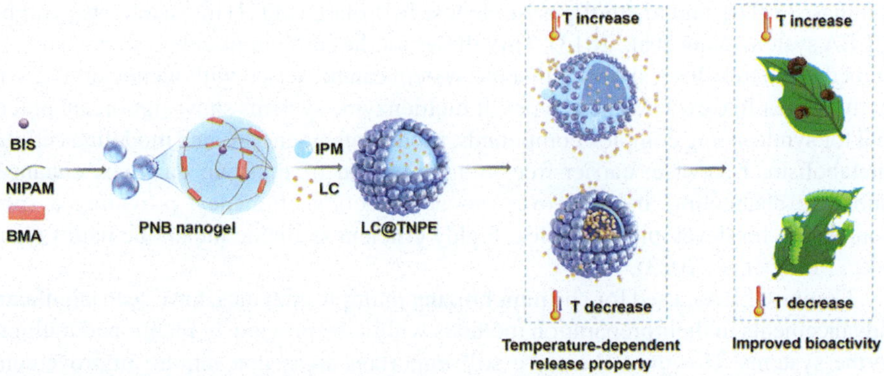


**FIGURE 10.5 Fabrication and temperature-responsive controlled release of the LC@TNPE delivery system.**

The preparation process of the LC@TNPE system involves the encapsulation of pesticides within PNIPAM-based temperature-responsive nanogels via a Pickering emulsion technique. This system exhibits volume phase transition behavior in response to temperature variations, facilitating intelligent, temperature-regulated delivery and management of agricultural pests.

*From Shen, Y., An, C., Jiang, J., Huang, B., Li, N., Sun, C., Wang, C., Zhan, S., Li, X., Gao, F., Zhao, X., Cui, H., Gooneratne, R., & Wang, Y. (2022). Temperature-dependent nanogel for pesticide smart delivery with improved foliar dispersion and bioactivity for efficient control of multiple pests.* ACS Nano, 16 *(12), 20622–20632. https://doi.org/10.1021/acsnano.2c07517.*

intelligence. Endowing immobilized enzymes with responsiveness to external stimuli such as pH, temperature, and light, enables the "on-demand switching" of catalytic activity and the precise recovery of enzyme resources. This fundamentally resolves the core contradiction in traditional immobilization of the incompatibility of stability and flexibility, paving a new pathway for high precision and efficiency industrial biomanufacturing (Liu et al., 2025).

### 10.3.3 Multienzyme carrier-free immobilization

The construction of multienzyme systems that imitate intracellular metabolic pathways is a leading area in carrier-free immobilization technology. By co-immobilizing multiple related enzymes within a single carrier-free system, substrate channel effects can be achieved, enhancing metabolic flux and product yields (Zhang et al., 2022). Key strategies include spatially organized co-assembly of enzymes for synergistic interactions, cell-like microdomains with optimized reaction conditions, and synthetic biology-designed interfaces for efficient substrate transfer. Talekar et al. (2013) co-immobilized alpha-amylase, glucoamylase, and pullulanase as a combination CLEA, developing a three-enzyme biocatalyst with one-pot starch hydrolysis activity (Talekar et al., 2013). Jung et al. (2013) employed a synergistic strategy of protecting GA cross-links with BSA to combine amylosucrase (AS), maltooligosyltrehalose synthase (MTS), and maltooligosyltrehalose hydrolase (MTH) into a one-step combi-CLEA system (Jung et al., 2013). This allows for the continuous conversion of sucrose directly into trehalose, showing that the system can be reused without any decrease in activity over five cycles. These types of multienzyme systems show significant potential for synthesizing complex compounds, producing bioenergy, and modeling cellular metabolism. Likewise, carrier-free immobilization of multiple enzymes enhances substrate channeling and improves cascade reaction efficiency, providing a solid foundation for developing complex, highly efficient synthetic metabolic pathways in vitro (Yang et al., 2023).

Combi-CLEAs, used for co-immobilizing multiple enzymes, have seen significant advancements in their preparation methods within the previously mentioned multienzyme systems. However, there are still important areas that require improvement, particularly in the coordinated control of their structure and function. Addressing these aspects would lay the groundwork for enhancing Combi-CLEAs' performance and expanding their use in specific applications. For example, creating targeted hierarchical porous structures within Combi-CLEAs can effectively reduce mass transfer resistance between substrates and intermediates in multienzyme cascade reactions, minimize mutual enzyme inhibition, and increase substrate availability for enzyme molecules. Additionally, the porous framework offers ample spatial flexibility for enzymes, reducing conformational rigidity during the cross-linking process (Sheldon, 2019). Periyasamy et al. (2016) developed carrier-free combi-CLEAs using triphasic partitioning precipitation (TPP) combined with GA cross-linking. This resulted in a coarse-grained, porous aggregate structure that effectively reduced substrate mass transfer resistance (Periyasamy et al., 2016). The loose and porous network within the cross-linked aggregate minimized diffusion resistance, enabling lignocellulosic substrates to reach the enzyme active sites more effectively. This design enabled the three enzymes to simultaneously interact with and hydrolyze cellulose, hemicellulose, and $\beta$-glucan, aligning with the core principle of the "Immobilized but not rigid" approach to carrier-free immobilization. This method offers a novel strategy for improving the performance of combi-CLEAs.

### 10.3.4 Interdisciplinary integration

The future advancement of carrier-free immobilization techniques will greatly benefit from integrating various fields. Material science can provide innovative smart materials and reactive polymers (Badzińska et al., 2025); nanotechnology offers precise nanofabrication and manipulation tools; synthetic biology enables the design of engineered enzymes with specific functions (Ge et al., 2022); and artificial intelligence (AI) can accelerate the optimization and design of immobilization methods. Therefore, upcoming research should include multiscale simulation methods, ranging from molecular dynamics and quantum chemical calculations to finite element analysis, to fully understand the relationship between structure and function in carrier-free immobilized enzymes (Luo et al., 2022). Additionally, real-time, on-site monitoring techniques should be developed to track dynamic changes in enzyme structure during immobilization, and predictive models must be established to determine the optimal immobilization strategies and conditions based on the enzyme's fundamental structure. Ge et al. (2022) had utilized a synthetic biology method to design a self-assembling "genetic tag" for the target enzyme (Ge et al., 2022). McGill et al. (2017) created sericin hydrogels using enzymatic cross-linking reactions and thoroughly investigated their cross-linking density, mechanical properties, and diffusion characteristics (McGill et al., 2017). Their findings demonstrated that by adjusting the concentrations of sericin and the cross-linking agent, the degree of cross-linking and diffusion behavior of the hydrogels could be precisely controlled. Additionally, they developed an analytical model spanning from molecular to macroscopic levels, providing a theoretical basis for designing biomaterials suitable for tissue engineering and drug delivery applications. Furthermore, integrating bionics with carrier-free immobilization techniques can leverage the principles of enzyme localization and assembly in biological systems to develop more efficient and intelligent immobilization methods (Decante et al., 2024).

### 10.3.5 Functional materials and green crosslinking agents

Conventional GA is known for its cytotoxic nature, so researchers have been exploring bio-based crosslinkers like natural polyphenols and genipin, which are emerging as a new research direction. These substances can effectively crosslink under gentle conditions without producing any harmful by-products, and at the same time, using biomimetic mineralization methods enables the creation of nanoscale protective layers on enzyme surfaces, which further enhances their stability. Martins et al. (2018) incorporated $Fe_2O_3$ nanoparticles and a glycerol imprinting method into rhamnose glycosidase CLEAs (mCLEAs@Rhmnase) to achieve magnetic separation and activity enhancement. Replacing conventional GA with polyethyleneimine (PEI) achieved up to 99% immobilization efficiency during a short 3-hour cross-linking process (Martins et al., 2018). Bilal et al. (2017) utilized a new mild cross-linking agent named ethylene glycol di (N-hydroxysuccinimide) carbonate (EG-NHS) to cross-linking horse radish peroxidase (HRP), and the resulting HRP-CLEAs had improved biocatalytic characteristics (Bilal et al., 2017). Cui has developed a novel mesoporous spherical CLEA and prepared a biosilicon shell using $CaCO_3$-based phenylalanine deaminase (Cui et al., 2017). This approach constructs CLEAs with regular morphology and a core-shell structure, effectively overcoming the issues of irregular morphology, mass transfer limitations, and poor stability inherent in traditional CLEAs. Furthermore, the incorporation of nanomaterials endows immobilized enzymes with novel and designable functionalities. For instance, magnetic nanoparticles enable rapid separation and recovery of immobilized enzymes via external magnetic fields, while other functional nanomaterials can generate synergistic catalytic effects with immobilized enzymes. Torabizadeh et al. prepared lysine-functionalized nanomagnetic composite CLEAs of Viscozyme, Alcalase, and Flavourzyme using ethanol as the aggregating agent, GA as the crosslinking agent, and ascorbic acid as the reducing agent (Torabizadeh et al., 2025). In contrast to using free mixed enzymes, this method allows for the simultaneous extraction of oil and protein hydrolysis from rice bran under gentle processing conditions in a single step. It eliminates solvent residues and reduces environmental pollution, making it an outstanding eco-friendly technology for oil extraction.

## 10.4 Conclusion

Carrier-free immobilization technology has reached a pivotal point in transitioning from laboratory research to industrial application, despite currently facing some challenges. It is anticipated that significant breakthroughs will be achieved within the next 5 to 10 years by exploring advanced interdisciplinary fields. The technology is expected to evolve towards greater precision, intelligence, and multifunctionality. With the aid of AI-driven design and automated experimental systems, customized and precise immobilization methods will become possible. Additionally, environmentally friendly and sustainable manufacturing practices will encourage the widespread use of bio-based cross-linking agents and functional materials. This technology will play a vital role across various sectors, providing essential technological support for global sustainable development goals. Ultimately, the goal of carrier-free immobilized enzymes is to create a new generation of biocatalytic systems characterized by excellent catalytic performance, stable operation, intelligent responsiveness, and multifunctional integration. This will enable precision, intelligence, and integration in biocatalysis, serving as a key driver for innovative progress in biotechnology.

## References

Abdul Wahab, M. K. H., El-Enshasy, H. A., Bakar, F. D. A., Murad, A. M. A., Jahim, J. M., & Illias, R. M. (2019). Improvement of cross-linking and stability on cross-linked enzyme aggregate (CLEA)-xylanase by protein surface engineering. *Process Biochemistry, 86*, 40–49. doi:10.1016/j.procbio.2019.07.017.

Akkas, T., Zakharyuta, A., Taralp, A., & Ow-Yang, C. W. (2020). Cross-linked enzyme lyophilisates (CLELs) of urease: A new method to immobilize ureases. *Enzyme and Microbial Technology, 132*, 109390. doi:10.1016/j.enzmictec.2019.109390.

Alves, N. R., Pereira, M. M., Giordano, R. L. C., Tardioli, P. W., Lima, Á. S., Soares, C. M. F., & Souza, R. L. (2021). Design for preparation of more active cross-linked enzyme aggregates of *Burkholderia cepacia* lipase using palm fiber residue. *Bioprocess and Biosystems Engineering, 44*(1), 57–66. doi:10.1007/s00449-020-02419-0.

Badzińska, W., Kołodziejczak-Radzimska, A., Zdarta, J., Piasecki, A., Siwińska-Ciesielczyk, K., Jesionowski, T., & Ciesielczyk, F. (2025). Novel protocol for design of the functional biocatalytic systems composed with laccase supported onto $Al_2O_3$-cellulose and $Al_2O_3$-chitosan modified electrospun materials. *International Journal of Biological Macromolecules, 333*(Pt 2), 148948. doi:10.1016/j.ijbiomac.2025.148948.

Barbosa, O., Ortiz, C., Berenguer-Murcia, Á., Torres, R., Rodrigues, R. C., & Fernandez-Lafuente, R. (2015). Strategies for the one-step immobilization-purification of enzymes as industrial biocatalysts. *Biotechnology Advances, 33*(5), 435–456. doi:10.1016/j.biotechadv.2015.03.006.

Bilal, M., Iqbal, H. M. N., Hu, H., Wang, W., & Zhang, X. (2017). Development of horseradish peroxidase-based cross-linked enzyme aggregates and their environmental exploitation for bioremediation purposes. *Journal of Environmental Management, 188*, 137–143. doi:10.1016/j.jenvman.2016.12.015.

Bilal, M., Iqbal, H. M. N., Guo, S., Hu, H., Wang, W., & Zhang, X. (2018). State-of-the-art protein engineering approaches using biological macromolecules: A review from immobilization to implementation view point. *International Journal of Biological Macromolecules, 108*, 893–901. doi:10.1016/j.ijbiomac.2017.10.182.

Bitterwolf, P., Gallus, S., Peschke, T., Mittmann, E., Oelschlaeger, C., Willenbacher, N., Rabe, K. S., & Niemeyer, C. M. (2019). Valency engineering of monomeric enzymes for self-assembling biocatalytic hydrogels. *Chemical Science, 10*(42), 9752–9757. doi:10.1039/c9sc04074a.

Bose, S., Vahabzadeh, S., & Bandyopadhyay, A. (2013). Bone tissue engineering using 3D printing. *Materials Today, 16*(12), 496–504. doi:10.1016/j.mattod.2013.11.017.

Bounegru, A. V., & Apetrei, C. (2023). Tyrosinase immobilization strategies for the development of electrochemical biosensors—A review. *Nanomaterials, 13*(4), 760. doi:10.3390/nano13040760.

Brem, J., Turcu, M. C., Paizs, C., Lundell, K., Toa, M. I., Irimie, F. D., & Kanerva, L. T. (2012). Immobilization to improve the properties of Pseudomonas fluorescens lipase for the kinetic resolution of 3-aryl-3-hydroxy esters. *Process Biochemistry, 47*(1), 119–126. doi:10.1016/j.procbio.2011.10.020.

Cao, L., Van Rantwijk, F., & Sheldon, R. A. (2000). Cross-linked enzyme aggregates: A simple and effective method for the immobilization of penicillin acylase. *Organic Letters, 2*(10), 1361–1364. doi:10.1021/ol005593x.

Cao, L., Langen, L., & Sheldon, R. A. (2003). Immobilised enzymes: Carrier-bound or carrier-free? *Current Opinion in Biotechnology, 14*(4), 387–394. doi:10.1016/s0958-1669(03)00096-x.

Chauhan, V., Kaushal, D., Dhiman, V. K., Kanwar, S. S., Singh, D., Dhiman, V. K., & Pandey, H. (2022). An insight in developing carrier-free immobilized enzymes. *Frontiers in Bioengineering and Biotechnology, 10*, 794411. doi:10.3389/fbioe.2022.794411.

Chen, Q., Qu, G., Li, X., Feng, M., Yang, F., Li, Y., Li, J., Tong, F., Song, S., Wang, Y., Sun, Z., & Luo, G. (2023). Active and stable alcohol dehydrogenase-assembled hydrogels via synergistic bridging of triazoles and metal ions. *Nature Communications, 14*(1), 2387. doi:10.1038/s41467-023-37921-y.

Cui, J., Zhao, Y., Feng, Y., Lin, T., Zhong, C., Tan, Z., & Jia, S. (2017). Encapsulation of spherical cross-linked phenylalanine ammonia lyase aggregates in mesoporous biosilica. *Journal of Agricultural and Food Chemistry, 65*(3), 618–625. doi:10.1021/acs.jafc.6b05003.

Díez, P., Sánchez, A., Gamella, M., Martínez-Ruíz, P., Aznar, E., De La Torre, C., Murguía, J. R., Martínez-Máñez, R., Villalonga, R., & Pingarrón, J. M. (2014). Toward the design of smart delivery systems controlled by integrated enzyme-based biocomputing ensembles. *Journal of the American Chemical Society, 136*(25), 9116–9123. doi:10.1021/ja503578b.

Du, Y., Jia, X., Zhong, L. e, Jiao, Y. i, Zhang, Z., Wang, Z., Feng, Y., Bilal, M., Cui, J., & Jia, S. (2022). Metal-organic frameworks with different dimensionalities: An ideal host platform for enzyme@MOF composites. *Coordination Chemistry Reviews, 454*, 214327. doi:10.1016/j.ccr.2021.214327.

Decante, G., Reis, R. L., Silva-Correia, J., & Oliveira, J. M. (2024). Enzymatic crosslinked hydrogels. In J. M. Oliveira, J. Silva-Correia, & R. L. Reis (Eds.), *Hydrogels for tissue engineering and regenerative medicine* (pp. 199–212). Academic Press.

Ge, H., Ge, Z., Mao, L., & Zhang, G. (2022). In vivo self-aggregation and efficient preparation of recombinant lichenase based on ferritin. *Shengwu Gongcheng Xuebao/Chinese Journal of Biotechnology, 38*(4), 1602–1611. doi:10.13345/j.cjb.210913.

Guauque Torres, M. P., Foresti, M. L., & Ferreira, M. L. (2014). CLEAs of Candida antarctica lipase B (CALB) with a bovine serum albumin (BSA) cofeeder core: Study of their catalytic activity. *Biochemical Engineering Journal, 90*, 36–43. doi:10.1016/j.bej.2014.05.004.

Hou, Q., Rooman, M., & Pucci, F. (2023). Enzyme stability-activity trade-off: New insights from protein stability weaknesses and evolutionary conservation. *Journal of Chemical Theory and Computation, 19*(12), 3664–3671. doi:10.1021/acs.jctc.3c00036.

Hudson, S., Cooney, J., & Magner, E. (2008). Proteins in mesoporous silicates. *Angewandte Chemie – International Edition, 47*(45), 8582–8594. doi:10.1002/anie.200705238.

Jung, D. H., Jung, J. H., Seo, D. H., Ha, S. J., Kweon, D. K., & Park, C. S. (2013). One-pot bioconversion of sucrose to trehalose using enzymatic sequential reactions in combined cross-linked enzyme aggregates. *Bioresource Technology, 130*, 801–804. doi:10.1016/j.biortech.2012.12.162.

Li, S., Zhang, A., Zatopek, K., Parvez, S., Gardner, A. F., Corrêa, I. R., Noren, C. J., & Xu, M. Q. (2018). Enhancing multistep DNA processing by solid-phase enzyme catalysis on polyethylene glycol coated beads. *Bioconjugate Chemistry, 29*(7), 2316–2324. doi:10.1021/acs.bioconjchem.8b00299.

Li, Y., Wang, Z., Xu, X., & Jin, L. (2015). A Ca-alginate particle co-immobilized with *Phanerochaete chrysosporium* cells and the combined cross-linked enzyme aggregates from *Trametes versicolor*. *Bioresource Technology, 198*, 464–469. doi:10.1016/j.biortech.2015.09.032.

Liu, X., Qiu, M., Zhang, Y., Pan, J., Zhang, W., Guo, L., Zhang, X., & Jiang, Y. (2025). Enzyme immobilization based on reticular framework materials: Strategy, food applications, and prospect. *Advances in Colloid and Interface Science, 344*, 103589. doi:10.1016/j.cis.2025.103589.

Lopez, S., Rondot, L., Leprêtre, C., Marchi-Delapierre, C., Ménage, S., & Cavazza, C. (2017). Cross-linked artificial enzyme crystals as heterogeneous catalysts for oxidation reactions. *Journal of the American Chemical Society, 139*(49), 17994–18002. doi:10.1021/jacs.7b09343.

Luo, S., Liu, L., Lyu, C. J., Sim, B., Liu, Y., Gong, H., Nie, Y., & Zhao, Y. L. (2022). Understanding the effectiveness of enzyme pre-reaction state by a quantum-based machine learning model. *Cell Reports Physical Science, 3*(11). doi:10.1016/j.xcrp.2022.101128.

Margolin, A. L., & Navia, M. A. (2001). Protein crystals as novel catalytic materials. *Angewandte Chemie (International ed. in English), 40*(12), 2204–2222.

Martins, S. L., Albuquerque, B. F., Nunes, M. A. P., & Ribeiro, M. H. L. (2018). Exploring magnetic and imprinted cross-linked enzyme aggregates of rhamnopyranosidase in microbioreactors. *Bioresource Technology, 249*, 704–712. doi:10.1016/j.biortech.2017.10.078.

McGill, M., Coburn, J. M., Partlow, B. P., Mu, X., & Kaplan, D. L. (2017). Molecular and macro-scale analysis of enzyme-crosslinked silk hydrogels for rational biomaterial design. *Acta Biomaterialia, 63*, 76–84. doi:10.1016/j.actbio.2017.09.020.

Noori, R., Perwez, M., & Sardar, M. (2019). Cross-linked enzyme aggregates: Current developments and applications. In M Sardar, M Perwez, & R Gupta (Eds.), *Biocatalysis:*

*Enzymatic basics and applications* (pp. 83–112). Springer International Publishing. doi:10.1007/978-3-030-25023-2_5.

Ottone, C., Romero, O., Aburto, C., Illanes, A., & Wilson, L. (2020). Biocatalysis in the winemaking industry: Challenges and opportunities for immobilized enzymes. *Comprehensive Reviews in Food Science and Food Safety, 19*(2), 595–621. doi:10.1111/1541-4337.12538.

Patti, S., Magrini Alunno, I., Pedroni, S., Riva, S., Ferrandi, E. E., & Monti, D. (2025). Advances and challenges in the development of immobilized enzymes for batch and flow biocatalyzed processes. *ChemSusChem., 18*(8), e202402007. doi:10.1002/cssc.202402007.

Periyasamy, K., Santhalembi, L., Mortha, G., Aurousseau, M., & Subramanian, S. (2016). Carrier-free co-immobilization of xylanase, cellulase and $\beta$-1,3-glucanase as combined cross-linked enzyme aggregates (combi-CLEAs) for one-pot saccharification of sugarcane bagasse. *RSC Advances, 6*(39), 32849–32857. doi:10.1039/c6ra00929h.

Reis, C., Sousa, E., Serpa, J., Oliveira, R., Oliveira, R., & Santos, J. (2019). Design of immobilized enzyme biocatalysts: Drawbacks and opportunities. *Química Nova, 42*(7), 768–783. doi:10.21577/0100-4042.20170381.

Roessl, U., Nahálka, J., & Nidetzky, B. (2010). Carrier-free immobilized enzymes for biocatalysis. *Biotechnology Letters, 32*(3), 341–350. doi:10.1007/s10529-009-0173-4.

Ruiz, A. I., Malavé, A. J., Felby, C., & Griebenow, K. (2000). Improved activity and stability of an immobilized recombinant laccase in organic solvents. *Biotechnology Letters, 22*(3), 229–233. doi:10.1023/A:1005698301681.

Sheldon, R. A. (2011). Characteristic features and biotechnological applications of cross-linked enzyme aggregates (CLEAs). *Applied Microbiology and Biotechnology, 92*(3), 467–477. doi:10.1007/s00253-011-3554-2.

Sheldon, R. A., Schoevaart, R., & Van Langen, L. M. (2005). Cross-linked enzyme aggregates (CLEAs): A novel and versatile method for enzyme immobilization (a review). *Biocatalysis and Biotransformation, 23*(3–4), 141–147. doi:10.1080/10242420500183378.

Sheldon, R. A., & Woodley, J. M. (2018). Role of biocatalysis in sustainable chemistry. *Chemical Reviews, 118*(2), 801–838. doi:10.1021/acs.chemrev.7b00203.

Sheldon, R. A. (2019). CLEAs, combi-CLEAs and "smart" magnetic CLEAs: Biocatalysis in a bio-based economy. *Catalysts, 9*(3), 261. doi:10.3390/catal9030261.

Shen, Y., An, C., Jiang, J., Huang, B., Li, N., Sun, C., Wang, C., Zhan, S., Li, X., Gao, F., Zhao, X., Cui, H., Gooneratne, R., & Wang, Y. (2022). Temperature-dependent nanogel for pesticide smart delivery with improved foliar dispersion and bioactivity for efficient control of multiple pests. *ACS Nano, 16*(12), 20622–20632. doi:10.1021/acsnano.2c07517.

Talekar, S., Pandharbale, A., Ladole, M., Nadar, S., Mulla, M., Japhalekar, K., Pattankude, K., & Arage, D. (2013). Carrier free co-immobilization of alpha amylase, glucoamylase and pullulanase as combined cross-linked enzyme aggregates (combi-cleas): A tri-enzyme biocatalyst with one pot starch hydrolytic activity. *Bioresource Technology, 147*, 269–275. doi:10.1016/j.biortech.2013.08.035.

Torabizadeh, H., Abbasi, S., & Tafaghodinia, B. (2025). Multi-immobilization of viscozyme, alcalase and flavourzyme by nanomagnetic combi-CLEAs method for oil and protein hydrolysates extraction from rice bran in aqueous phase. *International Journal of Biological Macromolecules, 311*(Pt 1), 143719. doi:10.1016/j.ijbiomac.2025.143719.

Wang, L., Ma, F., Li, Z., & Zhang, Y. (2024). Immobilizing amyloglucosidase on inorganic hybrid nanoflowers to prepare time–temperature integrators for chilled pork quality monitoring. *Food Chemistry, 437*(Pt 2), 137876. doi:10.1016/j.foodchem.2023.137876.

Wang, M., Jia, C., Qi, W., Yu, Q., Peng, X., Su, R., & He, Z. (2011). Porous-CLEAs of papain: Application to enzymatic hydrolysis of macromolecules. *Bioresource Technology, 102*(3), 3541–3545. doi:10.1016/j.biortech.2010.08.120.

Yang, C., Liu, W., Chen, S., Zong, X., Yuan, P., Chen, X., Li, X., Li, Y., Xue, W., & Dai, J. (2023). MOF-immobilized two-in-one engineered enzymes enhancing activity of biocatalytic cascade for tumor therapy. *Advanced Healthcare Materials, 12*(12), e2203035. doi:10.1002/adhm.202203035.

Zhai, C., Miao, L., Zhang, Y., Zhang, L., Li, H. E., & Zhang, S. (2022). An enzyme response-regulated colorimetric assay for pattern recognition sensing application using biomimetic inorganic-protein hybrid nanoflowers. *Chemical Engineering Journal, 431*, 134107. doi:10.1016/j.cej.2021.134107.

Zhang, X., Wang, S., Jiang, Z., Li, Y., & Jing, X. (2020). Boronic ester based vitrimers with enhanced stability via internal boron-nitrogen coordination. *Journal of the American Chemical Society, 142*(52), 21852–21860. doi:10.1021/jacs.0c10244.

Zhang, Y., Nie, N., & Zhang, Y. (2022). Toward modular construction of cell-free multienzyme systems. *Chinese Journal of Catalysis, 43*(7), 1749–1760. doi:10.1016/S1872-2067(21)64002-0.

# Index

Page numbers followed by "*f*" and "*t*" indicate, figures and tables respectively.

## P



## R



## S



## T



## U



## V



## X



## Z

Printed and bound by CPI Group (UK) Ltd, Croydon, CR0 4YY
24/09/2026
02222290-0005